Earth Observation, Remote Sensing and Geoscientific Ground Investigations for Archaeological and Heritage Research

Earth Observation, Remote Sensing and Geoscientific Ground Investigations for Archaeological and Heritage Research

Special Issue Editor

Deodato Tapete

MDPI • Basel • Beijing • Wuhan • Barcelona • Belgrade

MDPI

Special Issue Editor
Deodato Tapete
Italian Space Agency (ASI),
Italy

Editorial Office
MDPI
St. Alban-Anlage 66
4052 Basel, Switzerland

This is a reprint of articles from the Special Issue published online in the open access journal *Geosciences* (ISSN 2076-3263) from 2018 to 2019 (available at: https://www.mdpi.com/journal/geosciences/special_issues/EO_RS_Geo_Archaeological_Heritage_Research)

For citation purposes, cite each article independently as indicated on the article page online and as indicated below:

LastName, A.A.; LastName, B.B.; LastName, C.C. Article Title. *Journal Name* **Year**, *Article Number, Page Range.*

ISBN 978-3-03921-193-7 (Pbk)
ISBN 978-3-03921-194-4 (PDF)

Contents

About the Special Issue Editor

Deodato Tapete is a researcher in Earth observation in the Scientific Research Division of the Italian Space Agency (ASI). With more than 10 years of research experience in Earth sciences and remote sensing applied to the study and conservation of cultural heritage, Dr. Tapete is specialized in synthetic aperture radar imaging and interferometry for deformation monitoring, hazard assessment, and archaeological prospection. His publications in top-ranked journals (e.g., *Remote Sensing of Environment, Remote Sensing, and Journal of Archaeological Science*) are highly cited across the community. He has led research activities focused on advanced technologies to study renowned UNESCO World Heritage Sites, including the historic centres of Rome, Florence and Naples in Italy, Valletta in Malta, the Complex of Koguryo Tombs in the Democratic People's Republic of Korea, the Lines and Geoglyphs of Nasca and Palpa and the Historic Sanctuary of Machu Picchu in Peru, and heritage sites at risk in the Middle East and North Africa. He is a Fellow of the Higher Education Academy (FHEA) and an editor for various MDPI journals.

Preface to "Earth Observation, Remote Sensing and Geoscientific Ground Investigations for Archaeological and Heritage Research"

In the preface of the book "Remote Sensing and Geosciences for Archaeology" published one year ago, I expressed the wish that a series of MDPI books could be initiated starting from that publication. After one year, that wish came true with the publication of this second book that widens the scope by including not only archaeology, but more specifically, also cultural heritage, as well as Earth observation (EO) in addition to remote sensing (RS) and geoscientific ground investigations.

The fourteen papers (plus my editorial), published after rigorous peer review, provide a comprehensive overview of the capabilities, limitations, challenges, and perspectives of EO, RS, and geoscientific investigations (either in situ or in laboratory) with regard to:

(1) archaeological prospection with high-resolution satellite SAR and optical imagery;
(2) high-resolution documentation of archaeological features with drones;
(3) archaeological mapping with LiDAR towards automation;
(4) digital fieldwork using old and modern data;
(5) field and archaeometric investigations to corroborate archaeological hypotheses;
(6) new frontiers in archaeological research from space in contemporary Africa;
(7) education and capacity building in EO and RS for cultural heritage.

In the hope that readers will use these contributions to learn new methodologies and take inspiration for new research and applications, and that the questions left open by this book can find answers in a new book next year, I express my sincere gratitude to all the authors, editors, and reviewers for their commitment during this editorial project.

My special thanks go to Mr Richard Li, Geosciences Assistant Editor, for his dedication to this project and his valuable collaboration in the setup, promotion, and management of the Special Issue.

Deodato Tapete
Special Issue Editor

geosciences

MDPI

Editorial

Earth Observation, Remote Sensing, and Geoscientific Ground Investigations for Archaeological and Heritage Research

Deodato Tapete

Italian Space Agency (ASI), Via del Politecnico snc, 00133 Rome, Italy; deodato.tapete@asi.it

Received: 1 April 2019; Accepted: 4 April 2019; Published: 7 April 2019

Abstract: Building upon the positive outcomes and evidence of dissemination across the community of the first Special Issue *"Remote Sensing and Geosciences for Archaeology"*, the second edition of this Special Series of *Geosciences* dedicated to *"Earth Observation, Remote Sensing and Geoscientific Ground Investigations for Archaeological and Heritage Research"* collects a varied body of original scientific research contributions showcasing the technological, methodological, and interpretational advances that have been achieved in this field of archaeological and cultural heritage sciences over the last years. The fourteen papers, published after rigorous peer review, allowed the guest editor to make considerations on the capabilities, limitations, challenges, and perspectives of Earth observation (EO), remote sensing (RS), and geoscientific ground investigations with regard to: (1) archaeological prospection with high resolution satellite SAR and optical imagery; (2) high resolution documentation of archaeological features with drones; (3) archaeological mapping with LiDAR towards automation; (4) digital fieldwork using old and modern data; (5) field and archaeometric investigations to corroborate archaeological hypotheses; (6) new frontiers in archaeological research from space in contemporary Africa; and (7) education and capacity building in EO and RS for cultural heritage.

Keywords: Earth Observation; remote sensing; optical; SAR; drone; airborne LiDAR; GIS; OBIA; neutron diffraction; archaeological prospection; pattern recognition; archaeometry; geological mapping

---

1. Introduction

The first Special Issue on *"Remote Sensing and Geosciences for Archaeology"* that I was invited to lead as guest editor by the journal *Geosciences* in 2017, collected 21 high-quality peer-reviewed papers (plus the editorial) outlining the state-of-the-art of research in the fields of archaeological remote sensing and geosciences. The contributions published in that Special Issue provide a wide portfolio of methodologies, data, and techniques proving that remote sensing and geosciences for archaeology are currently vibrant research and practice domains, with expertise spread across the globe, and teams fully exploiting the capability of remote sensing to investigate sites and landscapes in different geographic, social, and environmental contexts [1].

After one year of publication, the metrics of the Special Issue summarized in Table 1 can be considered promising to assess the dissemination degree of these papers across the specialist community. We also need to account for the fact that the Special Issue was the first in *Geosciences* which was dedicated to remote sensing and archaeology, and the journal itself was not as much known to the specialist readership as it is nowadays. In particular, it is worth mentioning that two of the published papers, i.e., Traviglia & Torsello [2] and Agapiou et al. [3], have repeatedly been listed among the dynamic ranking of the 10 most-cited papers of *Geosciences* in the last 24 months.

Table 1. Article metrics of the papers published in the first edition of the Special Issue as of 01/04/2019 (source: *Geosciences*).

Authors	Views	Downloads	Citations
Agapiou et al. [8]	3278	1904	5
Agapiou et al. [3]	1708	970	8
Chyla [9]	1429	827	1
Comer et al. [5]	1570	1102	3
Corso et al. [12]	1753	1215	2
Danti et al. [10]	1943	1740	7
Drap et al. [13]	1571	1565	2
Gade et al. [6]	1751	1210	4
Garcia-Garcia et al. [14]	1282	995	1
Guidi et al. [15]	1719	976	3
Kalayci et al. [16]	1218	968	2
Křivánek [17]	1420	1174	2
Malinverni et al. [18]	1445	1345	1
Parcak et al. [11]	1486	1374	3
Poux et al. [19]	2306	2249	7
Rayne et al. [7]	1750	1500	5
Rutishauser et al. [4]	1784	1345	2
Sonnemann et al. [20]	1582	1046	2
Tapete [1]	1256	2313	3
Traviglia & Torsello [2]	1680	1311	9
Verhoeven [21]	1573	1399	6

Building upon the positive outcome achieved in 2017 and in order to continue this Special Series, in March 2018 I launched the call for papers for a second edition of the Special Issue with the title *"Earth Observation, Remote Sensing and Geoscientific Ground Investigations for Archaeological and Heritage Research"*.

Comparing the titles of the two editions of this Special Series, it clearly emerges that, in this second Special Issue, I intentionally:

(1) broadened the spectrum of the topics to include Earth Observation (EO), to acknowledge that satellite imagery is nowadays regarded by the archaeological and heritage communities as a resource of spatial and temporal information (see the majority of the papers published in the first edition of the Special Issue: [3–11]);

(2) cited "heritage" alongside "archaeology" to be more inclusive of the various disciplines and domains of geoscientific research focusing on cultural subjects;

(3) included geoscientific ground investigations, in the hope of receiving submissions highlighting not only new methods for ground-based surveying, archaeological prospection, and diagnostic investigation, but also validation of signals, parameters, features, and marks extracted from EO and remote sensing (RS) analyses with ground-truth data collected in the field.

The topics that I envisioned to cover for the submissions to this second edition included:

- archaeological prospection
- digital archaeological fieldwork
- condition assessment of heritage assets
- GIS analysis of spatial settlement patterns in modern landscapes
- assessment of natural or human-induced threats to conservation
- education and capacity building in EO and RS for archaeology

2. Facts and Figures of the Special Issue

A total of 21 submissions were received for consideration of publication in the Special Issue from April 2018 to January 2019. After rigorous editorial checks and the peer review process involving external and independent experts in the field, the acceptance rate was 67%. The published Special Issue therefore contains a collection of 14 research articles.

Figure 1a shows the countries where the study areas of the papers published in the Special Issue are located, while Figure 1b the spatial distribution of these study areas, distinguished between cultural landscapes and individual heritage sites. By comparison with Figure 1b published in [1], it is apparent that in this second edition the study areas are more widespread across the globe, while in the first edition the majority was concentrated in Europe and in the Middle East. The latter region, alongside Peru and Germany, is still of research interest. However, this time the archaeology of the Indian subcontinent and African continent gathered specific attention of the research community. It is also worth mentioning that one of the contributions [22] provides an overview of space law and space sciences for archaeological and heritage research in contemporary Africa. Thus, the African continent has been marked in grey in Figure 1a to signify the wider geographic focus of this paper.

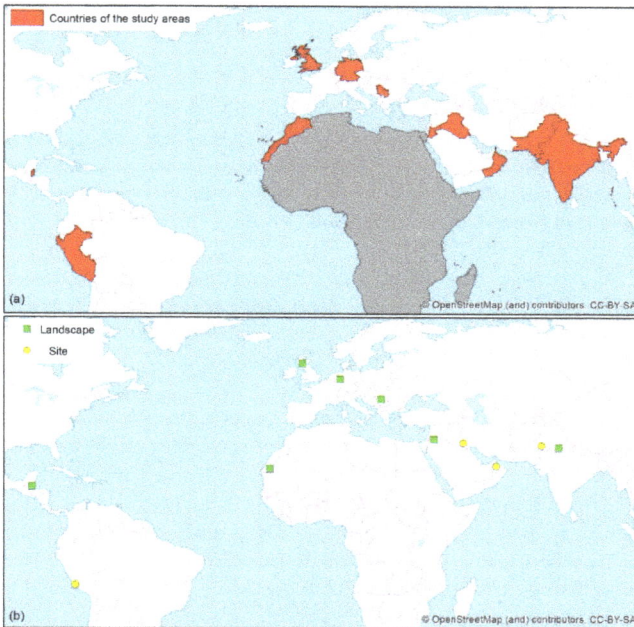

Figure 1. (**a**) Countries where the study areas of the papers published in the Special Issue are located; (**b**) geographic distribution of the study areas distinguished by typology ("landscape" in case of regional archaeological mapping and wide-area archaeological prospection; "site" in case of site-focused studies and investigations in single location). The African continent is marked in grey because one of the published papers [22] provides an overview of space law and space sciences for archaeological and heritage research in contemporary Africa.

This geographic distribution could not be predicted, was not intentional, and indeed, was the random result of the call for papers and following peer review. However, some considerations can be made.

The remote location and vastness of the study areas covered by the majority of the published papers once again prove the impact that EO and RS can generate in facilitating archaeological research, by making investigations more cost-effective and less risky for the operators.

Furthermore, it can be rightly said that with EO and RS there is no frontier for archaeological and heritage research. On the contrary, unexplored regions and areas with limited literature are ideal geographic locations for exercises of archaeological mapping and site discovery studies.

Finally, the predominance of landscape studies compared to site investigations (7 vs. 5; Figure 1b) highlights a growing interest in using EO and RS for regional and wide-area mapping. This trend has been recently observed and commented by several authors in the literature (e.g., [23,24]).

3. Overview of the Published Papers

As manuscripts were submitted and processed for peer review, it progressively became clear that this Special Issue was shaping not only along the topics that I had delineated in the call for papers (see Section 1), but also following other unexpected topics, including automation in archaeological prospection, methodological reflections on the use of old and new remote sensing data for digital fieldwork, and legal aspects of archaeological research. A summary of the published papers is reported in the following sections.

3.1. Archaeological Prospection with High Resolution Satellite SAR and Optical Imagery

The two papers published by Wiig et al. [25] and Zanni & De Rosa [26] respectively seem to contrast the controversial statements (sometimes written in the literature or claimed at conferences) that archaeologists are not familiar with satellite Synthetic Aperture Radar (SAR) imagery as a source of information for archaeological prospection due to difficulties with access, processing and interpretation of these data, and that high-resolution (HR) satellite optical imagery (i.e., 5–30 m) is of marginal usefulness in archaeology (see also [23,27]).

Wiig et al. [25] add a novel contribution to the still open discussion whether satellite SAR sensors operating at short wavelength (i.e., in C- and X-band, 5 to 3 cm wavelength) can penetrate through the subsurface in arid regions. The authors compared the observations made at the site of 'Uqdat al-Bakrah (Safah), Oman, with HR TanDEM-X bistatic and RADARSAT-2 images that were acquired at different incidence angles at scene center (from 27° to 53°) and polarization, and then processed to achieve pixel spacing of 0.87–1.14 m and 2.1–2.95 m, respectively. In particular, the authors' attention was concentrated on a subsurface paleo-channel that was not visible on the ground surface, but was first identified through Ground Penetrating Radar (GPR) survey and later verified by test excavations at a depth of 0.6–0.7 m. Although it is still unclear whether the microwaves are penetrating to the specific depth at which this paleo-channel was found, the findings are significant as this paper is one of the very few studies where features found in satellite SAR images were verified in the field.

Zanni & De Rosa [26] tested different combinations of the spectral information collected in the 13 bands of the Multispectral Instrument (MSI) onboard the satellite Sentinel-2A of the Copernicus programme, to investigate the capabilities of these satellite data for detection of buried features belonging to Roman roads. The experimental trials were run in the Srem District in Serbia, part of the original Roman itinerary between Aquileia (Italy) and Singidunum (Belgrade). Sentinel-2A images acquired in the summer season in 2016 were first carefully selected from the available catalogue and then processed to extract the Normalized Difference Vegetation Index (NDVI), Normalized Archaeological Index (NAI), the combination of Red and NIR (RN) and Crop Coefficient 3 (CC3). The visual assessment of the obtained maps and the comparison with the same processing outputs of a matching WorldView2 image led to the identification of 60 crop-marks in the portion of territory stretching from Sremska Mitrovica to Zemun. Of these, during the in-situ validation surveys, 13 were found to correspond to already known archaeological sites and stretches of the Roman road, whereas 47 crop-marks remained unmatched, thus highlighting the benefits and limitations of Sentinel-2 and WorldView2 observations.

3.2. High Resolution Documentation of Archaeological Features with Drones

In the current practice of archaeological remote sensing where small Unmanned Aerial Vehicles (UAV)/Remotely Piloted Aircraft System (RPAS) are increasingly used by archaeologists as data acquisition platforms and (semi-)autonomous measurement instrumentation, the paper published by Pavelka et al. [28] demonstrates the agility of this RS solution in arid environments and the opportunity that it can offer for fascinating discoveries while documenting cultural landscapes. The authors exploited very high resolution (VHR) satellite data and super resolution data from the drone to improve the digital documentation of the "Pista" geoglyph in Palpa, Peru, and refine the knowledge and interpretation of this geoglyph that had been researched several times by the archaeologists, but still poses some open questions. Through the description of the methodological workflow of data capture, processing and post-processing, the authors present the final vector map that they generated, achieving more detailed delineation of surviving archaeological features than older outputs based on satellite or old aerial data. The surveys also offered the opportunity to discover unknown geoglyphs (a bird, a guinea pig, and other small drawings), thus adding new information in an area of well-known geoglyphs. While dating these new geoglyphs remains a challenging task, the digital record of these newly found geoglyphs allowed the authors to observe similarity in the iconography compared with other well-known geoglyphs.

3.3. Archaeological Mapping with LiDAR towards Automation

There is no doubt about the great value of airborne LiDAR (Light Detection and Ranging) for archaeological mapping [29], as well as the high degree of appreciation that this technology finds across the archaeological community.

The contribution by Moyes & Montgomery [30] adds a further proof of the usefulness of this technology to explore Maya lowlands and other tropical regions, where dense vegetation usually prevents archaeologists from conducting extensive surveys or, at least, makes this type of archaeological survey less cost-effective. In particular, the authors describe a method for locating potential cave openings using local relief models that require only a working knowledge of relief visualization techniques. This method was exploited in Chiquibul Forest Reserve, a heavily forested area in western Belize, where caves were utilized by the ancient Maya people as ritual spaces. Almost all attempts to find caves using LiDAR data focused on locating sinkholes that lead to underground cave systems, but caves in Chiquibul can be entered in some cases by sinkholes, in others via vertical cliff faces or by dropping into small shafts. Therefore, the authors aimed to locate and investigate not only sinkholes but other types of cave entrances using point cloud modeling. Validation was undertaken through an opportunistic survey to verify selected caves identified on the LiDAR, and a systematic pedestrian survey that was completed over two six-week field seasons in the summers of 2017 and 2018 using two to three crews of three people each. The opportunistic survey led to 86% success rate with only three false positives, verifying 26 cave openings, and proved LiDAR to be expedient in meeting the project goals of locating and investigating unknown cave sites.

Regional and national LiDAR collections are increasingly made available by territorial administrations under open data policies for land management and scientific research purposes. Although these data are generally acquired in the context of flood or other hazard management, it is envisaged that their continuous release to the public will only further increase the impact of airborne LiDAR on archaeological research and heritage management [31]. While these initiatives are welcome as they provide an extraordinary source of spatial data, there is lively discussion about the impact that automation can bring to improve the operator's capabilities to handle huge quantities of LiDAR data for archaeological mapping of large regions. However, it cannot be neglected that the development of automation methods and approaches in archaeological prospection is still in its infancy.

Towards this direction, Meyer et al. [32] exploited the LiDAR datasets acquired between 2008 and 2010, and later in 2016, and made available to archaeologists in North Rhine-Westphalia, Germany, by the provincial government according to the Open Geodata principle, to assess the potential for automated classifications using Object-Based Image Analysis (OBIA). Three types of field monuments were considered: Ridge and Furrow areas (of early medieval fields), Burial Mounds (Bronze and Iron Ages), and Motte-and-Bailey castles. The latter two are not classified as binary, but in multiple classes, depending on their degree of erosion. After a detailed description of the methodology and processing workflow, the authors focus their results discussion around the challenge of discriminating between true and false positives in situations where the terrain becomes complex and a more anthropogenic influence is present. On the other side, the detection rate of field monuments with OBIA is ~90%, although this technique is vulnerable to distortions and frequently can be implemented in commercial software that may limit the accessibility to archaeologists due to fund constraints.

3.4. Digital Fieldwork and Reflections on Challenges of Archaeological Mapping with Old and Modern Data

One of the main objectives of this Special Issue was to capture the state-of-the-art of the methods of digital fieldwork in remote and inaccessible areas. The picture coming out from the collection of the papers described in this section is that archaeologists, from different countries, are making efforts to develop rigorous and robust methodologies for archaeological mapping which are at the same time systematic, accurate, reliable, and cost-effective. Digital fieldwork is undertaken as a desk-based task in the perspective of precisely planning ground-truth and validation surveys, to optimize resources and prioritize in-situ inspections in areas of higher archaeological potential.

In this regard, Nsanziyera et al. [33] present a predictive model based on GIS and remote sensing data to locate areas with high potential to be archaeological sites. The authors apply a multi-criteria decision making method—analytic hierarchy process (AHP)—that integrates archaeological data and environmental factors, geospatial analysis, and predictive modeling, to identify possible tumuli locations in Awsard (total study area of 980 km^2), southern Morocco. The results consist of a prediction map with a gain of 92.8%, in a scale where 1 means a high predictive model and 0 no a predictive model. Interestingly, 56.87% of all sites were found to be located in only 4.04% of the total study area. This method proves effective to prioritize areas for archaeological expeditions.

Smith & Chambrade [34] showcase the results of the systematic analysis of the arid "Black Desert" of north-eastern Jordan, which they conducted in the framework of the archaeological project Western Harra Survey (WHS), using the full VHR Google Earth coverage released in 2017, with further GeoEye and CNES/Airbus satellite imagery becoming available, as well as DigitalGlobe products appearing in Bing Maps. The high spatial resolution of such datasets enabled a more clear definition of structural differences between the types of prehistoric structures (e.g., enclosures, "wheels", "pendants", "kites", and meandering walls). The major benefit of this satellite digital fieldwork was the precise planning of ground surveys, with advanced knowledge of which sites were vehicle-accessible and how to efficiently visit a stratified sample of different site types. The fieldwork-derived data were then fed back into the satellite imagery survey, helping the authors to interpret what can be seen in remote sensing more accurately for future investigations.

However, the advent of new EO and RS data, visualization platforms, and processing technologies does not mean that archaeologists and heritage scientists disregard historical mapping resources. On the contrary, the community is working on bringing these old fashioned resources back to light, standardizing the methodology for their use and interpretation, and combining the information extracted with modern data, to achieve a diachronic and dynamic reconstruction of the cultural landscape evolution in time.

Petrie et al. [35] and Garcia et al. [36] are two interlinked papers that need to be read in conjunction, because they were conceived and published in the framework of TwoRains, WaMStrIn and Marginscapes projects. Petrie et al. [35] advocates the value and importance of the Survey of India 1" to 1-mile map series, an historical mapping resource which was under-utilized and, with this paper, gains the attention it deserves since it is a precious reservoir of spatial information of topographic features and elevated mounds visible at the time of the surveys, but which were either damaged or destroyed by the expansion of irrigation agriculture, and urbanism, and are no longer visible. The authors present a method for accurately georeferencing these maps and review the symbology that was used to represent elevated mound features that have the potential to be archaeological sites. Certainly, this method will be very useful to support further studies by other scholars willing to use this mapping resource alongside modern RS data, as it is well demonstrated by the accompanying Garcia et al. paper [36]. Within the latter paper, the authors investigate the historical inundation that hit the city of Dera Ghazi Kkan, in Punjab, Pakistan, in 1909. Historic news reports, books, and maps are used to undertake a regressive analysis to reconstruct the historical dynamics between the urban settlement and the river morphodynamics in the Indus alluvial plain. Declassified CORONA images, multispectral Landsat time series, and microtopographic data derived from ALOS Global Digital Surface Model "ALOS World 3D-30 m (AW3D30)" using the Multi-Scale Relief Model (MSRM), are combined to examine: (1) how historical hydrological dynamics are reflected in RS data; (2) the implications of river morphodynamics in the interpretation of settlement patterning; and (3) the documented socio-political responses to the geomorphological change of the local environment.

If old mapping data preserve an otherwise vanishing memory, they have to be handled carefully, especially if they have been collected by different operators and according to different study purposes. In this context, the feature paper by Banaszek et al. [37] will be, in the author's opinion, a reference piece of research, since it provides a practical discussion of the challenges that archaeologists need to deal with for creating systematic datasets of national-scale archaeological mapping, where the standards to which these datasets were created are explicit, and against which the reliability of the knowledge of the material remains of the past can be assessed. With the focus on Scotland, the authors start by acknowledging that the National Record of the Historic Environment (NRHE) is an inventory of what has been recorded over the years and it reflects the interests and recording policies of those who created it, with bias in content as a result. The lack of scalability in traditional approaches to large area mapping which rely heavily on human resources and field visits, is definitely a constraint to deal with. The authors use the Isle of Arran as an outdoor laboratory for scoping their approach to rapid large area mapping and test how airborne laser scanning derivatives and orthophotographs, supplemented by field observations, can help to increase the records of the known monuments. This exercise demonstrated the strengths and weaknesses of remotely sensed data acquired for general purpose, the variability of desk-based interpretation between individuals, and the necessity for targeted field observations in areas with poor data coverage and where background noise obscures the visibility of archaeological features in the visualizations derived from the airborne laser scanning surveys.

3.5. Field and Archaeometric Investigations to Corroborate Archaeological Hypotheses

In a multidisciplinary perspective, geoscientific ground investigations and laboratory analyses remain essential to achieve an insightful knowledge of the near surface in archaeological and heritage sites, as well as of objects and findings, that EO and RS alone could not be able to document or investigate. While most of the analytical techniques and research methodologies in geo-archaeology and archaeometry are well-established and standardized, there are always opportunities to employ advanced approaches and collect elements to support or modify existing archaeological hypotheses.

Festa et al. ref. [38] is an archaeometric paper presenting the results of non-destructive analyses carried out on 36 Sumerian pottery fragments found in the settlement of Abu Tbeirah (3rd millennium BC), southern Iraq. The analysis aimed to characterize the crystallographic composition of the ceramic material, to shed light on the ancient technology and manufacturing techniques. Combining non-invasive neutron diffraction (ND) with chemometrics such as Principal Component Analysis (PCA) and Cluster Analysis (CA), the authors observed a general uniformity of the raw materials and could suggest a local origin of the clay used for Sumerian vases, by comparison with modern clay collected from the canal near the excavated site. The secondary minerals found and their marker-temperature formation are compatible with two different ranges of firing temperature that never exceeded 1000 °C. In the absence of kiln traces in the archaeological site of Abu Tbeirah, it appears reasonable to hypothesize that the analyzed pottery was produced with pit-firing techniques and not kiln firing.

Because kilns have been documented in the Mesopotamian archaeological record for earlier periods, the finding of this research would suggest the coeval presence of different firing methodologies that has been neglected by archaeologists so far.

Delle Rose et al. [39] attempt to find stratigraphic evidence corroborating (or confuting) the hypothesis that the ceremonial center of Cahuachi, Rio Grande de Nazca, in southern Peru, was first severely damaged, then completely buried by catastrophic river floods as a result of two Mega El Niño events, which occurred around 600 Common Era (CE) and 1000 CE, respectively. The occurrence of such catastrophic events would be proved by the presence of a conglomerate layer in the stratigraphy. Therefore, during the 2012 archaeological excavation works at Cahuachi, the geological substratum close to the Piramide Sur was temporarily exposed and stratigraphic, grain-size distribution, and petrographic investigations were carried out. No fundamental discontinuity was found in the studied stratigraphic interval which instead, due to the lithological features, matches with common regional successions (i.e., Changuillo or Changuillo–Canete Formations) of the pampa of Nazca rather than the deposits related to El Niño–Southern Oscillation (ENSO) events.

3.6. New Frontiers in Archaeological Research from Space in Contemporary Africa

As recalled in Figure 1a, the last paper published in the Special Issue [22] provides an overview of space law and space sciences for archaeological and heritage research in contemporary Africa, which could become a new frontier for activities of discovery and preservation in this continent. This paper also reminds the reader that there are far more diverse categories of heritage and archaeological features than those commonly studied with EO and RS. Indeed Oduntan [22] articulates a series of insightful reflections on the legal aspects of EO and RS, trying to answer questions about the impact that these aspects of space law and space sciences have in relation to: (a) international boundaries disputes and demarcation activities; (b) management and preservation of the African heritage; (c) disaster and conservation management. In particular, the paper tests the hypothesis that it is crucial for the development of the African continent that states should sustain and increase investment in the following areas: archaeological prospection, condition assessment of heritage assets; Geographic Information System (GIS) analysis of spatial settlement patterns in modern landscapes, and assessment of natural or human-induced threats to conservation. Through a critical, comparative, and socio-legal methodology, the author focuses on the space active African states and the emergent patterns in African domestic space-related policies and space-dedicated legislation. The connection with the EO and RS practice of archaeological and heritage research lies in the area of the reconstruction of African territories from space, the demarcation of boundaries, and geodetic ground investigations, not only to resolve disputes but also to preserve state boundaries and ancient African "relict boundaries". The latter term refers to antecedent boundaries which were abandoned for political purposes but are still evident in the cultural landscape and, as such, manifest themselves in space by, among other features, direct border remains such as border stones, mounds, ancient walls, border roads, clearings, customs houses, and watch-towers. The latter are among the less known African heritage and treasures that EO and RS can help to unveil, document, and preserve within national and international legal frameworks and space policies.

3.7. Education and Capacity Building in EO and RS for Cultural Heritage

All the papers summarized above were published by expert scientists and researchers who are extremely familiar with and competent in EO, RS, geoscientific ground investigations, and laboratory analytical techniques. The knowledge transfer and the capacity building to heritage stakeholders and early beginners are still challenging tasks, and require a specialist educational preparation that is not obvious. Showcasing the ability of a technology to support a specific operational task (e.g., condition assessment of heritage sites) does not mean that the potential users of that technology will be able to use it themselves or, after training, will recognize the value of that technology and will search for it in their daily duties. In the current context where more work is definitely required to reach the

Geosciences **2019**, *9*, 161

users and stakeholders and generate real impact on archaeological and heritage practice, the paper by Matusch et al. [40] is proof that some initiatives are ongoing. The authors present the e-learning module Space2Place that they developed in the framework of the project "Space4Geography" carried out between 2013 and 2017, with the aim to empower UNESCO site stakeholders to incorporate EO into their working routines. This e-learning module is contextualized in the current situation of knowledge gaps by the user, limited technical and financial facilities, or the lack of ready-to-use data, despite the abundance of satellite data and user-oriented services made available by EO programs such as the European Commission Copernicus. Space2Place is therefore a capacity building initiative to enable heritage stakeholders obtain a substantial introduction into EO and overcome the knowledge barriers that may exist. One of the key features of this paper is the discussion of the results collected after an expert survey that the authors ran with the participation of 11 experts coming from various institutions. The survey provides insights into the main barriers and expected benefits that stakeholders perceive in the use of EO to address specific threats to conservation of cultural heritage (e.g., climate change, natural hazards, intentional destruction, and warfare). Of all the interesting elements emerging from this direct feedback, two are worthy of mention. First, not all EO data are appropriate for each task, thus stakeholders need to be able to choose themselves the appropriate EO sensor(s) with regard to their specific needs, the study time, and the size and location of the site to observe. This approach will make the stakeholders aware and become critical users of these technologies. Second, there is a clear demand for up-to-date information with high cost-efficiency, that can be used in support of daily and routine tasks such as detection of impacts, evaluation of interventions, and early detection of critical changes in heritage sites. However, accessibility in terms of finance, infrastructure, and human resources remains a constraint.

Funding: This research received no external funding.

Acknowledgments: The Guest Editor thanks all the authors, Geosciences' editors, and reviewers for their great contributions and commitment to this Special Issue. Special thanks go to Richard Li, Geosciences' Assistant Editor, for his dedication to this project and his valuable collaboration in the setup, promotion, and management of the Special Issue.

Conflicts of Interest: The author declares no conflict of interest.

References

1. Tapete, D. Remote sensing and geosciences for archaeology. *Geosciences* **2018**, *8*, 41. [CrossRef]
2. Traviglia, A.; Torsello, A. Landscape pattern detection in archaeological remote sensing. *Geosciences* **2017**, *7*, 128. [CrossRef]
3. Agapiou, A.; Lysandrou, V.; Hadjimitsis, D. Optical remote sensing potentials for looting detection. *Geosciences* **2017**, *7*, 98. [CrossRef]
4. Rutishauser, S.; Erasmi, S.; Rosenbauer, R.; Buchbach, R. SARchaeology—Detecting palaeochannels based on high resolution radar data and their impact of changes in the settlement pattern in Cilicia (Turkey). *Geosciences* **2017**, *7*, 109. [CrossRef]
5. Comer, D.; Chapman, B.; Comer, J. Detecting landscape disturbance at the Nasca Lines using SAR data collected from airborne and satellite platforms. *Geosciences* **2017**, *7*, 106. [CrossRef]
6. Gade, M.; Kohlus, J.; Kost, C. SAR imaging of archaeological sites on Intertidal Flats in the German Wadden Sea. *Geosciences* **2017**, *7*, 105. [CrossRef]
7. Rayne, L.; Bradbury, J.; Mattingly, D.; Philip, G.; Bewley, R.; Wilson, A. From above and on the ground: Geospatial methods for recording endangered archaeology in the Middle East and North Africa. *Geosciences* **2017**, *7*, 100. [CrossRef]
8. Agapiou, A.; Lysandrou, V.; Sarris, A.; Papadopoulos, N.; Hadjimitsis, D. Fusion of satellite multispectral images based on Ground-Penetrating Radar (GPR) data for the investigation of buried concealed archaeological remains. *Geosciences* **2017**, *7*, 40. [CrossRef]
9. Chyla, J. How can remote sensing help in detecting the threats to archaeological sites in Upper Egypt? *Geosciences* **2017**, *7*, 97. [CrossRef]

10. Danti, M.; Branting, S.; Penacho, S. The American schools of oriental research cultural heritage initiatives: Monitoring cultural heritage in Syria and Northern Iraq by geospatial imagery. *Geosciences* **2017**, *7*, 95. [CrossRef]

11. Parcak, S.; Mumford, G.; Childs, C. Using open access satellite data alongside ground based remote sensing: An assessment, with case studies from Egypt's delta. *Geosciences* **2017**, *7*, 94. [CrossRef]

12. Corso, J.; Roca, J.; Buill, F. Geometric analysis on stone façades with terrestrial laser scanner technology. *Geosciences* **2017**, *7*, 103. [CrossRef]

13. Drap, P.; Papini, O.; Pruno, E.; Nucciotti, M.; Vannini, G. Ontology-Based photogrammetry survey for medieval archaeology: Toward a 3D geographic information system (GIS). *Geosciences* **2017**, *7*, 93. [CrossRef]

14. Garcia-Garcia, E.; Andrews, J.; Iriarte, E.; Sala, R.; Aranburu, A.; Hill, J.; Agirre-Mauleon, J. Geoarchaeological core prospection as a tool to validate archaeological interpretation based on geophysical data at the Roman Settlement of Auritz/Burguete and Aurizberri/Espinal (Navarre) †. *Geosciences* **2017**, *7*, 104. [CrossRef]

15. Guidi, G.; Gonizzi Barsanti, S.; Micoli, L.; Malik, U. Accurate reconstruction of the Roman circus in Milan by georeferencing heterogeneous data sources with GIS. *Geosciences* **2017**, *7*, 91. [CrossRef]

16. Kalayci, T.; Simon, F.-X.; Sarris, A. A manifold approach for the investigation of early and middle Neolithic settlements in Thessaly, Greece. *Geosciences* **2017**, *7*, 79. [CrossRef]

17. Křivánek, R. Comparison study to the use of geophysical methods at archaeological sites observed by various remote sensing techniques in the Czech Republic. *Geosciences* **2017**, *7*, 81. [CrossRef]

18. Malinverni, E.; Pierdicca, R.; Bozzi, C.; Colosi, F.; Orazi, R. Analysis and processing of Nadir and Stereo VHR Pleiadés images for 3d mapping and planning the land of Nineveh, Iraqi Kurdistan. *Geosciences* **2017**, *7*, 80. [CrossRef]

19. Poux, F.; Neuville, R.; Van Wersch, L.; Nys, G.-A.; Billen, R. 3D point clouds in archaeology: Advances in acquisition, processing and knowledge integration applied to quasi-planar objects. *Geosciences* **2017**, *7*, 96. [CrossRef]

20. Sonnemann, T.; Comer, D.; Patsolic, J.; Megarry, W.; Herrera Malatesta, E.; Hofman, C. Semi-Automatic detection of indigenous settlement features on Hispaniola through remote sensing data. *Geosciences* **2017**, *7*, 127. [CrossRef]

21. Verhoeven, G. Are We There Yet? A Review and assessment of archaeological passive airborne optical imaging approaches in the light of landscape archaeology. *Geosciences* **2017**, *7*, 86. [CrossRef]

22. Oduntan, G. Geospatial sciences and space law: Legal aspects of earth observation, remote sensing and geoscientific ground investigations in Africa. *Geosciences* **2019**, *9*, 149. [CrossRef]

23. Tapete, D.; Cigna, F. Appraisal of opportunities and perspectives for the systematic condition assessment of heritage sites with copernicus sentinel-2 high-resolution multispectral imagery. *Remote Sens.* **2018**, *10*, 561. [CrossRef]

24. Casana, J.; Laugier, E.J. Satellite imagery-based monitoring of archaeological site damage in the Syrian civil war. *PLoS ONE* **2017**, *12*, e0188589. [CrossRef]

25. Wiig, F.; Harrower, M.J.; Braun, A.; Nathan, S.; Lehner, J.W.; Simon, K.M.; Sturm, J.O.; Trinder, J.; Dumitru, I.A.; Hensley, S.; et al. Mapping a subsurface water channel with x-band and c-band synthetic aperture radar at the Iron Age archaeological site of 'Uqdat al-Bakrah (Safah), Oman. *Geosciences* **2018**, *8*, 334. [CrossRef]

26. Zanni, S.; De Rosa, A. remote sensing analyses on sentinel-2 images: Looking for Roman roads in Srem region (Serbia). *Geosciences* **2019**, *9*, 25. [CrossRef]

27. Tapete, D.; Cigna, F. Trends and perspectives of space-borne SAR remote sensing for archaeological landscape and cultural heritage applications. *J. Archaeol. Sci. Reports* **2016**, *14*, 716–726. [CrossRef]

28. Pavelka, K.; Šedina, J.; Matoušková, E. High resolution drone surveying of the Pista Geoglyph in Palpa, Peru. *Geosciences* **2018**, *8*, 479. [CrossRef]

29. Chase, A.S.Z.; Chase, D.Z.; Chase, A.F. LiDAR for archaeological research and the study of historical landscapes. In *Sensing the Past: From Artifact to Historical Site*; Masini, N., Soldovieri, F., Eds.; Springer International Publishing: Cham, Switzerland, 2017; pp. 89–100.

30. Moyes, H.; Montgomery, S. Locating cave entrances using lidar-derived local relief modeling. *Geosciences* **2019**, *9*, 98. [CrossRef]

31. Opitz, R.; Herrmann, J. Recent trends and long-standing problems in archaeological remote sensing. *J. Comput. Appl. Archaeol.* **2018**, *1*, 19–41. [CrossRef]

32. Meyer, M.F.; Pfeffer, I.; Jürgens, C. Automated detection of field monuments in digital terrain models of Westphalia using OBIA. *Geosciences* **2019**, *9*, 109. [CrossRef]

33. Nsanziyera, A.F.; Rhinane, H.; Oujaa, A.; Mubea, K. GIS and remote-sensing application in archaeological site mapping in the Awsard Area (Morocco). *Geosciences* **2018**, *8*, 207. [CrossRef]

34. Smith, S.L.; Chambrade, M.-L. The application of freely-available satellite imagery for informing and complementing archaeological fieldwork in the "Black Desert" of North-Eastern Jordan. *Geosciences* **2018**, *8*, 491. [CrossRef]

35. Petrie, C.A.; Orengo, H.A.; Green, A.S.; Walker, J.R.; Garcia, A.; Conesa, F.; Knox, J.R.; Singh, R.N. Mapping archaeology while mapping an empire: Using historical maps to reconstruct ancient settlement landscapes in modern India and Pakistan. *Geosciences* **2018**, *9*, 11. [CrossRef]

36. Garcia, A.; Orengo, H.A.; Conesa, F.C.; Green, A.S.; Petrie, C.A. Remote sensing and historical morphodynamics of alluvial plains. The 1909 Indus flood and the city of Dera Ghazi Khan (province of Punjab, Pakistan). *Geosciences* **2018**, *9*, 21. [CrossRef]

37. Banaszek, .; Cowley, D.C.; Middleton, M. Towards national archaeological mapping. Assessing source data and methodology—A case study from Scotland. *Geosciences* **2018**, *8*, 272. [CrossRef]

38. Festa, G.; Andreani, C.; D'Agostino, F.; Forte, V.; Nardini, M.; Scherillo, A.; Scatigno, C.; Senesi, R.; Romano, L. Sumerian pottery technology studied through neutron diffraction and chemometrics at Abu Tbeirah (Iraq). *Geosciences* **2019**, *9*, 74. [CrossRef]

39. Delle Rose, M.; Mattioli, M.; Capuano, N.; Renzulli, A. Stratigraphy, petrography and grain-size distribution of sedimentary lithologies at Cahuachi (South Peru): ENSO-Related deposits or a common regional succession? *Geosciences* **2019**, *9*, 80. [CrossRef]

40. Matusch, T.; Schneibel, A.; Dannwolf, L.; Siegmund, A. Implementing a modern e-learning strategy in an interdisciplinary environment—Empowering UNESCO stakeholders to use earth observation. *Geosciences* **2018**, *8*, 432. [CrossRef]

geosciences

MDPI

Article

Mapping a Subsurface Water Channel with X-Band and C-Band Synthetic Aperture Radar at the Iron Age Archaeological Site of 'Uqdat al-Bakrah (Safah), Oman

Frances Wiig [1,*], Michael J. Harrower [2], Alexander Braun [3], Smiti Nathan [4], Joseph W. Lehner [5], Katie M. Simon [6], Jennie O. Sturm [6], John Trinder [1], Ioana A. Dumitru [2], Scott Hensley [7] and Terence Clark [8]

[1] School of Civil and Environmental Engineering, The University of New South Wales, UNSW SYDNEY, Kingsford, NSW 2052, Australia; j.trinder@unsw.edu.au
[2] Department of Near Eastern Studies, Johns Hopkins University, Baltimore, MD 21218, USA; mharrower@jhu.edu (M.J.H.); idumitr1@jhu.edu (I.A.D.)
[3] Department of Geological Sciences and Geological Engineering, Queen's University, Kingston, ON K7L 3N6 Canada; braun@queensu.ca
[4] Johns Hopkins University, Sheridan Libraries and Museums, Baltimore, MD 21218, USA; smiti.nathan@jhu.edu
[5] Department of Archaeology, The University of Sydney, Sydney, NSW 2006, Australia; joseph.lehner@sydney.edu.au
[6] Center for Advanced Spatial Technologies, University of Arkansas, Fayetteville, AR 72701, USA; katie@cast.uark.edu (K.M.S.); jsturm@tagrsi.com (J.O.S.)
[7] Jet Propulsion Laboratory, Pasadena, CA 91109 USA; scott.hensley@jpl.nasa.gov
[8] Department of Archaeology and Anthropology, University of Saskatchewan, Saskatoon, SK S7N 5C9 Canada; terence.clark@usask.ca
* Correspondence: f.wiig@student.unsw.edu.au; Tel.: +61-413-712-100

Received: 10 July 2018; Accepted: 29 August 2018; Published: 5 September 2018

Abstract: Subsurface imaging in arid regions is a well-known application of satellite Synthetic Aperture Radar (SAR). Archaeological prospection has often focused on L-band SAR sensors, given the ability of longer wavelengths to penetrate more deeply into sand. In contrast, this study demonstrates capabilities of shorter-wavelength, but higher spatial resolution, C-band and X-band SAR sensors in archaeological subsurface imaging at the site of 'Uqdat al-Bakrah (Safah), Oman. Despite having varying parameters and acquisitions, both the X-band and C-band images analyzed were able to identify a subsurface paleo-channel that is not visible on the ground surface. This feature was first identified through Ground Penetrating Radar (GPR) survey, then recognized in the SAR imagery and further verified by test excavations. Both the GPR and the excavations reveal the base of the paleo-channel at a depth of 0.6 m–0.7 m. Hence, both X-band and C-band wavelengths are appropriate for subsurface archaeological prospection in suitable (dry silt and sand) conditions with specific acquisition parameters. Moreover, these results offer important new insights into the paleo-environmental context of ancient metal-working at 'Uqdat al-Bakrah and demonstrate surface water flow roughly contemporary with the site's occupation.

Keywords: synthetic aperture radar; subsurface imaging; microwave penetration; archaeology; arid environments; remote sensing; Oman

1. Introduction

1.1. Context of Research

The use of Synthetic Aperture Radar (SAR) as a tool for archaeological prospection has a limited history, commencing in the 1980s when NASA's (National Aeronautics and Space Administration) airborne L-Band sensor detected Mayan irrigation channels and cultivated wetlands in the Yucatán peninsula [1–4] and the SIR-A (Shuttle Imaging Radar-A) sensor identified subsurface paleo-channels in North Africa [5–7]. These early examples present an alternative to optical imagery as they exploit the ability of SAR microwaves to penetrate through different media, whether tropical foliage in the Yucatán Peninsula or aeolian sands in the Sahara Desert. Because of this capability, SAR is now being used for prospection of archaeological sites and/or paleo-environmental features that are not discernable in the visible or infrared portions of the electromagnetic spectrum used by multi-spectral satellites [8–10].

A SAR system transmits electromagnetic pulses to illuminate a portion of the earth's surface and subsurface and then receives the backscattered returning pulse, which provides information about the surface and subsurface characteristics in the illuminated scene [10,11]. Subsurface imaging is dependent on having a fine-grained (relative to the radar wavelength), physically homogenous medium through which the microwaves can propagate, with the target providing a contrasting surface that allows the microwaves to reveal a change in scattering processes. In addition to wavelength and grain size, the interaction between radar waves and subsurface materials is further governed by physical parameters, such as the soil's dielectric permittivity and conductivity (directly related to soil moisture), incidence angle and polarization [6,10–13].

Research into microwave propagation in arid environments has been undertaken with varying results. Early theoretical work proposed that longer wavelengths (L-band) were able to penetrate deeper than 5 m in dry sand [6,12,14] while later investigations supported more conservative penetration depths of 0.05–0.3 m for X-band, 0.1–0.5 m for C-band and 0.4–2.0 m for L-band in the silica blow sand and alluvium of Egypt's Western desert [15]. Because of their ability to penetrate further, longer wavelengths such as P-band (270–430 MHz frequency or 80–110 cm wavelength) and L-Band (1–2 GHz frequency or 15–30 cm wavelength) are often chosen for archaeological subsurface prospection in these environments [8,13,16–18]. However, C-band (4–8 GHz frequency or 3.75–7.5 cm wavelength) [17,19] and X-band (8–12.5 GHz frequency or 2.5–3.75 cm wavelength) [20,21] have also been used.

Further parameters that affect subsurface imaging include the look direction from the sensor and angle from the sensor to the ground (incidence angle), as targets are more likely to be visible if they have a strong profile that is perpendicular to the direction of radar propagation [17]. Microwave sensors are also configured to transmit and receive electromagnetic waves with specific polarizations, the simplest and most common being the single polarizations: Horizontal (HH) or vertical (VV) linear, in which the same polarization is transmitted and received. Different polarizations can provide additional information about a target and are another advantage of SAR imaging [11], although multi-polarization observations are often not available at the same fine resolution as single polarized data.

Despite the promising capabilities of SAR, archaeological applications have been hindered by the relatively low spatial resolution of early sensors, the limited availability and high costs of SAR data and software, as well as difficulties involved in processing and interpreting SAR images compared to optical imagery. Over the past few years many of these obstacles have diminished. There are an increasing number of higher spatial resolution C-band and X-band satellite missions (TerraSAR-X, TanDEM-X, COSMO-SkyMed, Sentinel series, RADARSAT-2) that acquire imagery in different modes (e.g., strip-map, spotlight) with different spatial resolutions, and different polarizations (single-pol, dual-pol, quad-pol). The Sentinel SAR data are freely available to the general public, while TerraSAR-X, TanDEM-X, COSMO-SkyMed and RADARSAT-2 data are available free of charge for research purposes from the respective space agencies upon successful application to specific Announcements of Opportunities. These data are greatly complemented by user-friendly open-source software (e.g., the SNAP toolbox from the European Space Agency). Additionally, more accessible historical data archives, expanding research, and forums on image interpretation [11,22] are making C-band and X-band SAR increasingly valuable tools for archaeological prospection.

1.2. The Archaeological Site of 'Uqdat al-Bakrah

The recently discovered (2012) Iron Age archaeological site of 'Uqdat al-Bakrah (also known as Safah) is situated on the eastern border of the Rub al-Khali Desert in Oman, approximately 50 km west of the town of Dhank. This location is at the periphery of the (ancient and contemporary) Wadi Bakrah alluvial fan and the fringe of the desert with its overlying aeolian sand veneer (Figure 1). The climate in this area is hyper-arid with an average rainfall of less than 100 mm/year [23].

Figure 1. Regional map showing the contemporary town of Dhank. The exact location of 'Uqdat al-Bakrah is not indicated given its sensitivity and need for cultural heritage protection.

'Uqdat al-Bakrah has yielded hundreds of bronze objects and pits that could have been used for producing charcoal or as furnaces for melting and finishing/recycling bronze objects [24]. In 2013, excavations of a small number of pits undertaken by an Italian team sponsored by the Sultanate of Oman Ministry of Heritage and Culture demonstrated that they were buried under a shallow layer of sand at depths ranging from 0.4 to 1.5 m [25].

In January 2017, investigations of the Archaeological Water Histories of Oman (ArWHO) Project at 'Uqdat al-Bakrah incorporated a geophysical survey, which included Ground Penetrating Radar (GPR). In addition to discovering a large number of new subsurface pit features, the survey also led to the identification of a shallowly buried channel-like feature with a northeast/southwest trajectory [26]. The results of this geophysical survey and excavations at 'Uqdat al-Bakrah will be published in greater detail elsewhere; this paper specifically assesses C-band and X-band SAR subsurface imaging.

GPR is commonly used for archaeological prospection and is analyzed in conjunction with SAR data as it can provide complementary information and/or be used to verify SAR interpretation. There are examples of this in the tropical environment of Angkor Wat [27], as well as in Egypt's Western Desert [13].

Located in dry aeolian and alluvial deposits, the shallowly buried features at 'Uqdat al-Bakrah provide a valuable opportunity for evaluating and clarifying the proficiency of SAR subsurface imaging. The identification of subsurface features at 'Uqdat al-Bakrah with SAR is also valuable in revealing details about human activities at the site, its paleo-environmental context, and is helpful in directing future research (remote sensing and excavation). In addition to clarifying the capabilities of C-band and X-band SAR, the discovery and mapping of a subsurface channel is highly significant as it shows water flow that may have supported vegetation. Many of the hundreds of pits at 'Uqdat al-Bakrah are thought to have been used for producing charcoal, which would have required large amounts of wood as fuel. If woody vegetation was available near the site (a possibility we are working to evaluate), this might help explain the presence of the site and hundreds of valuable metal objects in such a remote and otherwise hyper-arid desert location. These observations and resultant hypotheses to be tested by future archaeobotanical and archaeometallurgical research are also significant in considering the similarly remote and hyper-arid context of other recently discovered desert metal-working sites in southeast Arabia, including the impressive finds at Saruq al-Hadid, UAE [28].

2. Materials and Methods

2.1. Data

The data used for this research included SAR products, a Digital Elevation Model (DEM) product, and multispectral satellite imagery. Details of these data are outlined in Tables 1–3.

Table 1. Product Specifications of TanDEM-X bistatic acquisitions (German Aerospace Center (DLR)) used in the analysis. All scenes were acquired in the 300 MHz High Resolution spotlight mode, right looking, with a range and azimuth resolution of 0.6 m × 1.1 m, resulting in a processed pixel spacing of 0.87 m–1.14 m (dependent on the incidence angle).

Image Acquisition Date Range	Incidence Angle at Scene Centre (Degrees)	Number of Scenes	Sensor Mode (Polarization)	Orbit (Ascending or Descending)	Channel Visible in Image
15 April and 29 May 2017	38–52	6	HV and VH	A and D	No
9 June 25 September and 20 June 2017	38	3	HH (× 2), VV	A	Yes
7 May and 14 August 2017	39	5	VV	D	Yes
18 May and 25 August 2017	39	5	HH	D	Yes
9 June 2017	40	1	VV	D	Yes
3 June and 14 June 2017	52	2	HH and VV	A	No
13 May and 20 August 2017	53	5	VV	D	No
2 May and 9 August 2017	53	6	HH	D	No

Table 2. Product Specifications of RADARSAT-2 acquisitions (Canadian Space Agency (CSA)) used in the analysis. All scenes were acquired in the Ultrafine mode, right looking, with a range and azimuth resolution of 1.3 m × 2.1 m, resulting in a processed pixel spacing of 2.1 m–2.95 m (dependent on the incidence angle).

Image Acquisition Date(s)	Incidence Angle at Scene Centre (Degrees)	Number of Scenes	Sensor Mode (Polarization)	Orbit (Ascending or Descending)	Channel Visible in Image
17 July 2017	39	1	HH	A	No
2 July 2017	33	1	HH	D	Yes
30 October 2017	33	1	VV	D	Yes
14 April and 8 May 2017	27	2	HH	D	No
26 April 2017	27	1	HH	A	No
16 November 2017	27	1	VV	D	Yes

Table 3. Product Specifications of Digital Elevation Model (DEM) product and optical imagery used in the analysis.

Sensor	Acquisition Date	Resolution
SRTM	February 2000	30 m
Worldview-3	22 April 2016	1.2 m multispectral and 0.3 m panchromatic

2.2. Processing

Both the TanDEM-X bistatic products and RADARSAT-2 products were provided as Single Look Complex data, in which the product is minimally processed to maintain the complex information required for specific types of processing as well as the optimum resolution [29]. The difference for the TanDEM-X products was their bistatic acquisition from the TanDEM-X and TerraSAR-X satellites orbiting in tandem and acquiring image pairs. This meant that the Coregistered Single look Slant Range Complex (CoSSC) data were already processed so that the image pairs could be coregistered and did not require further calibration [30]. This radiometric correction step is required in order to interpret the data quantitatively (for comparing against other SAR images) as the calibration ensures that the pixel values correctly represent radar backscatter of the scene [29].

The processing of the X-band and C-band data was undertaken with a two-pronged approach during which the single data products were processed differently than the multi-temporal products. This allowed for all products to be assessed individually but also took advantage of the multi-temporal data to be coregistered and stacked. Figure 2 details the processing chain used with the Sentinel 1 toolbox software (SNAP—European Space Agency (ESA) Sentinel Application Platform v6.0) to produce comparative and geocoded images.

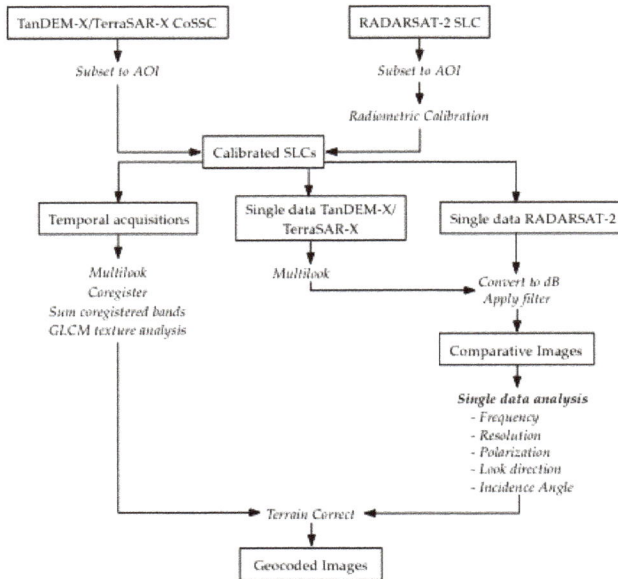

Figure 2. Flowchart of processing chain.

For the single data processing chain, all products were first subsetted for the area of interest (AOI). Calibration was applied to the RADARSAT-2 products so that the images were comparable. Then the TanDEM-X bistatic complex products were detected and multilooked. Since the pixel dimensions were already nearly square, this step converted the data from CoSSC products to real valued and interpretable intensity images by computing the modulus squared of the complex value. Both the TanDEM-X and RADARSAT-2 intensity images were then converted to decibel (dB), thereby reducing the dynamic range between the brightest and darkest pixels and making the images more interpretable. A low pass filter was applied to reduce speckle noise level, with the 3 × 3 pixel window size to preserve texture and enhance the subsurface channel, thus better facilitating identification of subsurface features [17,29]. These processed images could then be assessed in terms of radar frequency, spatial resolution, polarization, look direction and incidence angle. Terrain correction was applied to geocode the images to the Universal Transverse Mercator (UTM) projection (Zone 40 North WGS1984) using the Shuttle Radar Topographic Mission (SRTM) [31] DEM version 3, at 1 arc second (30 m) resolution.

The multi-temporal products (with the same acquisition parameters) were also subsetted for the area of interest and multilooked to produce detected intensity images. The sets of images (HH and VV) were then coregistered into two stacks. For the VV images, the 1 July 2017 scene was used as the master and the remaining four bistatic pairs were resampled to the master using the cubic convolution method. For the HH product, the 3 August 2017 scene was the master with two other pairs as slaves. The bands in each stack were summed to reduce image speckle and improve the signal-to-noise ratio thus enhancing subtle features [16,17,20]. The Gray Level Co-occurrence Matrix (GLCM) texture analysis (with a 5 × 5 pixel window, utilizing all angles, for 32 quantization levels and with a probabilistic quantizer) was then applied to the summed images. This analysis measures the pattern of intensity variations in an image based on the probability of occurrence of two gray levels at a given distance in specific direction(s) [29,32]. These measurements are then categorized into contrast, orderliness and statistics groups [29]. As with the single data images, the stacks were terrain corrected in the same manner.

The WorldView-3 (WV-3) product was not processed, as it was provided as a geocoded image, with georeferencing accuracy of 5 m [33]. In conjunction with field investigations, the high spatial resolution panchromatic band of WV-3 (0.3 m) was used to pansharpen other WV-3 bands and evaluate if any features identified in the GPR and SAR imagery were visible on the surface.

3. Results

3.1. SAR Analysis and Results

As shown in Figure 3a, no drainage channels are visible in the WV-3 image within the area surveyed by GPR in 2017. However, there are contemporary northeast/southwest drainage channels visible on the desert surface ~800 m to the northeast of the GPR survey area (Figure 3b).

Delineation of a northeast/southwest trending linear feature first identified by GPR was most evident in the TanDEM-X bistatic image multi-channel stacks (Figure 4), although it is also detectable in many (Figure 5), but not all (Figure 6) of the single data TanDEM-X bistatic images as well as some of the RADARSAT-2 images (Figure 7).

This linear feature is very similar in appearance to the drainage channels occasionally visible on the surface in areas surrounding 'Uqdat al-Bakrah (Figure 3b). However, during repeated visits to the site over multiple years there were no discernable differences in color, texture, or surface topography that would indicate a subsurface linear feature at 'Uqdat al-Bakrah in this location (Figure 3a). Due to its sinewy appearance and backscatter properties, this feature was interpreted as a natural subsurface paleo-channel, which was later confirmed by excavation.

The channel is visible in all the co-polarized TanDEM-X bistatic images that have an incidence angle of 30 to 40 degrees, across different linear polarizations and look directions (Figures 4 and 5 and Table 1). In contrast, the co-polarized images with incidence angles of 52 or 53 degrees (Figure 6a) changed the backscatter behavior between the channel and its surroundings to such a degree that the feature could not be distinguished. These images were similar in appearance to the VH and HV images

(Figure 6b) with their high speckle, suggesting a comparable low signal-to-noise ratio, which provides poor imaging for archaeological prospection [8,10,17].

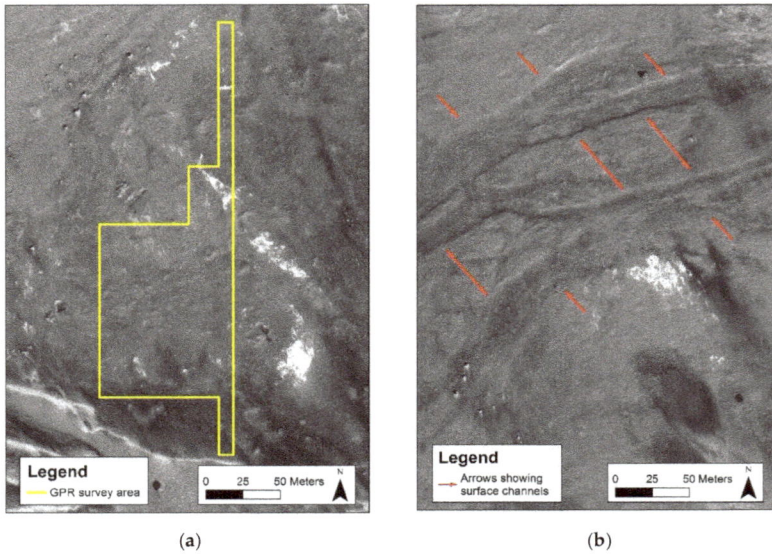

(a)   (b)

Figure 3. WorldView 3 panchromatic images. (**a**) No channels are visible within the GPR survey area. (**b**) Surface channels visible ~800 m to the northeast of the survey area.

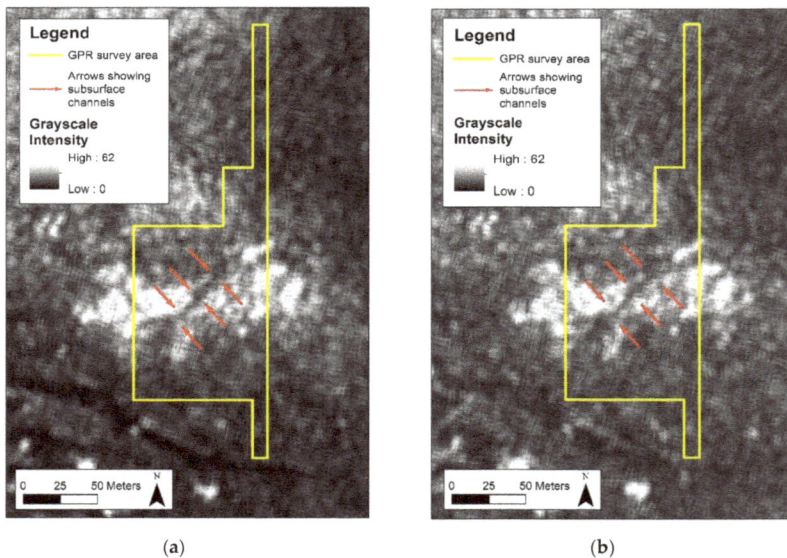

(a)   (b)

Figure 4. TanDEM-X bistatic image coregistered stacks (grayscale intensity images processed with Gray Level Co-occurrence Matrix ((GLCM)) variance texture analysis), with black representing low intensity values and white representing high intensity values. (**a**) Summed stack of 10 VV images. (**b**) Summed stack of 6 HH images.

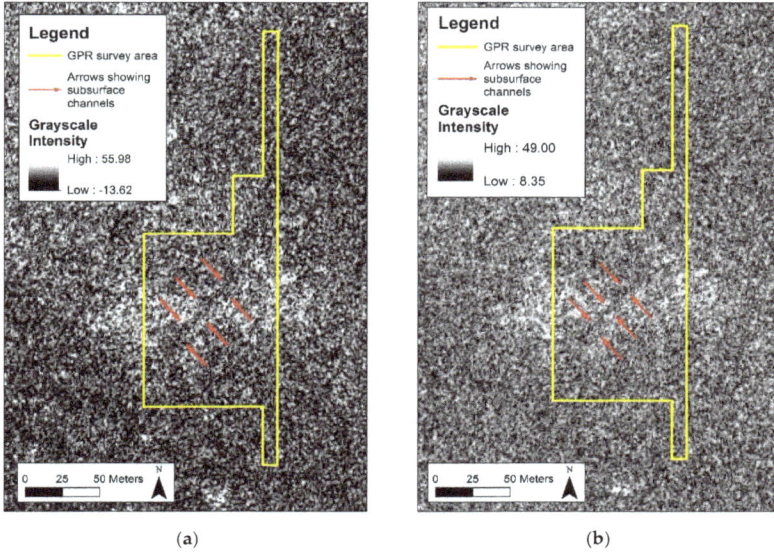

Figure 5. TanDEM-X bistatic grayscale intensity images (dB) with black representing low intensity values and white representing high intensity values. (**a**) 23 July 2017 VV image. (**b**) 25 August 2017 HH image.

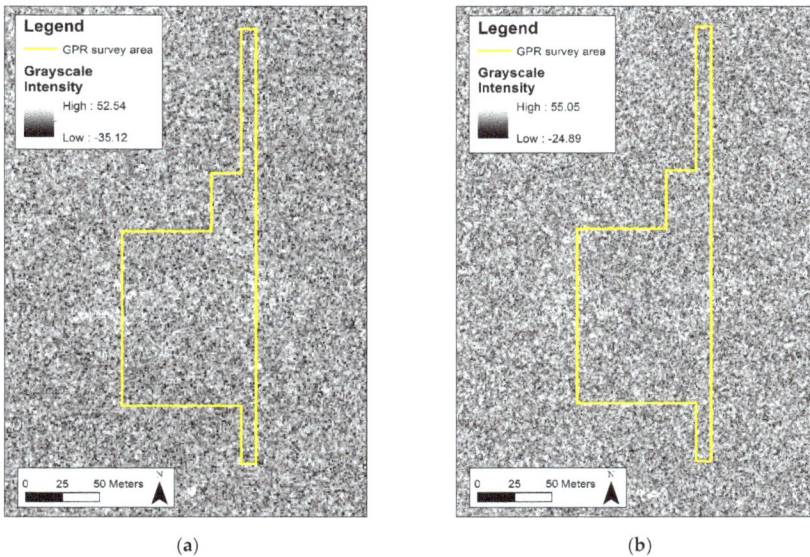

Figure 6. TanDEM-X bistatic grayscale intensity images (dB) with black representing low intensity values and white representing high intensity values. (**a**) 13 May 2017 VV image. (**b**) 15 April 2017 HV image.

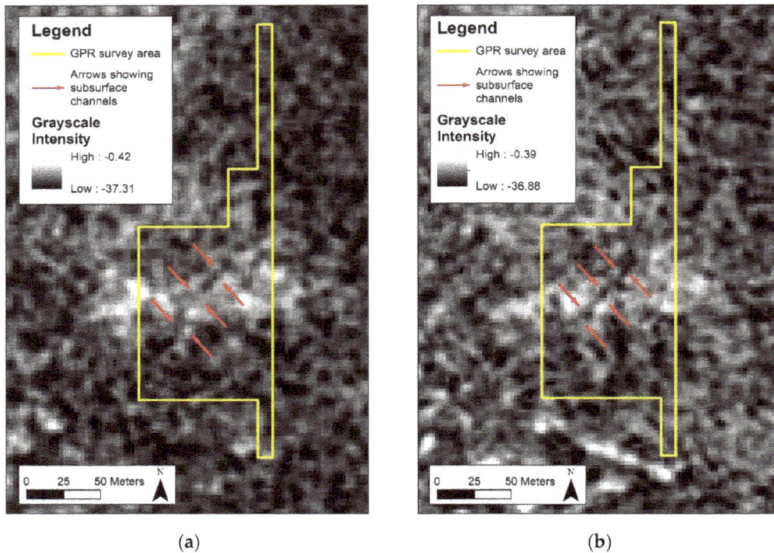

Figure 7. RADARSAT-2 processed grayscale intensity images (db) (low pass 3 × 3 speckle filter applied) with black representing low intensity values and white representing high intensity values. (a) 30 October 2017 VV image. (b) 2 July 2017 HH image.

While the paleo-channel is visible in the individual TanDEM-X bistatic processed images (Figure 5), it becomes more discernable with the GLCM mean variance texture analyses on the stacked images (Figure 4) due to the improved signal-to-noise ratio achieved with the coregistration and summing of a temporal series [8,20].

Although not as clearly delineated, this channel is also visible in three of the seven analyzed RADARSAT-2 images. It is best imaged in the HH and VV descending images at a 33-degree incidence angle (Figure 7) but is also visible in the VV polarized image with the 27-degree incidence angle. It was not discernable in the HH polarization images with the 27-degree or 39-degree incidence angle as the lower signal-to-noise ratio in these images obscured any identification of this subsurface feature. In contrast to the TanDEM-X bistatic images, the backscatter behavior that allows identification of the channel is limited to smaller incidence angles (27 to 33 degrees) with the VV polarization also affecting identification. Due to the slightly coarser resolution, the paleo-channel is better displayed with the low pass 3 × 3 pixel window filter rather than the GLCM texture analysis.

This analysis demonstrates that the identification of this subsurface channel in both the TerraSAR-X bistatic and RADARSAT-2 images is highly dependent on a low radar incidence angle. However, despite the positive identification of this subsurface channel, it is unclear what exactly is responsible for the changed scattering mechanism: remnant moisture in the stratigraphy, differences in the geometric size of the pebbles in the channel base relative to the radar wavelengths, or other chemical/physical properties of the soils in the stratigraphy that provide a contrast against the surrounding medium. In contrast, the loss of sensitivity to the subsurface feature in the higher and lower incidence angle images is likely a result of a decreasing signal-to-noise ratio (whether from wave attenuation [34] or increased surface roughness due to the change in viewing geometry [10,12,17]), which does not allow differentiation of the feature from its surroundings [8,10,11,17].

Although successful in identifying the subsurface channel, neither dataset could identify the pits at the site, likely due to their small size (~0.8 m–3 m), relative to either SAR mode resolution. While some pits have a pebble base or lining, many appear to be degraded, leaving an insubstantial base and charcoal layer, thus providing only a subtle contrast to the background medium of alluvial, aeolian and calcrete sands.

3.2. Ground Verification

The subsurface linear feature described above was first identified in GPR data (Figures 8 and 9) collected at the site of 'Uqdat al-Bakrah. This geophysical survey was undertaken in January 2017, during which 620 GPR profiles were acquired with an average spacing of 0.50 m. A GSSI SIR-3000 GPR system (Geophysical Survey Systems, Nashua, NH, USA) was used with a 400 MHz antenna. Confirmatory identification of the feature in SAR prompted heightened scrutiny of the GPR data, which were processed using GPR Slice (version 7.0, Geophysical Archaeometry Laboratory Inc., Woodland Hills, CA, USA). Velocity analysis for the site revealed an average relative dielectric permittivity of 4, which converts to a depth of approximately 0.75 m/ns.

Due to the nature of the GPR processing in north/south transects, the subsurface channel is displayed as approximately 8 m wide in the radargram profile as it is not perpendicular to the channel like the excavated trench. Additionally, the depth is slightly shallower (approximately 0.6 m) in Figure 9. Of the ten radargrams produced along this profile the subsurface channel depth varies from 0.6 m to 0.7 m.

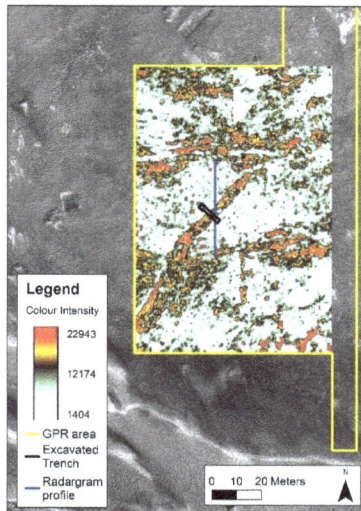

Figure 8. Ground Penetrating Radar (GPR) time slice color intensity image (with red representing high intensity and white representing low intensity) at 6.2–12.2 ns/44.5–89.1 cm depth with location of excavated trench and radargram profile.

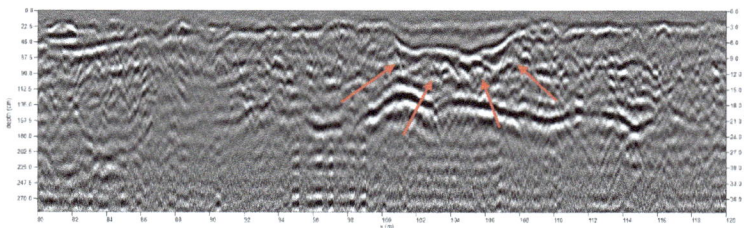

Figure 9. GPR radargram showing the subsurface channel in vertical profile.

Excavations conducted by the ArWHO Project in January 2018 included a trench dug perpendicular to the linear feature identified in GPR and SAR, confirming its interpretation as a natural subsurface paleo-channel. The channel cuts into a hardened unit of concreted pebbles and was covered in deposits of calcrete, compact and loose windblown sand, and cut-fill sedimentary units over a loose pebble-layer bed. The depth of the channel is approximately 0.7 m below ground surface (Figure 10).

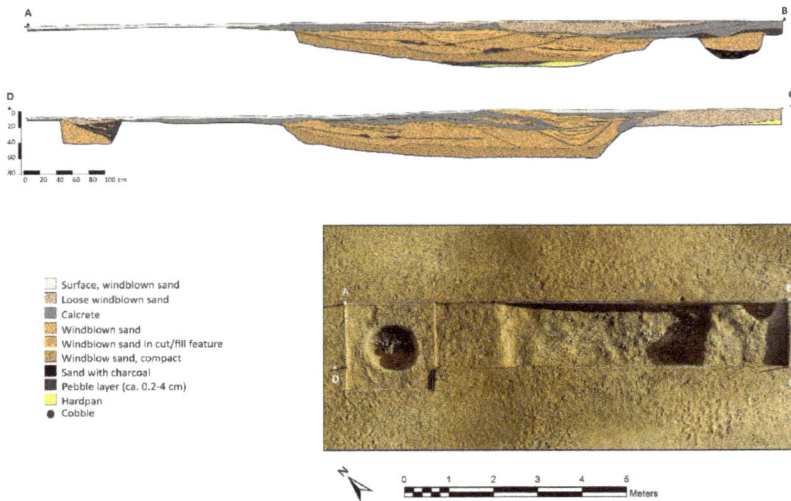

Figure 10. Illustrated profile and plan view photo of trench dug perpendicular to channel with adjacent pit. Base of channel approximately 4–5 m wide in center of trench (deepest part).

Subsurface imaging at 'Uqdat al-Bakrah indicates that both X-band and C-band microwaves are able to identify this channel, the base of which has been measured to a depth of 0.6 m–0.7 m below ground surface, as validated through GPR survey and excavation. In this case, the lower frequency wavelength provided the best subsurface image. The look direction does not seem to affect the interpretability of the subsurface feature, likely because the channel is a sinewy shaped feature with indistinct edges rather than a solid feature that would create a strong profile from the sensor. Although the VV polarization displays a slightly clearer image, the HH polarization is also adequate for imaging this channel. In addition to the polarization, the incidence angle seems to be the deciding factor for imaging the subsurface channel. For both X-band and C-band, the feature was visible in images with incidence angles between 33 and 40 degrees (except for the RADARSAT-2 VV image at 27 degrees).

4. Discussion

Penetration depth of X-band microwaves in arid environments has not been extensively studied or verified with quantitative fieldwork. The foundational work on subsurface penetration focused on the Mojave Desert with the SEASAT sensor and the Sahara with the SIR-A sensor, both of which provided measured L-band penetration depths of up to 2 m in arid environments [14,35]. Ongoing study in the Sahara with the SIR-C/X sensor further substantiated Schaber's [15] calculated imaging depths of 0.4–2 m for L-band, 0.1–0.5 m for C-band and 0.05–0.3 m for X-band, but through comparative analysis only, with later investigations in this region using GPR data to confirm similar imaging depths for both C-band and L-band [36].

Based on this foundational work, identifying larger subsurface features with SAR data has become relatively common in arid environments. However, verification for the depths of penetration has often only been explained comparatively (versus other SAR sensor imaging penetration depths or in comparison to optical imagery) rather than empirically measured [18,37–40]. This lack of verification is especially evident regarding shorter wavelengths and in archaeological contexts. One exception to this would be a recent investigation at the Roman fortress Qreiye in Syria where the authors claimed an X-band penetration depth of ca. 25 cm [20,21]. Unfortunately, other recent C-band archaeological investigations have not been verified due to political tensions in subject regions [19], lack of confirmatory fieldwork [17] or lack of success in identifying subsurface features due to the limits in ground resolution of the available sensor [41]. Hence, while the depths of penetration into desert sands have been calculated for different wavelengths, empirical testing of these depths is limited, especially for shorter wavelengths.

The discovery and verification of a subsurface paleochannel at 'Uqdat al-Bakrah is significant as it demonstrates the ability of shorter wavelengths for subsurface imaging in arid environments. However, although the depth of the channel has been measured at 0.6 to 0.7 m (in the GPR and the excavation), it is unclear whether the microwaves are penetrating to this specific depth. In attempting to determine the subsurface interface that will help us understand the depth of microwave penetration there are a few possibilities that require further investigation and will be addressed in future work. Surface/subsurface moisture and dielectric permittivity could be affecting the penetration depth and will be measured during upcoming field seasons. The relationship between this channel or other potential subsurface features with the ubiquitous calcrete soils at the site will be further considered as this type of soil is known to have properties that affect microwave backscattering [14,15]. The effect of the incidence angle from refraction of the microwave into the soil will also be considered, as this factor may have enhanced the subsurface backscatter [42].

Despite the continued research required in order to understand how exactly the X-band and C-band microwaves are interacting with this subsurface feature, it is still clear that these sensors can be useful for subsurface imaging in archaeological applications of arid environments. This work also contributes to the lack of investigation regarding microwave penetration of shorter wavelengths.

5. Conclusions

Our results show that X-band and C-band data are suitable for subsurface archaeological mapping of small hydrological features in arid contexts. While the subsurface channel is visible in the TanDEM-X bistatic individual images, the sum of these images increased the signal-to-noise ratio and allowed a better representation of the area [20]. Subsequently applying the GLCM texture analysis further reduced the speckle and better articulated the channel. Single data images (both TanDEM-X bistatic and RADARSAT-2) display the channel best with a low pass 3 × 3 pixel window filter to reduce the speckle.

The identification of a paleo-channel at the Iron Age site of Uqdat al-Bakrah is integral to the understanding of water resources in arid environments of the Arabian Peninsula. Water availability, including small paleo-channels, were crucial to past human activity and are therefore important targets of archaeological prospection. The assessment of data with varying acquisition parameters has provided informative results, with VV polarization and incidence angles of 30 to 40 degrees being the most successful for subsurface imaging of this channel. Ideally, the successful results of this investigation will be replicated in similar environments, providing archaeologists with more useful prospection tools.

Further work on this site will include the use of TerraSAR-X data; the staring spotlight mode offered by this data is the highest resolution satellite SAR data available. This imagery has been used successfully in archaeological applications for detecting remains of historical land-use on intertidal flats on the German North Sea Coast [43] as well as monitoring heritage looting over time at Apamea in western Syria [44]. Our work will expand on this repertoire of case studies with subsurface prospection at 'Uqdat al-Bakrah. We expect that a stack of these products will improve the signal-to-noise ratio and provide a higher quality image [8,16,20] that will allow further subsurface imaging of features at the site. At this stage it is difficult to trace the path of the paleo-channel, but an improved image may support a more precise delineation. In addition, it is a primary goal of this further work to identify the small pits or other possible features. Despite the fact that many of these pits are degraded, the staring spotlight mode may be sensitive enough to reveal changes in the backscatter behavior that will differentiate some of the pits from their surroundings if they have a solid pebble base and/or walls or are spatially clustered. Multi-polarized products also potentially offer additional subsurface information if their resolution is fine enough for the scale of the features at this site. Ongoing excavations will continue to be integral to interpreting GPR and SAR results.

Author Contributions: F.W. conceptualized the research and methodology, carried out all the SAR data processing and prepared the original draft. Technical editing was provided by M.J.H., A.B., S.H., J.T., J.O.S. and K.M.S., M.J.H., S.N., J.W.L., I.A.D. and T.C. contributed archaeological expertise. Field validation was conducted by F.W., M.J.H., S.N., J.W.L., K.M.S. and J.O.S. Supervision was provided by M.J.H., J.T. and S.H.

Funding: The SAR data for this research was granted by the German Aerospace Center (DLR) Science Program, (Proposal ID: Other7038) and the SOAR-E (Science and Operational Applications Research—Education Initiative) of the Canada Space Agency (Project #5410). RADARSAT-2 Data and Products © MacDonald Dettwiler and Associates Ltd. (2017)—All Rights Reserved. RADARSAT is an official trademark of the Canadian Space Agency. Funding for fieldwork and analysis included a NASA ROSES (Research Opportunities in Space and Earth Sciences) Grant (#NNX13AO48G), a Johns Hopkins University Catalyst Grant, a Space@Hopkins Grant, an Australian Research Council Discovery Early Career Researcher Award (Project ID: DE180101288), and a grant from the University of Arkansas, Centre for Advanced Spatial Technologies, Spatial Archaeometry Research Collaborations (CAST/SPARC) Program.

Acknowledgments: We are very grateful to the Sultanate of Oman, Ministry of Heritage and Culture for permission and collaborative support for our research, in particular His Excellency Salim M. Almahruqi, Sultan Al-Bakri, Khamis Al-Asmi, Mohammed Al-Waili, and Suleiman Al-Jabri deserve special thanks for their professionalism, support, and collegiality. Our investigations also rely on the gracious hospitality of the general public in Oman, including Shafi and Mutaab Al-Shukri who have long contributed crucial logistical assistance to our team. CSA is gratefully acknowledged for providing the RADARSAT-2 data as is the German Space Agency (DLR) for providing the TanDEM-X bistatic products.

Conflicts of Interest: The authors declare no conflicts of interest.

References

1. Adams, R.E.W. Swamps, canals, and the locations of ancient Maya cities. *Antiquity* **1980**, *54*, 10. [CrossRef]
2. Adams, R.E.W.; Brown, W.E.J.; Culbert, P.T. Radar mapping, archaeology and ancient maya land use. *Science* **1981**, *213*, 6. [CrossRef] [PubMed]
3. Adams, R.E.W.; Culbert, P.T.; Brown, W.E.J. News and short contributions: Rebuttal to pope and dahlin. *J. Field Archaeol.* **1990**, *17*, 4. [CrossRef]
4. Pope, K.O.; Dahlin, B.H. Ancient maya wetland agriculture: New insights from ecological and remote sensing research. *J. Field Archaeol.* **1989**, *16*, 19.
5. Elachi, C. Seeing under the sahara: Spaceborne imaging radar. *Eng. Sci.* **1983**, *47*, 4–8.
6. McCauley, J.F.; Schaber, G.G.; Breed, C.S.; Grolier, M.J.; Haynes, C.V.; Issawi, B.; Elachi, C.; Blom, R.G. Subsurface valleys and geoarchaeology of the eastern sahara revealed by shuttle radar. *Science* **1982**, *218*, 1004–1020. [CrossRef] [PubMed]
7. McCauley, J.F.; Breed, C.S.; Schaber, G.G.; McHugh, P.W.; Issawi, B.; Haynes, C.V.; Grolier, M.J.; Kilani, A.E. Paleodrainages of the Eastern Sahara—The radar rivers revisited (SIR-A/B implications for a mid-tertiary trans—African drainage system). *IEEE Trans. Geosci. Remote Sens.* **1986**, *24*, 24.
8. Stewart, C.; Montanaro, R.; Sala, M.; Riccardi, P. Feature extraction in the North Sinai Desert using spaceborne synthetic aperture radar: Potential archaeological applications. *Remote Sens.* **2016**, *8*, 825. [CrossRef]
9. Tapete, D.; Cigna, F. Trends and perspectives of space-borne SAR remote sensing for archaeological landscape and cultural heritage applications. *J. Archaeol. Sci.* **2017**, *14*, 11. [CrossRef]

10. Chen, F.; Lasaponara, R.; Masini, N. An overview of satellite synthetic aperture radar remote sensing in archaeology: From site detection to monitoring. *J. Cult. Herit.* **2017**, *21*, 7. [CrossRef]

11. Lasaponara, R.; Masini, N. Satellite synthetic aperture radar in archaeology and cultural landscape: An overview. *Archaeol. Prospect.* **2013**, *20*, 71–78. [CrossRef]

12. Elachi, C.; Granger, J. Spaceborne imaging radars probein depth. *IEEE Spectr.* **1982**, *19*, 24–29. [CrossRef]

13. Gaber, A.; Koch, M.; Griesh, M.H.; Sato, M.; El-Baz, F. Near-surface imaging of a buried foundation in the Western Desert, Egypt, using space-borne and ground penetrating radar. *J. Archaeol. Sci.* **2013**, *40*, 1946–1955. [CrossRef]

14. Schaber, G.G.; McCauley, J.F.; Breed, C.S.; Olhoeft, G.R. Shuttle imaging radar: Physical controls on signal penetration and subsurface scattering in the Eastern Sahara. *IEEE Trans. Geosci. Remote Sens.* **1986**, *24*, 603–623. [CrossRef]

15. Schaber, G.G.; McCauley, J.F.; Breed, C.S. The use of multifrequency and polarimetric SIR-C/X-SAR Data in geologicstudies of Bir Safsaf, Egypt. *Remote Sens. Environ.* **1997**, *59*, 337–363. [CrossRef]

16. Stewart, C.; Lasaponara, R.; Schiavon, G. ALOS PALSAR nalysis of the Archaeological Site of Pelusium. *Archaeol. Prospect.* **2013**, *20*, 109–116. [CrossRef]

17. Chen, F.; Masini, N.; Liu, J.; You, J.; Lasaponara, R. Multi-frequency satellite radar imaging of cultural heritage: the case studies of the Yumen Frontier Pass and Niya ruins in the Western Regions of the Silk Road Corridor. *Int. J. Digit. Earth* **2016**, *9*, 19. [CrossRef]

18. Paillou, P. Mapping paleohydrography in deserts: Contributions from space-borneimaging radar. *Water* **2017**, *9*, 194. [CrossRef]

19. Patruno, J.; Dore, N.; Crespi, M.; Pottier, E. Polarimetric multifrequency and multi-incidence SAR sensors analyis for archaeological purposes. *Archaeol. Prospect.* **2013**, *20*, 89–96. [CrossRef]

20. Linck, R.; Busche, T.; Buckreuss, S.; Fassbinder, J.W.E.; Seren, S. Possibilities of archaeological prospection by high-resolution X-band Satellite Radar—A case study from Syria. *Archaeol. Prospect.* **2013**, *20*, 97–108. [CrossRef]

21. Linck, R.; Busche, T.; Buckreuss, S. Possibilities of TerraSAR-X Data for the Prospection of Archaeological Sites by SAR. In Proceedings of the 5th TerraSAR-X/4th TanDEM-X Science Team Meeting, Oberpfaffenhofen, Germany, 10–14 June 2013; German Aerospace Centre (DLR): Oberpfaffenhofen, Germany, 2013; p. 4.

22. Tapete, D. Remote sensing and geosciences for archaeology. *Geosciences* **2018**, *8*, 41. [CrossRef]

23. Kwarteng, A.Y.; Dorvlo, A.S.; Kumar, G.T.V. Analysis of a 27-year rainfall data (1977–2003) in the Sultanate of Oman. *Inter. J. Climatol.* **2009**, *29*, 13. [CrossRef]

24. Yule, P.A.; Gernez, G. *Early Iron Age Metal-Working Workshop in the Empty Quarter, al-Zahira Province, Sultanate of Oman*; Habelt-Verlag: Bonn, Germany, 2018.

25. Genchi, F.; Giardino, C. The field work. In *Early Iron Age Metal-Working Workshop in the Empty Quarter, al-Zahira Province, Sultanate of Oman*; Yule, P.A., Gernez, G., Eds.; Habelt-Verlag: Bonn, Germany, 2018; pp. 11–31.

26. Harrower, M.J.; Dumitru, I.A.; Wiig, F.; David-Cuny, H.; Taylor, S.P.; Sivitskis, A.J.; Simon, K.M.; Sturm, J.O.; Mazzariello, J.C.; Lehner, J.W.; et al. *Archaeological Water Histories of Oman (ArWHO) Project*; Field Report for the Ministry of Heritage and Culture; Sultanate of Oman: Muscat, Oman, 2017; p. 151.

27. Sonnemann, T.F. Spatial configurations of water management at an early angkorian capital—combining GPR and TerraSAR-X data complement an archaeological map. *Archaeol. Prospect.* **2015**, *22*, 11. [CrossRef]

28. Weeks, L.; Cable, C.; Franke, K.; Newton, C.; Karacic, S.; Roberts, J.; Stepanov, I.; David-Cuny, H.; Price, D.; Bukhash, R.M.; et al. Recent archaeological research at Saruq al-Hadid, Dubai, UAE. *Arab. Archaeol. Epigr.* **2017**, *28*, 30. [CrossRef]

29. ESA. *SNAP_ESA Sentinel Application Platform Help*; Eurpoean Space Agency: Paris, France, 2018.

30. Institute, R.S.T. *TanDEM-X Payload Ground Segment: CoSSC Generation and Interferometric Considerations*; German Aerospace Centre (DLR): Cologne, German, 2012; p. 31.

31. Farr, T.G.; Rosen, P.A.; Caro, E.; Crippen, R.; Duren, R.; Hensley, S.; Kobrick, M.; Paller, M.; Rodriguez, E.; Roth, L.E.; et al. The shuttle radar topography mission. *Rev. Geophys.* **2007**, *45*, 33. [CrossRef]

32. Haralick, R.M.; Shanmugam, K.; Its'Hak, D. Textural features for image classification. *IEEE Trans. Syst. Man Cybern.* **1973**, *SMC-3*, 21. [CrossRef]

33. Digital Globe. *Accuracy of WorldView Products*; Digital Globe: Westminster, CO, USA, 2016; p. 11.

34. O'Grady, D.; LeBlanc, M.; Gillieson, D. Relationship of local incidence angle with satellite radar backscatter for different surface conditions. *Inter. J. Appl. Earth Obs. Geoinf.* **2013**, *24*, 12. [CrossRef]

35. Blom, R.G.; Cripper, R.E.; Elachi, C. Detection of subsurface features in SEASAT radar images of Means Valley, Mojave Desert, California. *Geology* **1984**, *12*, 346–349. [CrossRef]

36. Paillou, P.; Grandjean, G.; Baghdadi, N.; Heggy, E.; August-Bernex, T.; Achache, J. Subsurface imaging in South-Central Egypt using low-frequency radar: Bir safsaf revisited. *IEEE Trans. Geosci. Remote Sens.* **2003**, *41*, 1672–1684. [CrossRef]

37. Ghoneim, E.; El-Baz, F. The application of radar topographic data to mapping of a mega-paleodrainage in the Eastern Sahara. *J. Arid Environ.* **2007**, *69*, 658–675. [CrossRef]

38. Paillou, P.; Schuster, M.; Tooth, S.; Farr, T.; Rosenqvist, A.; Lopez, S.; Malezieux, J.-M. Mapping of a major paleodrainage system in eastern Libya using oribital imaging radar: The Kufrah River. *Earth Planetary Sci. Lett.* **2009**, *277*, 327–333. [CrossRef]

39. Robinson, C.A.; El-Baz, F.; Al-Saud, T.S.M.; Jeon, S.B. Use of radar data to delineate paleodrainage leading to the Kufra Oasis in the eastern Sahara. *J. Afr. Earth Sci.* **2006**, *44*, 229–240. [CrossRef]

40. Dabbagh, A.E.; Al-Hinai, K.G.; Asif Khan, M. Detection of sand-covered geologic features in the Arabian peninsula using SIR-C/X-SAR data. *Remote Sens. Environ.* **1997**, *59*, 375–382. [CrossRef]

41. Tapete, D.; Cigna, F.; Masini, N.; Lasaponara, R. Prospection and monitoring of the archaeological heritage of Nasca, Peru, with ENVISAT ASAR. *Archaeol. Prospect.* **2013**, *20*, 15. [CrossRef]

42. Elachi, C.; Roth, L.E.; Schaber, G.G. Spaceborne radar subsurface imaging in hyperarid regions. *IEEE Transact. Geosci. Remote Sens.* **1984**, *22*, 383–388. [CrossRef]

43. Gade, M.; Kohlus, J.; Kost, C. SAR imaging of archaeological sites on intertidal flats in the German Wadden Sea. *Geosciences* **2017**, *7*, 105. [CrossRef]

44. Tapete, D.; Cigna, F.; Donoghue, D.N.M. Looting marks in space-borne SAR imagery: Measuring rates of archaeological looting in Apamea (Syria) with TerraSAR-X Staring Spotlight. *Remote Sens. Environ.* **2016**, *178*, 17. [CrossRef]

geosciences

MDPI

Article

Remote Sensing Analyses on Sentinel-2 Images: Looking for Roman Roads in Srem Region (Serbia)

Sara Zanni [1],* and Alessandro De Rosa [2]

[1] Domaine Universitaire, Maison de l'Archéologie, Institut Ausonius (UMR 5607),
Université Bordeaux Montaigne, 8 Esplanade des Antilles, 33600 Pessac, France
[2] Independent Researcher, via XXV Aprile 16, 87053 Celico CS, Italy; aderosa77@gmail.com
* Correspondence: zanni.sara@gmail.com

Received: 25 November 2018; Accepted: 28 December 2018; Published: 5 January 2019

Abstract: The present research is part of the project "From Aquileia to Singidunum: reconstructing the paths of the Roman travelers—RecRoad", developed at the Université Bordeaux Montaigne, thanks to a Marie Skłodowska-Curie fellowship. One of the goals of the project was to detect and reconstruct the Roman viability between the Roman cities of *Aquileia* (Aquileia, Italy) and *Singidunum* (Belgrade, Serbia), using different sources and methods, one of which is satellite remote sensing. The research project analyzed and combined several data, including images produced by the Sentinel-2 mission, funded by the European Commission Earth Observation Programme Copernicus, in which satellites were launched between 2015 and 2017. These images are freely available for scientific and commercial purposes, and constitute a constantly updated gallery of the whole planet, with a revisit time of five days at the Equator. The technical specifications of the satellites' sensors are particularly suitable for archaeological mapping purposes, and their capacities in this field still need to be fully explored. The project provided a useful testbed for the use of Sentinel-2 images in the archaeological field. The study compares traditional Vegetation Indices with experimental trials on Sentinel images applied to the Srem District in Serbia. The paper also compares the results obtained from the analysis of the Sentinel-2 images with WorldView-2 multispectral images. The obtained results were verified through an archaeological surface survey.

Keywords: remote sensing; satellite; Sentinel-2; surface survey; Roman archaeology

---

1. Introduction

This paper aims to present the research methodology developed within the "RecRoad— Reconstructing the Paths of the Roman Travelers from *Aquileia* to *Singidunum*" project, funded through a Marie Skłodowska-Curie Individual Fellowship at the Université Bordeaux Montaigne. The project, started in February 2016 and ended in January 2018, aimed to retrace, with the highest possible reliability level, the Roman itinerary between *Aquileia* (Italy) and *Singidunum* (Belgrade, Serbia), following the course of the Sava River.

Between the 2nd century BC and the 4th century AD, *Aquileia* was an important military base and the main port of the Northern Adriatic basin, particularly for its relationships with the people living in Hystria and in the basin of the Danube (Strab. V, 1, 8, 214 C). *Singidunum*, established at the confluence of the Sava and Danube Rivers, where Belgrade is now located, was an important city and one of the main military camps in the province of *Moesia Superior*. The Romans traced several itineraries to connect Northern Italy to the Danube area: the travelers could choose the one they preferred according to their personal needs. These routes are described in the itinerary sources, namely the *Itinerarium Antonini* [1] (pp. 1–85), [2,3], the *Itinerarium Burdigalense* [1] (pp. 86–190), [4,5] and the *Tabula Peutingeriana* [6–11].

According to these resources, two main routes led from *Aquileia* to *Singidunum*; both crossed the Alps at the *Ad Pirum* pass (Hrušica, Slovenia), in the Julian Alps, to reach *Emona* (Ljubljana). The road, as reported in the *Itinerarium Antonini* and *Itinerarium Burdigalense*, subsequently headed north-east towards *Celeia* (Celje) and *Poetovio* (Ptuj), where it started following the valley of the Drava River across Croatia, to reach its confluence with the Danube. On the other hand, the *Tabula Peutingeriana* shows another itinerary, passing by *Emona* and turning to south-east, in the direction of the Sava River, that is reached at *Siscia* (Sisak). Then, it follows the course of the Sava until it flows into the Danube in front of *Singidunum*. This latter is the itinerary that was mapped within the RecRoad Project: and it was the first to be traced by the consul P. Cornelius Lupus in BC 156, in its attempt to reach *Segestica* (the Celtic settlement nowadays covered by the modern Sisak), as Appian (*Illyr.* 22 and 135) and Polibius (fr. 64) have reported [12] (pp. 437–438). Octavian's armies took the same direction in BC 36–35, when he decided to conquer *Segestica* and to temporarily take control of the *Iapodes* [13] (pp. 29,30).

Notwithstanding the importance of this itinerary, its topographical layout and the location of its remains are partially unknown, and a precise mapping of its archaeological traces has never been accomplished in detail along its whole extension. Due to the length of the route (about 650 km), it was necessary to design a methodology to integrate different sources of information and different techniques, with a strong use of digital methods and the development of a GIS platform to manage the whole dataset. Satellite remote sensing techniques played a central role, enabling the scanning and analysis of very large areas, and to identify the buried remains of the Roman road and other nearby archaeological sites within a distance of about 2 km from the road itself. Among the data used, we decided to compare the results obtained from the images produced by the Sentinel-2 mission for the detection and identification of archaeological remains, to the analysis outcomes for other types of images. We then performed a final reliability assessment of the hypothetical reconstruction of the road mapping, through an archaeological surface surveys.

This paper focuses on the results obtained in the region of Srem, in Serbia, where today's road network has completely changed its layout in comparison to the Roman one, so that the Roman itinerary currently lays under the cultivated fields: this is the best condition to ensure good visibility conditions. Otherwise, in other regions involved in this project, the modern roads lay on top of the Roman ones, preventing an effective detection of the archaeological remains. More specifically, the paper is focused on the territories depending on the settlements of Šašinci, Voganj, Ruma, Kraljevci, Dobrinci, Donji Petrovci, Popinci, Golubinci, and Vojka (Figure 1). The analysis of multi-spectral Sentinel-2 images led to the identification of sixty crop-marks possibly related to the presence of buried archaeological remains in this area.

1.1. Geographic and Historical Framing

The Srem District is one of the seven administrative districts of the autonomous province of Vojvodina. Srem is the western part of the province and its name derives from the Roman city of *Sirmium*, that stood in the location of the modern city of Sremska Mitrovica. Vojvodina is bound by three main rivers: the Drava at the north, the Danube at the east, and the Sava at the south. It is a part of the Pannonian plain and is a very fertile region, where 70% of crops are cereals [14]. In the northern part of the Srem District, the Fruška Gora mountain is a part of the National Park that includes 35 orthodox monasteries. Before the Roman conquest at the end of the 1st century BC, the region was inhabited by Illyrians. The fortress of *Sirmium* was built beside the Sava River, and played an important role in the Great Illyrian Revolt in AD 6–9. When *Pannonia* was finally conquered, *Sirmium* became its economic capital, thanks to its strategic position. In AD 293, when Diocletian established the Tetrarchy, *Sirmium* became one of the four capitals of the Empire.

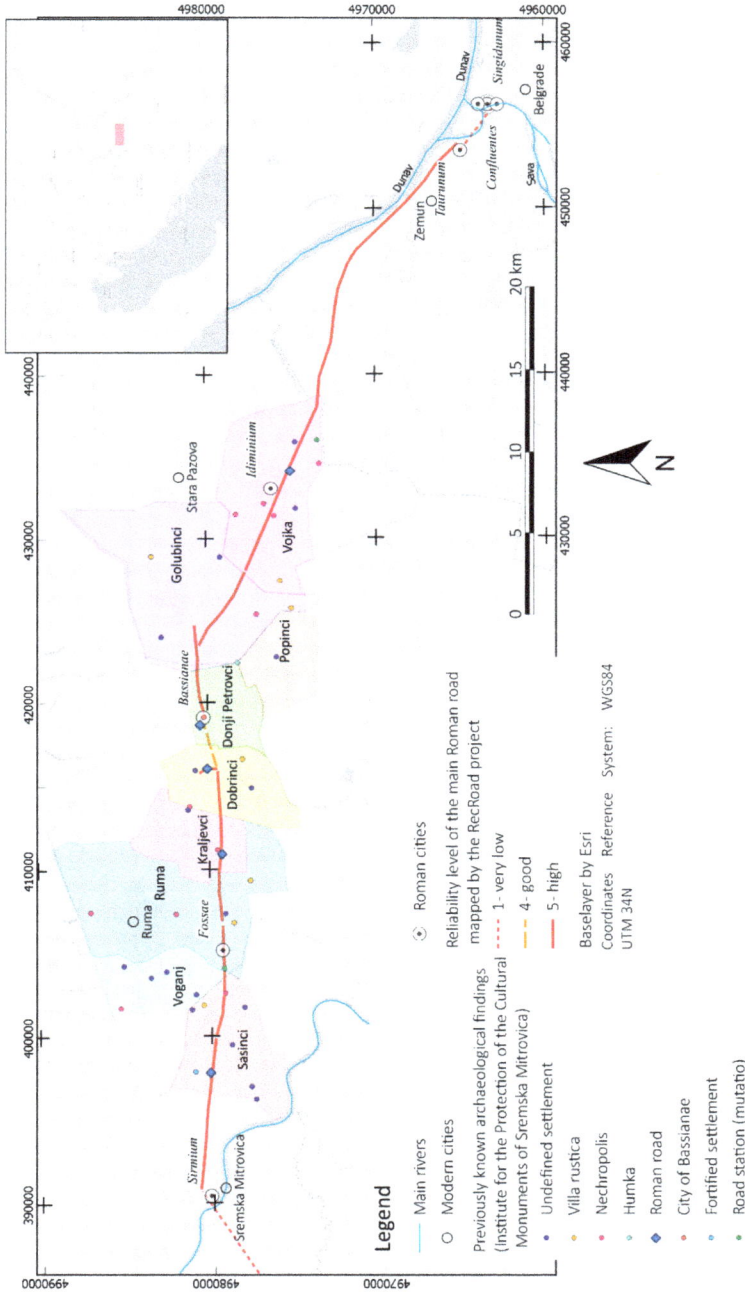

Figure 1. Map of the area of interest in the presented research. The five blue squares identify the remains of the Roman road already known before the beginning of the RecRoad project (courtesy of the Institute for the Protection of the Cultural Monuments of Sremska Mitrovica).

The Roman province of *Pannonia* had a great strategic relevance and commercial importance. Its north-eastern border was the River Danube, that became the eastern frontier of the Roman Empire, protected by an imposing number of military camps and fortresses connected by a necessary communication network. Among these communication lines, the itinerary from *Aquileia* to *Singidunum*, had a central role, linking Italy to the *limes*, following the course of the Sava River. The region is crossed by some of the major European waterways that were integrated within the land road network. The city of *Aquileia*, located in the Augustan *X Regio*, had a key role in spreading the Roman culture eastward, and counted among the Empire's main ports and crossroads [15,16]. The communication network of the province was primarily designed in response to the military and economical needs of Rome as the political capital of the Empire [17] (pp. 7–12).

Historical sources report the existence of two land routes linking *Aquileia* to the Danube River, and to the city of *Singidunum*. They followed the courses of the two main rivers flowing from the west towards the east: one ran along the Drava River, while the second along the Sava River. Both the *Itinerarium Antonini* (It. Ant. 128-6–132.1) and *Itinerarium Burdigalense* (It. Burd. 559.11–563.14) report the stages of the Drava itinerary, which passes by *Poetovio* (Ptuj, Slovenia) and *Celeia* (Celje, Slovenia). A third source, the *Tabula Peutingeriana*, provides information about the Sava itinerary. This true road map of the Roman Empire shows the main cities and towns crossed by the route between *Aquileia* and *Singidunum*, among which we can count *Neviodunum, Siscia, Marsonia, Sirmium*, and *Bassianae*, the main cities on the course of the Sava river [18] (pp. 150–152).

1.2. State of the Art

At the beginning of the RecRoad project, it was possible to map the already known rests of the Roman main road in the territories of the towns of Šašinci, Voganj, Ruma, Kraljevci, Dobrinci, Donji Petrovci, Popinci, Golubinci, and Vojka, under the jurisdiction of the Institute for the Protection of Cultural Monuments of Sremska Mitrovica. The analysis of the Institute's archive enabled the identification of five locations where the existence of the Roman road had been previously assessed during archaeological surface surveys. The extension of the area of interest of the RecRoad project in the Srem district, and the location of the precedent findings can be seen in Figure 1. The presence of the road was documented in the Šašinci, Kraljevci, Dobrinci, Donji Petrovci, and Vojka territories. Even if this information enabled a general knowledge of its direction, it was not possible to map the layout of the Roman road in full detail, especially concerning its path at the exit from the site of the Roman city *Bassianae* (located to the north of Donji Petrovci modern settlement) eastward.

In the 1960s, a survey of the eastern section of the road, from *Taurunum* (Klisina) to Vojka was performed by the Regional Museum of Zemun [19–21]. This research was closely followed, in 1969–1971, by the prospections of the Yugoslav–American "Sirmium Project" [22–24], that identified on the ground the road in Tovarnik, close to the border between Croatia and Serbia, to Zemun. It was documented that the road was trackable thanks to surface linear dispersions of pebbles of an approximately 15-m-wide surface, mostly even with the ground level and only in some places slightly elevated at 0.5–1 m from the surrounding ground level [23] (p. 261). The reported width of the Roman road's trace encouraged the use of Sentinel-2 images for the remote sensing analysis, since their spatial resolution can be sufficient to detect a mark 15-m-wide. Within the same project, a first excavation of the Roman road was carried out between August and September 1969 [23].

More recent scientific works and research projects focused on the road from *Aquileia* to the Danube River, with a specific focus on the segment from *Sirmium* to *Singidunum*. Some work published by Petar Milošević in 1988 [25] reported important information for the recognition of the road's segment linking *Sirmium* to *Bassianae*: the focus of the paper was on the topographical evidence of the two roads connecting *Sirmium* with *Mutatio Fossis*—which is the initial part of the road *Sirmium–Singidunum*—and with *Bononia*. The author discussed the previously affirmed theory according to which the Roman road followed the course of the Sava River down to its crossing with the Jarčina channel in Jarak. This interpretation had already been questioned by the finding of two milestones in the field of Crepovac, between Sremska Mitrovica and Šašinci. By the end of the 19th century, Ignat Jung—representative of the National Museum of Zagreb and teacher in Mitrovica—had remarked the correct direction of the Roman road, through a surface survey, but it was finally accepted in the literature only much later [22] (p. 102). Only in 1981 did a second archaeological excavation document the Roman road in Rumska Petlja [26] (p. 96), [27] (p. 187), with good results especially for the lower layers of the road's foundation, while the upper levels had been destroyed by the agricultural works.

In 2010, Hrvoje Gračanin made a broad analysis of the historical and itinerary Latin sources about the Roman road network in *Pannonia*, identifying the stations mentioned by the sources and the relative distances [28]. Two years later, a new survey campaign was performed by the Institute for the Protection of the Cultural Monuments of Sremska Mitrovica within the "Study for the protection of archaeological sites for the needs of the South Stream pipeline construction" [29]: this led to the identification of the traces of the Roman road in the area of Međice, south of Kraljevci, over a length of about 2 km. The trace, 10-m-wide, was marked by circular stones and pebbles [29] (pp. 39,40). The results of the survey were reported in the scientific deliverable of the "ARCHEST Project: Developing archaeological audiences along the Roman route *Aquileia–Emona–Sirmium–Viminacium*" (www.archest.eu).

In 2017, one of the most recent efforts in the analysis of the epigraphic and geographic knowledge about the Roman itinerary from *Aquileia* to *Singidunum* was undertaken by Florin Fodorean. He published two studies focused on the analysis of the epigraphic and geographic sources referring to the road passing by *Emona, Siscia, Sirmium,* and *Taurunum* [30], and analysis of the information about the road *Emona–Singidunum* in the *Tabula Peutingeriana, Itinerarium Antonini,* and *Itinerarium Burdigalense* [31].

It is clear that the most recent works regarding the Roman road from *Sirmium* to *Singidunum* are focused on two different trends: the first one is analysis of the information contained in the historical sources, while the other remains much more connected to the topographical research. Even if, in comparison with other regions interested by the RecRoad project, the Srem district presented a much detailed and established knowledge of the road path, there were still some questions that required an answer and a need for the precise mapping of the main road's traces, in order to better plan the monument's protection in case of future public works in the area.

First of all, it was necessary to understand whether the Roman road maintained a straight direction from Šašinci to Kraljevci, the two locations where it had been identified and documented at the west of *Bassianae*. The second problem that needed addressing was, of course, the exact behavior of the road between *Bassianae* and Vojka, where its presence had already been verified. Thanks to the collaboration with the Institute for the Protection of the Cultural Monuments of Sremska Mitrovica, it was possible to identify and map the whole extent of the Roman road from the outskirts of Sremska Mitrovica to the entrance of Zemun, through the analysis of satellite images and a focused archaeological surface survey.

2. Materials and Methods

The RecRoad project raised from its beginning several issues regarding the collection of the necessary geographical data to accomplish the foreseen analysis, and complete the mapping and the reconstruction of the Roman itinerary from *Aquileia* to *Singidunum*. The itinerary, extended on an overall distance of 650 km, crossed five different countries: Italy, Slovenia, Croatia, Bosnia Herzegovina, and Serbia. Providing an exhaustive documentation of aerial images has raised, on its own, several issues in terms of delivery time of the materials and of their availability to the public, together with the problem of the dispersion of the documentation in different institutes and countries with different policies for the use and publication of the data.

The beginning of the RecRoad project temporally followed only a few months from the launch of the Sentinel-2A satellite (23 June 2015), the first of the twin satellites making up the Sentinel-2 constellation. Sentinel-2B was launched on 7 March 2017. The images produced by the Sentinel-2 mission immediately attracted our attention, because of their free availability through the Copernicus

Open Access Hub (https://scihub.copernicus.eu/), and whole planet coverage. This was especially relevant for the project, due to the extension of the area of interest. The revisit time of the Sentinel-2 mission of five days enabled a continuous coverage over the different stages of the crops' growth. The two satellites carry a wide swath high-resolution multispectral imager with 13 spectral bands (see Table 1 for detailed spectral characteristics): four bands, corresponding to RGB and NIR, with a resolution of 10 m; six bands corresponding to Vegetation Red Edge (VRE) and SWIR, with a resolution of 20 m; and three bands with a resolution of 60 m (Coastal aerosol, Water Vapour and SWIR—Cirrus). The spectral characteristics of the imagers, resumed in the following table, had been assessed as especially suitable for the identification of crop-marks due to the presence of buried archaeological remains, before the launch of the satellites [32].

Table 1. Sentinel-2 bands with the corresponding central wavelengths and spatial resolution.

Sentinel-2 Bands	Central Wavelength (µm)	Resolution (m)
Band 1—Coastal aerosol	0.443	60
Band 2—Blue	0.490	10
Band 3—Green	0.560	10
Band 4—Red	0.665	10
Band 5—Vegetation Red Edge	0.705	20
Band 6—Vegetation Red Edge	0.740	20
Band 7—Vegetation Red Edge	0.783	20
Band 8—NIR	0.842	10
Band 8A—Vegetation Red Edge	0.865	20
Band 9—Water Vapour	0.945	60
Band 10—SWIR—Cirrus	1.375	60
Band 11—SWIR	1.610	20
Band 12—SWIR	2.190	20

Since the images produced by the Sentinel-2 satellites had only been recently released for public use, the research team decided to test their capabilities in the archaeological research field. This was the first testbed for the Sentinel-2 images, since the previous publication record only tried to assess the theoretical archaeological potential of the images based on their spectral characteristics released by the European Commission Earth Observation Programme Copernicus to the scientific community before the launch [32]. Other applications in the cultural heritage field have recently been experimented, with a special focus on the monitoring and protection of archaeological sites [33].

In the Srem District, the satellite remote sensing analysis focused on the area where, according to past research activities, the Roman road would most likely lie, within a strip 10-km-wide. Generally speaking, the research methodology included four steps:

1. Satellite remote sensing analysis, resulting in the identification of 60 crop-marks in the Srem District;
2. Examination of the data stored in the archive of the Institute for the Protection of the Cultural Monuments of Sremska Mitrovica, that led to the mapping of 50 archaeological sites previously identified during field research activities;
3. Comparison of the data obtained during Steps 1 and 2: 13 archaeological sites already filed in the archive corresponded with the anomalies identified on the satellite images;
4. Archaeological surface survey to assess the reliability of the data obtained in Step 1.

In some cases, the evidence documented during previous archaeological research could easily be integrated with the stretches of the Roman road visible on the satellite images. Consequently, the archaeological evidence already established increased the reliability level of the satellite remote sensing anomalies. The surface survey, organized in collaboration with the Institute for the Protection of the Cultural Monuments of Sremska Mitrovica, confirmed this hypothesis, allowing the collection of pottery fragments and other autoptic materials that testified to the exact path of the Roman road over the 70 km from *Sirmium* to the confluence of the Danube and the Sava Rivers, across the city of *Bassianae*.

2.1. Satellite Remote Sensing Analysis

2.1.1. Processing of Sentinel-2 Images

There are several techniques to map buried archaeological sites through the use of different typologies of satellite images, which have been broadly explained in many different publication that constitute a solid scientific literature [34–38]. Agriculture is the main economic activity in Srem and cereals are the main crop, covering around 70% of its extension [14]: for this reason, the research team considered that analyzing multi-spectral satellite images would have ensured better results in comparison with other kinds of data. To investigate the presence of buried archaeological remains, the researchers mainly used Vegetation Indices to measure crops' health, which indirectly indicated the presence of buried structures, highlighted by the presence of anomalies in the plants' growth and health [39]. The analysis of multi-spectral images would have enabled the identification of crop-marks, possibly due to the presence of buried archaeological features: their presence is connected to the concurrency of different factors, among which the soil composition and the vegetation phenological and biophysical cycles play a central role [39] (pp. 114–116). Indeed, crop-marks formation is a dynamic process that depends on vegetal species and on their growth stage: this characteristic was especially taken into account when we decided to use the Sentinel-2 images within the research project. Their revisit time of five days at the Equator actually permitted monitoring crop-marks evolution over the time [40]. Furthermore, the main goal of the Sentinel mission is to provide information for agricultural and forestry practices, with special attention to food security. Their imagers are especially designed to produce images suitable for the calculation of indices such as leaf area chlorophyll and water content indices [41].

Remote sensing analyses were performed over four images generated in 2016:

1. S2A_OPER_MSI_L1C_TL_MPS__20160616T131744_A005138_T34TDQ, recorded on 16 June,
2. S2A_OPER_MSI_L1C_TL_SGS__20160629T150301_A005324_T34TDQ recorded on 29 June,
3. S2A_OPER_MSI_L1C_TL_MPS__20160407T131800_A004137_T34TDQ recorded on 4 July,
4. S2A_OPER_MSI_L1C_TL_EPA_20160815T031336_A000426_T34TDQ recorded on 15 August.

Even if the images were all processed with the same techniques, the results obtained from the analysis of the image recorded on 16 June was selected for this publication, because it gave the best results and it could be compared to a WorldView2 image of the same area acquired thanks to an agreement with the European Space Agency (ESA) for the supplying of third party commercial missions' data (Restrained Dataset Project n. 36166).

After exploring the potentiality of Sentinel-2 images for the first time, we decided to test the efficacy of different Vegetation Indices designed to enhance and exploit the contrast between different regions of the electromagnetic spectrum for archaeological purposes [39] (p. 119). Among those indices, the Normalized Difference Vegetation Index (NDVI) is often used to assess crops' health through the comparison of Infra-Red and Red bands [35] (pp. 92–94), with the following formula:

$$NDVI = (NIR - Red)/(NIR + Red). \tag{1}$$

The index measures the spectral absorption of light performed by chlorophyll, in the Red band, and the reflection of light in the NIR, depending on the leaves' structure. The resulting values are always less than one: the vegetation will be healthier where values are closer to 1, the ground will be less vegetated where values are closer to 0.

In Figure 2, we observe details of the NDVI index applied to the Sentinel-2 image, in the territory of Golubinci. The magenta polygons mark the location of the crop-marks that enabled the probable identification of the Roman road. The data show that even if some anomalies were visible, it was not possible to map the road through this image. This could be due to the spatial resolution of the original images: since the width of the Roman road has been measured between 10 and 15 m, it might be too small to be visible in an image in which pixel size is 10 m.

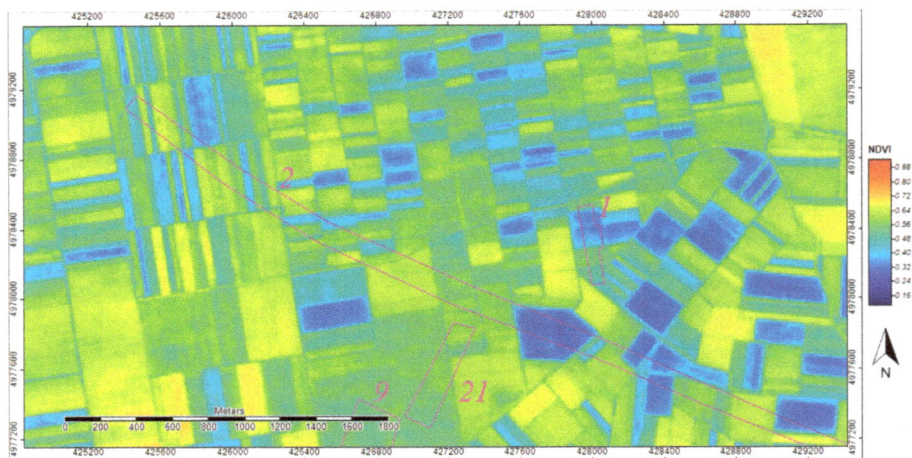

Figure 2. Detail of the output of the NDVI application to the Sentinel-2 image recorded on 16 June 2016. The areas including the Roman road and other buried archaeological remains are marked in magenta, but it was not possible to detect traces of the Roman road in this image. For a geographical reference, see the crop-marks identification numbers and locations in Figure 9.

To take a step further and to get better results in the identification of crop-marks, we tried to apply different algorithms for processing the same image. The major issue in the use of the Sentinel-2 images to detect buried archaeological features was the spatial resolution of the bands in comparison with the dimensions of the features themselves. Agapiou et al. remarked that the most efficient reflectance region to detect archaeological crop-marks is included between 700 µm and 800 µm, and consequently elaborated the Normalized Archaeological Index to exploit the spectral characteristics of the Sentinel-2 imagers [32] (pp. 2183–2185). The algorithm thus elaborated applies the formula:

$$\text{NAI} = (800\ \mu m - 700\ \mu m)/(800\ \mu m + 700\ \mu m), \tag{2}$$

where 700 µm and 800 µm respectively stand for band 5 (VRE) and 7 (VRE). The NAI is a reformulation of the NDVI, based on the specific spectral characteristics of the Sentinel-2 bands in relationship with the optimum archaeological spectral region (700 µm and 800 µm), accordingly with the analysis of Agapiou et al. [32] (pp. 2183–2185). The application of this equation did not give meaningful results, probably due to the low spatial resolution of bands 5 and 7: the output was an image with 20 m pixel size.

Trying to get most out of the data from the images, we tried a simpler combination of bands 4 (Red) and 8 (NIR), exploiting at once their spectral characteristics and their higher spatial resolution, obtaining a 10 m pixel size image after processing with the Semi-Automatic Classification Plugin [42] developed for QGIS [43]:

$$\text{RN} = \text{Red} + \text{NIR}. \tag{3}$$

The combination of Red and NIR gave good results, as can be seen in Figure 3, where the crop-marks were better defined than in Figure 9. It was even possible to map additional crop-marks, not visible in the NDVI output. The magenta arrows mark the presence of a crop-mark indicating the presence of the Roman road (see a better detail in Figure 4). The visibility of the crop-marks did not depend only on the pixel size of the images, but varied based on the different processes of band manipulation, like in this case.

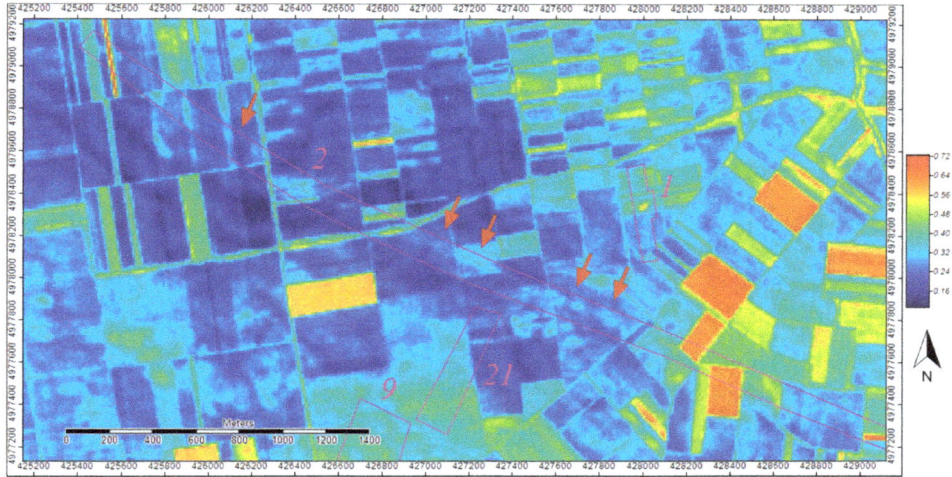

Figure 3. Detail of the output of RN application to the Sentinel-2 image recorded on 16 June 2016. The areas including the Roman road and other buried archaeological remains are marked in magenta. The arrows indicate the visible traces of the road. For a geographical reference, see the crop-marks identification numbers and locations in Figure 9.

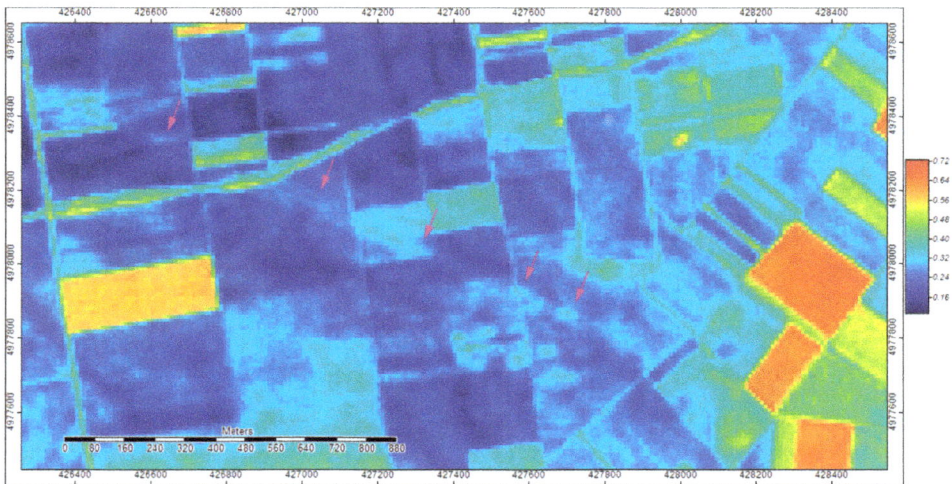

Figure 4. Details of the output image of RN application to the Sentinel-2 image. The crop-mark n.2, identified as the Roman road, is marked by the magenta arrows.

Finally, we attempted a test using orthogonal equations, developed from the Tasseled Cap transformation [44–46], and designed to detect the presence of archaeological remains [47]. The equations, among which the Crop Coefficient 3 (CC3) has a specific relevance, had already been tested on the images recorded by third party missions (Quickbird, IKONOS, Worldview-2, GeoEye-1, ASTER, Landsat 4TM, Landsat 5TM, and Landsat 7ETM), with good results in the archaeological field [45]. Nevertheless, the original formula [47] (pp. 6568,6569) needed adaptation to the new Sentinel-2 images to test its reliability with the new data:

$$CC3 = 0.19 \times \rho\text{Band 1TM} + 0.56 \times \rho\text{Band 2TM} - 0.81 \times \rho\text{Band 3TM} - 0.04 \times \rho\text{Band 4TM}. \quad (4)$$

In Figure 5, we can observe the resulting image from the application of this formula. Indeed, it was impossible to distinguish any meaningful crop-marks. Subsequently, we tried to further improve the results by introducing a variant to the formula:

$$CC3\_DR = (0.19 \times \rho\text{Band 1TM} + 0.56 \times \rho\text{Band 2TM})/(-0.81 \times \rho\text{Band 3TM} - 0.04 \times \rho\text{Band 4TM}). \quad (5)$$

The image resulting from the application of the Equation (5) is shown in Figure 6. The variant changed the ratio among the components of the algorithm's structure, producing an output with a higher definition of the crop-marks and ground principal components. The CC3_DR algorithm, applied on images recorded by other sensors (GeoEye-1, Worldview-2, Quickbird) produced better outputs in the archaeological perspective, while the Sentinel-2 images showed once again their main limit: a lower spatial resolution that prevented the improvement of the results.

Once again, Figure 6 does not show any clear improvement in comparison to previous tests.

To conclude, the only algorithm that really enabled the identification of crop-marks generated by the presence of the Roman road was the combination of Red and NIR bands. Nevertheless, also in that case, the low spatial resolution of the Sentinel-2 images did not allow the detection of many anomalies that are clearly visible on other images, as we exemplify by comparing these results with the elaboration of a WorldView-2 image of the same area, but it was enough to map the Roman road.

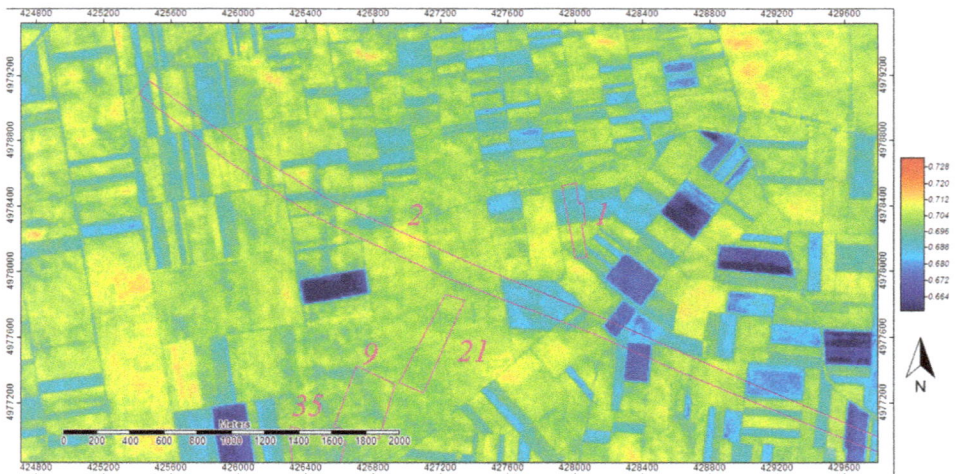

Figure 5. Output details of CC3 algorithm application to the Sentinel-2 image. The areas including the Roman road and other buried archaeological remains are marked in magenta, but it was not possible to detect them through this image. For a geographical reference, see the crop-marks identification numbers and locations in Figure 9.

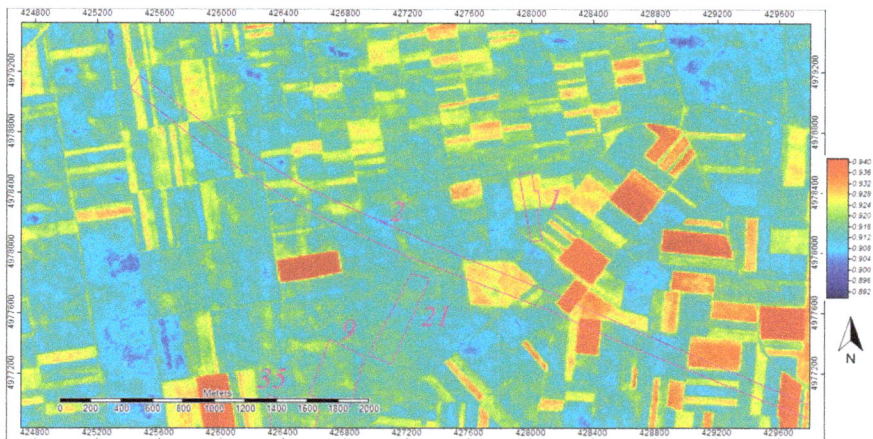

Figure 6. Output details for CC3_DR algorithm application to the Sentinel-2 image. The areas including the Roman road and other buried archaeological remains are marked in magenta, but it was not possible do identify the Roman road in this image. For a geographical reference, see the crop-marks identification numbers and locations in Figure 9.

2.1.2. Comparison with the Results Obtained with a WorldView-2 Image

Together, with the assessment of the efficacy of the Sentinel-2 images in the archaeological perspective, the research still needed to deal with mapping the traces and remains of the Roman road from *Sirmium* to *Singidunum*. Thanks to an agreement with ESA for the acquisition of a WorldView-2 multi-spectral image, it was possible to map the path of the road to the East of *Bassianae*, where its layout was completely under the question.

The 15JUL23095803-P3DS-056453190020_01_P001 WorldView-2 image was recorded on 15 July 2016 (for its spectral and spatial characteristics, see: Table 2), and it was processed using both the CC3 (Figure 7) and the CC3_DR (Figure 8) algorithms. It can be argued from the comparison of the two images, that the advantage of the CC3_DR algorithm consists in generating a much better detailed output. Thanks to the latter, it was possible to determine the direction and path of the Roman road between *Bassianae* and *Singidunum*.

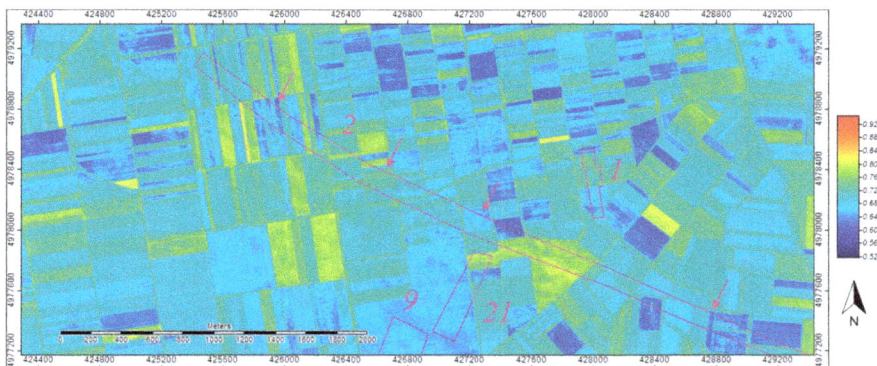

Figure 7. Output details for CC3 algorithm application to the WorldView2 image. The crop-marks identifying the presence of the Roman road and of other buried archaeological remains are marked in magenta. Data provided by the European Space Agency. The arrows mark the visible traces of the road. For a geographical reference, see the crop-marks identification numbers and locations in Figure 9.

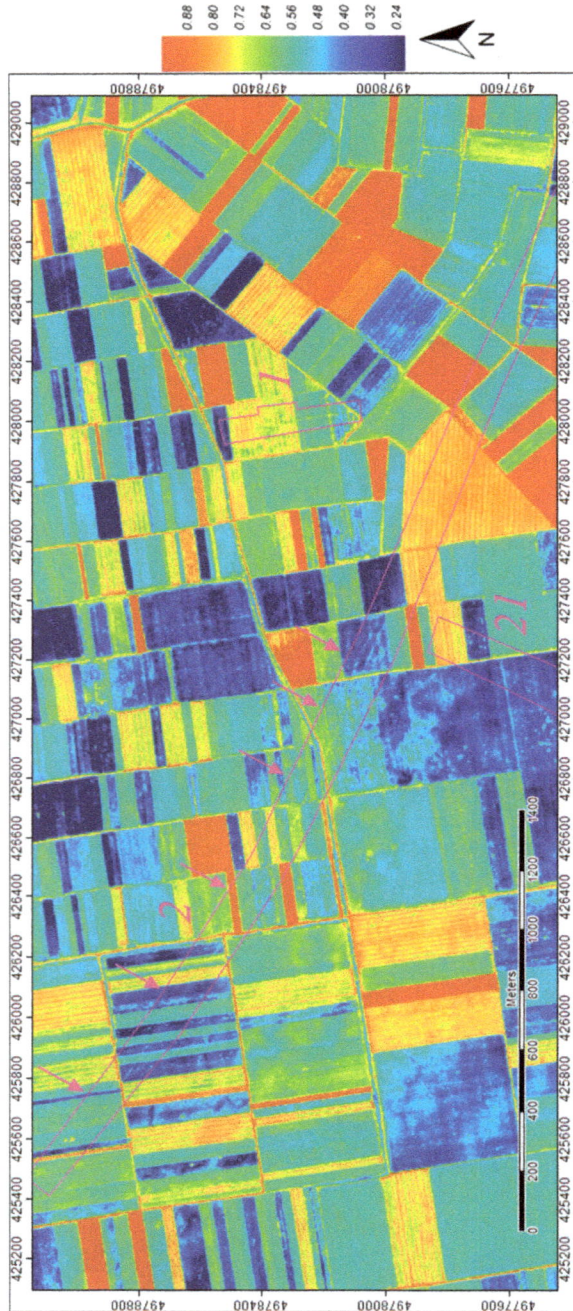

Figure 8. Output details for CC3_DR application to the WorldView2 image. The crop-marks identifying the presence of the Roman road and other buried archaeological remains are marked in magenta. The arrows mark the best-defined traces of the road. Data provided by the European Space Agency. For a geographical reference, see the crop-marks identification numbers and locations in Figure 9.

Table 2. Spectral characteristics and spatial resolution of the WorldView-2 spectral images.

WorldView-2 Bands	Wavelength (μm)	Spatial Resolution (m)	
		at nadir	20° off-nadir
Panchromatic	450–800	0.46	0.52
Coastal	400–450	1.85	2.07
Blue	450–510	1.85	2.07
Green	510–580	1.85	2.07
Yellow	585–625	1.85	2.07

3. Results

Thanks to the specific distribution of modern settlements in the area of interest, the Srem district revealed to be particularly favorable for the application of satellite remote sensing to detect Roman roads and buried archaeological remains.

The combined analysis of Sentinel-2 and commercial missions images led to the identification of 60 crop-marks, hypothetically due to buried archaeological features in the portion of territory stretching from Sremska Mitrovica to Zemun (Figure 1). Figure 9 shows that some of these anomalies were linear, possibly connected to the Roman road or to other historical branches of the road network, while others were polygonal. It was necessary, once we established this fact, to verify the reliability of these data and to assign a more precise chronology to the remains that originated the crop-marks. To this end, an archaeological surface survey campaign was organized in collaboration with the Institute for the Protection of the Cultural Monuments of Sremska Mitrovica.

The survey was planned after combining the information derived from previous research, stored in the Institute's archive, with the results from remote sensing analysis. A procedure integrating mobile mapping technologies was established in order to optimize the available resources in terms of funding and time.

During the field research, it was possible to map several archaeological sites dating to different chronological periods, falling within the area of interest, in addition to the Roman road from *Sirmium* to *Singidunum*. More precisely, 13 crop-marks correspond to already known archaeological sites and stretches of the Roman road, whereas, 47 crop-marks remain unmatched. Some of them, of course, can be due to the presence of artifacts of geo-pedological or artificial nature. On the other hand, 36 files from the Institute's archive did not correspond to any anomaly. This is probably due to the small dimensions of the sites and their inconsistency in terms of visibility (e.g., isolated graves, hearth remains, ...). Nevertheless, integrating the results coming from the remote sensing analysis with the archive information, allowed the researchers to focus on the areas where it was most probable to identify archaeological, and particularly Roman sites, evidence of the Roman road and sites connected with the road itself.

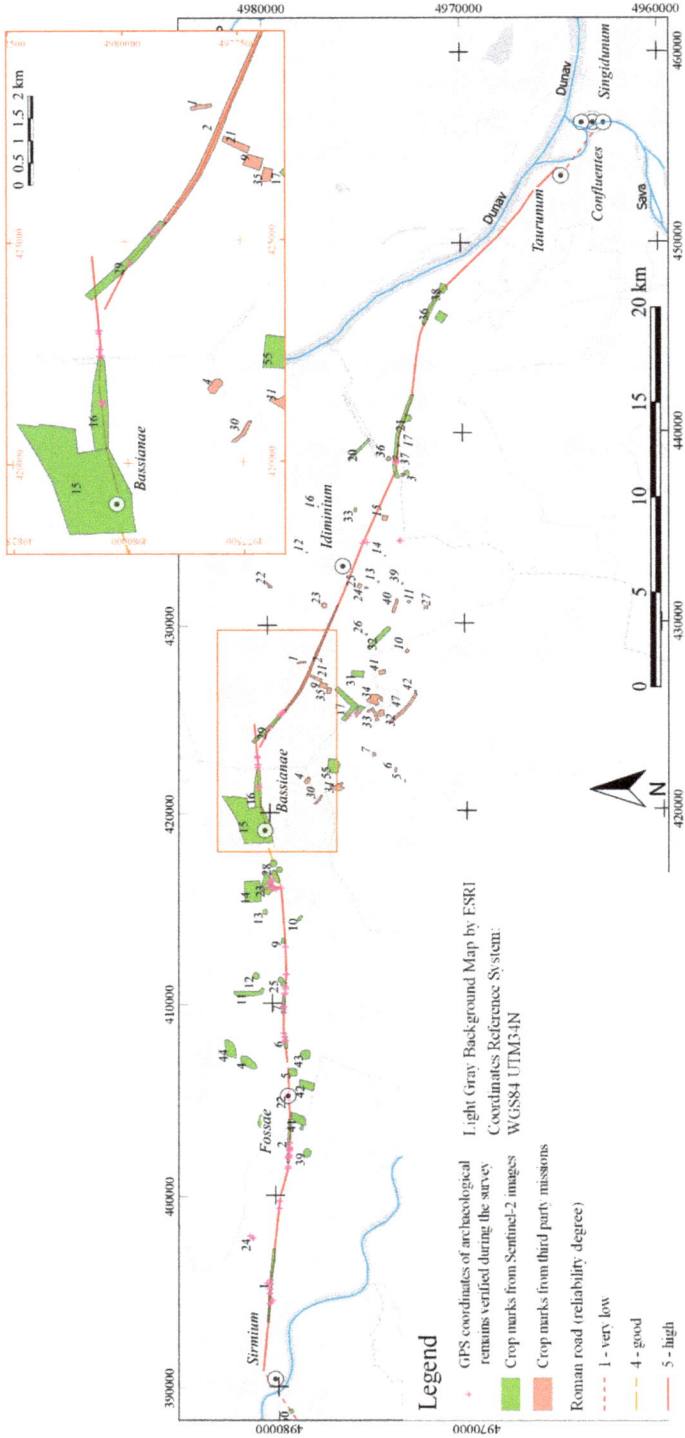

Figure 9. Map representing the results obtained during the archaeological survey in Srem. In the focus is the segment of the Roman road between *Bassianae* and *Singidunum*, mapped for the first time within the RecRoad project.

4. Discussion

The District of Srem is well known from an archaeological point of view, also due to the relevance held by the city of *Sirmium* during Roman times. Several field research projects have been developed before the beginning of the RecRoad project in 2016, and they generated general knowledge of the layout of the main Roman road from *Sirmium* to *Singidunum*, passing through the city of *Bassianae* [19–24]. Nevertheless, some relevant questions remained unanswered: indeed, the Roman road path was documented only in five points and there was no certainty about its precise layout. This was an issue especially for the protection of the archaeological remains.

Satellite remote sensing techniques, applied to Sentinel-2 images and to other kinds of images (WorldView-2, OrbView, QuickBird), enabled mapping of 70 km of the Roman road and the development of an integrated survey methodology. It is desirable to include in further research new kinds of multi-spectral images with higher pixel resolution, such as PlanetScope images. Thanks to the integration of remote sensing data and mobile mapping technologies, the researchers were able to verify the layout of the Roman road and to identify buried archaeological sites along its 70 km path in only six days of effective field work. Furthermore, the survey enabled the collection of archaeological materials that are proving to be very useful for the interpretation of crop-marks. An important issue regarding archaeological remote sensing must be noted: where remote sensing techniques can prove their usefulness for archaeological research, they cannot provide any information about the chronology, cultural pertinence, and function of the buried archaeological remains. Only the direct prospect of the sites, analysis of pottery fragments, and other archaeological materials can lead to an interpretation of the remains.

Concerning the efficacy of Sentinel-2 images for archaeological research, especially in the research of Roman roads, as it has been shown in this paper, the spatial resolution of the images is a major limitation. This is due to the average width of the Roman road in this area, since it is comprehended between 9 and 15 m [29]. Nevertheless, Sentinel-2 images have proven to be very useful in the detection of different kinds of archaeological remains in the same area, enabling the identification of previously undocumented archaeological sites [48,49]. The recent introduction of the Sentinel-2 images, has led to the testing of several algorithms to process them. The ones that gave the best results were the combination of Red and NIR bands and the CC3_DR Equation (4). The combination of Red and NIR bands enabled a better exploitation of both the spectral characteristics and the spatial resolution of this kind of images, enhancing their capabilities of showing crop-marks and allowed the mapping of the remains of the Roman road.

On the other hand, in comparison with the CC3 formula, in the CC3_DR formula, the contribution of the NIR band results were almost irrelevant, since it was weighted on a minor coefficient than the others. Thus, we obtained the weighted sum of 1TM and 2TM bands, normalized by the 3TM band. This practically eliminated the effect of band 3TM in the image. These elements increased the capabilities of showing the presence of buried archaeological features. Furthermore, the introduction of the ratio strengthened the formula at the variation of the coefficients. Nevertheless, it is true to be said that the potentialities of the CC3_DR formula have given better results when applied to higher resolution images (Figure 8).

To conclude, the integration of data coming from the systematic archaeological survey and research in the Srem District [50] was an opportunity to test the capabilities of Sentinel-2 images in this field. The combined methodology also improved the positioning of the segments of the main Roman route that connected *Sirmium* with *Bassianae* and *Singidunum*, and gave hints for future exploration of the Roman road network in the area. It would be especially interesting to investigate the connection of the main road from *Bassianae* to *Singidunum* with the *limes* road that functioned as main communication tool among the military camps and forts along the Danube border. The first segment of this connection was already documented during the survey, shown in Figure 9, but the remaining part requires mapping.

The data obtained through the analysis of multi-spectral images were also important for identifying broader archaeological site areas, proving that the combination of field survey and remote sensing analysis often gives a limited understanding in the case of poorly known sites. New perspectives of research could be imagined integrating the analysis of more high-resolution multi-spectral images, recorded by satellite imagers or by drones equipped with multi-spectral cameras, improving also the data detail.

Author Contributions: Conceptualization, methodology, software, and validation, S.Z. and A.D.R.; writing—original draft preparation, S.Z.; writing—review and editing, A.D.R.; supervision and project administration, S.Z.; funding acquisition, S.Z.

Funding: This work received funding from the European Union's Horizon 2020 research and innovation programme under the Marie Skłodowska-Curie grant agreement No 660763. Also with the support of the Marie Curie Alumni Association.

Acknowledgments: The authors acknowledge Biljana Lučić and the Institute for the Protection of the Cultural Monuments of Sremska Mitrovica, for the archive data and archaeological material analysis; the European Space Agency for providing WorldView-2 images; and Antonio Augimeri for the discussing the images processing methodology.

Conflicts of Interest: The authors declare no conflicts of interest. The funders had no role in: the design of the study; in the collection, analyses, or interpretation of data; in the writing of the manuscript, or in the decision to publish the results.

References

1. Cuntz, O. *Itineraria Romana*, 1990th ed.; Vieweg+Teubner Verlag: Wiesbaden, Germany, 1929; Volume 1.
2. Arnaud, P. L'Itinéraire d'Antonin: Un témoin de la littérature itinéraire du Bas Empire. *Geogr. Antiq.* **1993**, *2*, 33–50.
3. Calzolari, M. *Introduzione allo Studio Della Rete Stradale dell'Italia Romana: l'Itinerarium Antonini*; Atti della Accademia Nazionale dei Lincei: Roma, Italy, 1996; Volume 7.4.
4. Douglass, L. A new look at the Itinerarium Burdigalense. *J. Early Christ. Stud.* **1996**, *4*, 313–333. [CrossRef]
5. Calzolari, M. Ricerche sugli Itinerari Romani. L'Itinerarium Burdigalense. *Studi onore Nereo Alfieri. Atti dell'Accademia Scienze Ferrara* **1997**, *74*, 125–189.
6. Miller, K. *Die Weltkarte des Castorius genannt die Peutingerische Tafel*; Nabu Press: Charleston, SC, USA, 1887.
7. Cuntz, O. Die Grundlagen der Tabula Peutingeriana. *Hermes* **1895**, *29*, 586–594.
8. Miller, K. *Itineraria Romana*; HATHI TRUST Digital Library: Stuttgart, Germany, 1916.
9. Weber, E. *Tabula Peutingeriana (Faksimileausgabe)*; Akademische Druck- und Verlagsanstalt: Graz, Austria, 1976.
10. Pontrera, F. (Ed.) *Tabula Peutingeriana: Le Antiche vie del Mondo*; Olschki: Firenze, Italy, 2003.
11. Talbert, R. *Rome's World: The Peutinger Map Reconsidered*; Cambridge University Press: Cambridge, UK, 2010.
12. De Sanctis, G. *Storia dei Romani*; Milano Bocca: Torino, Italy, 1923; Volume IV.1.
13. Pavan, M. Aquileia città di frontiera. *Antich. Altoadriatiche* **1987**, *29*, 17–55.
14. Mihailović, B.; Cvijanović, D.; Milojević, I.; Filipović, M. The role of irrigation in development of agriculture in Srem district. *Econ. Agric.* **2014**, *61*, 989–1004. [CrossRef]
15. Zaccaria, C. Il ruolo di Aquileia e dell'Istria nel processo di romanizzazione della Pannonia. In Proceedings of the Atti del Convegno La Pannonia e l'Impero Rom, Roma, Italy, 13–16 January 1994; pp. 51–70.
16. Cencig, D. *Elementi Topografici Notevoli sulle vie di Accesso di Aquileia Romana e Sull'antica Viabilità sud Orientale del Friuli Venezia Giulia*; Archeomedia-Rivista di Archeologia Online a cura di Mediares: Torino, Italy, 2018.
17. Burghardt, A.F. The origin of the road and city network of Roman Pannonia. *J. Hist. Geogr.* **1979**, *5*, 1–20. [CrossRef]
18. Zanni, S. La route d'Aquileia à Singidunum: Aspects méthodologiques. Du terrain à la publication et à la mise en valeur. In Proceedings of the Table Ronde Internationale La Route Antique et Médiévale: Nouvelles Approche, Nouveaux Outils, Bordeaux, France, 15 November 2016; pp. 145–164.
19. Dimitrijević, D. Nekoliko podataka o rimskom limesu u istočnom Sremu. In *Limes u Jugoslaviji*; Arheolosko drustvo Jugoslavije Beograd: Belgrade, Serbia, 1961; Volume I.
20. Dimitrijević, D. Vojka–Ugrinovci, Stara Pazova–Zemun, lok. Brestove Mede. *Arheol. Pregl.* **1966**, *8*, 124–127.

21. Dimitrijević, D. Istraživanja rimskog limesa u isstočnom Sremu s posebnim osvrtom na pitanje komunikacija. *Osiječki Zb.* **1969**, *12*, 87–112.

22. Popović, D. Glavna antička komunikacija u Sremu u svetlu arheoloških istraživanja, Putevi i komunikacije u antici. *Materijali* **1980**, *XVII*, 101–108.

23. Popović, D.; Vasiljević, M. Rekognosciranje I sondiranje rimskog puta Sirmium-Bassianae. *Arheol. Pregl.* **1969**, *11*, 261–262.

24. Popović, D.; Vasiljević, M. Rekognosciranje u Sremu. *Arheol. Pregl.* **1970**, *12*, 193–194.

25. Milošević, P. O traci puta Sirmium-Fossis i Sirmium-Bononia/Sur le trace de la route Sirmium-Fossis et Sirmium-Bononia. *Starinar* **1988**, *39*, 117–123.

26. Brukner, O. Žirovac, Ruma—Rimski Put. *Arheol. Pregl.* **1982**, *23*, 95–97.

27. Brukner, O. Rimski put Sirmium—Singidunum (11). *Archaeological Investigations along the Highway Route in Srem*. Available online: http://archest.eu/wp-content/uploads/2016/06/Archest-WP3-Task-3.1-%E2%80%93-Historiographic-research-update-on-the-Roman-route_28.7.2016.pdf (accessed on 25 November 2018).

28. Gračanin, H. Rimske prometnice i komunikacije u kasnoantičkoj južnoj Panoniji. *Scr. Slavon.* **2010**, *10*, 9–69.

29. Lučić, B. Contribution to the research of the main Roman road through Srem. In *Archest—Developing Archaeological Audiences along the Roman Route Aquileia-Emona-Sirmium-Viminacium*; Historiographic Research Update on the Roman Route; Muzej in galerije mesta Ljubljane: Ljubljana, Slovenia, 2016; pp. 37–44.

30. Fodorean, F. Praetorium and the Emona–Siscia–Sirmium–Tauruno road in the ancient geographical and epigraphic sources. *Arheol. Vestn.* **2017**, *68*, 337–348.

31. Fodorean, F. Listing settlements and distances: The Emona-Singidunum road in Tabula Peutingeriana, Itinerarium Antonini and Itinerarium Burdigalense. *Starinar* **2017**, 95–108. [CrossRef]

32. Agapiou, A.; Alexakis, D.D.; Sarris, A.; Hadjimitsis, D.G. Evaluating the Potentials of Sentinel-2 for Archaeological Perspective. *Remote Sens.* **2014**, *4*, 2176–2194. [CrossRef]

33. Tapete, D.; Cigna, F. Appraisal of Opportunities and Perspectives for the Systematic Condition Assessment of Heritage Sites with Copernicus Sentinel-2 High-Resolution Multispectral Imagery. *Remote Sens.* **2018**, *10*, 561. [CrossRef]

34. Wiseman, J.R.; El-Baz, F. (Eds.) *Remote Sensing in Archaeology*; Springer: Berlin, Germany, 2007.

35. Parcak, S.H. *Satellite Remote Sensing for Archaeology*; Routledge: London, UK, 2009.

36. Lasaponara, R.; Masini, N. (Eds.) *Satellite Remote Sensing—A new tool for Archaeology*; Springer Netherlands: Dordrecht, The Netherlands, 2012.

37. Silver, M.; Törmä, M.; Silver, K.; Okkonen, J.; Nuñez, M. Remote sensing, landscape and archaeology tracing ancient tracks and roads between Palmyra and the Euphrates in Syria. *ISPRS Ann. Photogramm. Remote Sens. Spat. Inf. Sci.* **2015**, *II-5/W3*, 279–285. [CrossRef]

38. Traviglia, A.; Torsello, A. Landscape Pattern Detection in Archaeological Remote Sensing. *Geosciences* **2017**, *7*, 128. [CrossRef]

39. De Guio, A. Cropping for a better future: Vegetation Indices in Archaeology. In *Detecting and Understanding Historic Landscapes*; Chavarria Arnau, A., Reynolds, A., Eds.; SAP: Società Archeologica srl: Mantova, Italy, 2015; pp. 109–152.

40. Malenovský, Z.; Rott, H.; Cihlar, J.; Schaepman, M.E.; García-Santos, G.; Fernandes, R.; Berger, M. Sentinels for science: Potential of Sentinel-1, -2, and -3 missions for scientific observations of ocean, cryosphere, and land. *Remote Sens. Environ.* **2012**, *120*, 91–101. [CrossRef]

41. Sentinel-2. Available online: http://www.esa.int/Our_Activities/Observing_the_Earth/Copernicus/Sentinel-2/Introducing_Sentinel-2 (accessed on 25 November 2018).

42. Congedo, L. Semi-Automatic Classification Plugin Documentation. Available online: https://media.readthedocs.org/pdf/semiautomaticclassificationmanual-v3/latest/semiautomaticclassificationmanual-v3.pdf (accessed on 25 November 2018).

43. Quantum GIS Development Team. *Quantum GIS Geographic Information System. Open Source Geospatial Foundation Project*. Available online: https://www.scirp.org/%28S%28vtj3fa45qm1ean45vvffcz55%29%29/reference/ReferencesPapers.aspx?ReferenceID=1515458 (accessed on 25 November 2018).

44. Kauth, R.J.; Thomas, G.S. The Tasseled Cap—A Graphic Description of the Spectral-Temporal Development of Agricultural Crops as Seen by LANDSAT. *LARS Symp.* **1976**, *1*, 159.

45. Agapiou, A.; Alexakis, D.D.; Sarris, A.; Hadjimitsis, D.G. Orthogonal Equations of Multi-Spectral Satellite Imagery for the Identification of Un-Excavated Archaeological Sites. *Remote Sens.* **2013**, *5*, 6560–6586. [CrossRef]

46. Yarbrough, L.D.; Navulur, K.; Ravi, R. Presentation of the Kauth-Thomas transform for WorldView-2 reflectance data. *Remote Sens. Lett.* **2014**, *5*, 131–138. [CrossRef]

47. Agapiou, A. Orthogonal equations for the detection of hidden archaeological remains de-mystified. *J. Archaeol. Sci. Rep.* **2017**, *14*, 792–799. [CrossRef]

48. Zanni, S.; Lučić, B.; De Rosa, A. From the sky to the ground. A spatial approach to the archaeological research in the Srem region (Serbia). The case study of Pusta Dreispitz. In Proceedings of the Cultural Heritage and New Technology Conference (CHNT), Wien, Germany, 8–10 November 2017.

49. Zanni, S. Seek and Ye Shall Find. A Spatial Approach to Mapping Roman Roads and Buried Archaeological Sites in the Srem Region. The Case Study of Tapavice Site 2018. Available online: https://www.researchgate.net/publication/327623205_Seek_and_Ye_Shall_Find_A_spatial_approach_to_mapping_Roman_roads_and_buried_archaeological_sites_in_the_Srem_region_The_case_study_of_Tapavice_site_XXIV_LIMES_Congress/download (accessed on 25 November 2018).

50. Lučić, B. Roman viability through the Srem Region—Territory of the town of Sirmium. In Proceedings of the Actes du Colloque International, Bordeaux, France, 11–12 September 2017.

geosciences

MDPI

Article

High Resolution Drone Surveying of the Pista Geoglyph in Palpa, Peru

Karel Pavelka *, Jaroslav Šedina and Eva Matoušková

Department of Geomatics, Faculty of Civil Engineering, Czech Technical University in Prague, Thakurova 7, Prague 6 16629, Czech Republic; jaroslav.sedina@fsv.cvut.cz (J.Š.); eva.matouskova@fsv.cvut.cz (E.M.)
* Correspondence: pavelka@fsv.cvut.cz; Tel.: +42-22435-3865

Received: 3 November 2018; Accepted: 10 December 2018; Published: 13 December 2018

Abstract: Currently, satellite images can be used to document historical or archaeological sites in areas that are distant, dangerous, or expensive to visit, and they can be used instead of basic fieldwork in several cases. Nowadays, they have final resolution on 35–50 cm, which can be limited for searching of fine structures. Results using the analysis of very high resolution (VHR) satellite data and super resolution data from drone on an object nearby Palpa, Peru are discussed in this article. This study is a part of Nasca project focused on using satellite data for documentation and the analysis of the famous geoglyphs in Peru near Palpa and Nasca, and partially on the documentation of other historical objects. The use of drone shows advantages of this technology to achieve high resolution object documentation and analysis, which provide new details. The documented site was the "Pista" geoglyph. Discovering of unknown geoglyphs (a bird, a guinea pig, and other small drawings) was quite significant in the area of the well-known geoglyph. The new data shows many other details, unseen from the surface or from the satellite imagery, and provides the basis for updating current knowledge and theories about the use and construction of geoglyphs.

Keywords: photogrammetry; RPAS; UAV; Peru; geoglyph Pista; mapping; drones

1. Introduction

Remote sensing and Google Earth using satellite images became an instrument for archaeological documentation and research at the beginning of the new millennium [1]. New types of multispectral and stereoscopic data with very high resolution are an important source of data; in many projects, Google Earth can be used as a tool for the searching of new potential archaeological sites. The long-term usage of an aerial archaeological survey is very popular. RPAS (remotely piloted aircraft system) also known as UAV (unmanned aircraft vehicle) or drones, are nowadays gaining importance [2]. Very high resolution (VHR) satellite data with a ground sampling distance (GSD) better than 1 m have been at disposal for civilian purposes since 1999. They support a new analytical possibility in archaeology and historical object documentation. Current civilian satellite images with GSD 0.35–1.0 m in panchromatic or pan-sharpened images serve as a perfect source for the documentation of known objects or finding possible newly detected archaeological sites or objects [3]. Satellite imagery can be used in some cases instead of basic fieldwork for searching for new objects and generally for documentation or mapping of an area of interest. Our possibility to detect a new object from satellite data ends in the case of linear objects by their thickness at 30–50 cm (0.7–1.0 pixels), in the case of 3D or point objects, some pixels (depend on object, but at least four pixels) are necessary for statistically appropriate detection. This means that linear objects with a thickness smaller than 30 cm and a point object smaller than 50 cm in diameter are invisible to us and for this reason unknown. Outputs with a better GSD can be obtained using aerial methods; however, they are not at their disposal or they are too expensive for small projects in many countries. There is a gap between remotely sensed data with GSD 50 cm and GSD 5 cm; however, the better resolution can be important in special cases—e.g., searching of fine details in orthophoto or in shaded digital surface model.

This problem can solve use of RPAS (remotely piloted aircraft system or simple "drones"), they typically provide data with a GSD better than 5 cm. It has been possible to use data from RPAS since 2010. However, it is not easy to use it in some cases, because of a meteorological, safety, or bureaucratic point of view. However, surface large archaeological objects are very well visible from height such as subsurface objects based on crop marks [3–7].

In particular, they are objects that are not overshadowed by growing vegetation—i.e., object in desert or semi-desert areas. Such an area is Peru, where more advanced civilizations existed in the dry coastal areas before the conquest.

1.1. Geoglyphs and Lines: State-of-the-Art

Near Nasca city, approximately 400 km southeast from the Peruvian capital city Lima (Figure 1), there is a famous archaeological area, which contains thousands of world-known geoglyphs and lines. The Nasca city is located 40 km from the Pacific coast in a dry landscape, interwoven with fertile valleys with occasional rivers. The first scientific information about geoglyphs and lines in Peruvian desert was brought by P. Kosok, who later collaborated with M. Reiche. Dr. Maria Reiche, a charismatic scientist born in Dresden/Germany, worked in the Peruvian desert near Nasca city to document and explain interesting geoglyphs and lines in the city neighbourhood since WW II. Based on her work, the Nasca lines and geoglyphs were added to the UNESCO (United Nations Educational, Scientific and Cultural Organization) world heritage list in 1995 and they are preserved by Peruvian law [8–11]. Other areas were not under auspices of law and they are in very bad condition, such as cases [10,11]. The origin of geoglyphs and lines is not clear, with many scientists and laics trying to explain the origin and reason [11,12]. The main hypothesis, which attempts to explain the origin of geoglyphs and lines, talks about the astronomical reason (calendar) [8,9,12], the designation of water resources [13], and religious rituals [9,14–16]. There are also technical theories that they were areas for weaving or spinning long ropes [17–19]. There are also bizarre theories about extra-terrestrials, pre-Columbian Olympia games, art, etc. [11,20].

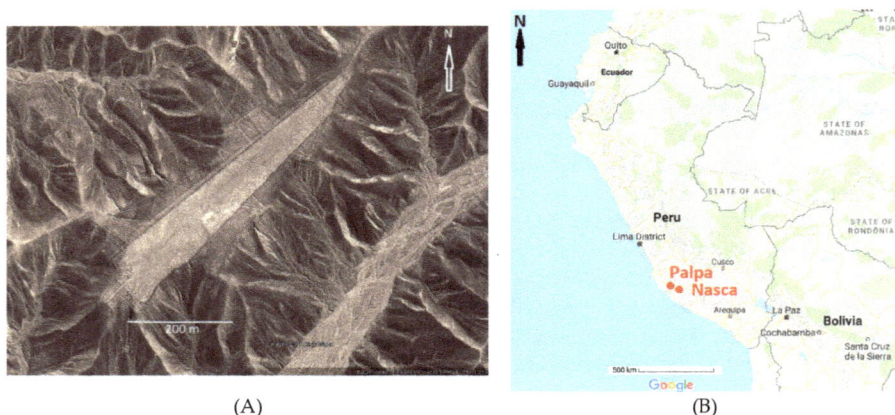

Figure 1. (**A**) Geoglyph Pista near Palpa, (Google Earth, satellite image from 2013) and (**B**) localization of cities Palpa and Nasca in Peru (Google maps).

Some lines and objects were analysed as astronomically oriented [8,9,11,12,20–23], with many lines or geoglyphs neither of known origin nor original use. It is not the goal to argue about the origin or purposes of geoglyphs; however, the result of detailed documentation may serve other professionals [16]. Unfortunately, simply explained theories are always sought. Apparently, there is no uniform use of geoglyphs and lines, which is due to long-term land use. It should be noted that the geoglyphs are of different types and they could have been created by different civilizations, which used geoglyphs and lines differently. From both, the satellite imagery and the ground survey, it is obvious that they previously covered a much larger area, which was later destroyed by the episodes of El Niño; line remnants or erosion scars are visible in the area. Thanks to the still intact area that resists El Niño, the geoglyphs and lines remain visible in the Nasca neighbourhood, but they are still threatened [11,24].

Thanks to the scientific work of M. Reiche [9,10] and the popular books of E. von Däniken [20], geoglyphs and lines around Nasca and Palpa have become world-renowned. They have come to the forefront of the interest of historians, archaeologists, cartographers, and experts from other disciplines. A major breakthrough in knowledge occurred after 2000, when commercially available satellite imagery with sub-meter resolution was already available [25].

Geoglyphs and lines have been investigated by all available methods. Significant is the "Nasca GIS" project at the Dresden University of Applied Sciences [21] or documentation of Palpa geoglyphs and lines from aerial images [26]. Many other projects have used optical VHR satellites in the form of orthophoto or DSM (digital surface model [27,28], it can be created from at least two different image data sets based on image correlation technique) [29]. Other projects have focused on the use of radar recordings from satellites and creating of digital relief model (DRM) using InSAR technology (an interferometric analysis of SAR data) [30–32].

Of course, ground research was also done. The German (M. Reindl and others) [33], Italian (G. Orefici, N. Massini, R. Lasaponara, and others) [14] and American (Ch. Conley and others) [34] archaeological missions have been working in this locality in cooperation with Peruvian archaeologists (J. Isla, director of Management for the Territory of Nasca and Palpa and others) [35] and conservation experts from INC (Instituto Nacional de Cultura del Perú). However, the archaeologists are more concerned with excavations (objects, cities, pyramids, and graves), and not geoglyphs and lines, which are especially surface structures that cannot be routinely excavated and examined by classical archaeological methods (perhaps only the possibility of surface collections of potsherds that are almost everywhere). For this reason, conservators, geographers, cartographers, and, in general, other scientists are focused more on geoglyphs and lines; of course, cooperation with archaeologists is necessary. There are hypotheses that connect geoglyphs and lines with excavations (e.g., the Nazca culture center Cahuachi near Nasca city) [14].

Terrestrial research has focused on documentation and protection of geoglyphs and lines, but also on the technology of their creation and associated with their problematic dating. Some projects have also used geophysical methods. In the last decade, there was a great boom in the use of unmanned aerial vehicles (drones) that allow safe and low-cost mapping of individual parts in this area. Their official use is often problematic in terms of different permissions and subordination [3,5,36]. There is a general problem connected with scientific documentation of geoglyphs: what is old and original and what is new (falsification, graffiti, etc.). Graffiti (names, new themes) are, of course, recognizable. A big problem is dating of geoglyphs and lines. It is not simple, as they were made from purely natural materials; simply by removing the upper eroded and oxidized material thereby a light background was exposed. It is even harder to attribute the geoglyphs to a historical epoch. The situation, when it is necessary to define a historical epoch, is even more complicated (Nasca and Palpa region was inhabited by several civilizations). Typically, the iconography—likeness of desert drawings with drawings on dated ceramic or textile findings is used for basic dating. Based on different sources, the Nazca culture is dated from 200 BC to about AD 800, the Paracas culture approximately between 1200 BC and 100 BC [11,35,37]. Physical dating using archaeological findings (ceramics, wooden parts, etc.) on geoglyphs and lines is not certain—these findings may have no connection with geoglyphs and lines. Even archaeological finds on geoglyph such as tombs or constructions cannot define the epoch of geoglyphs because they can be from a different time. However, geoglyphs and lines may even come from different periods (exact dating of all geoglyphs and lines is not possible until now). The recently presented method—the use of thermoluminescence—might accurately determine the age of some geoglyphs. Especially optically stimulated thermoluminescence technology, that was founded by the University of Heidelberg, may be effective [38]. Based on the analysis of samples taken directly from geoglyphs, their relative age was calculated. The principle is the time of exposure of individual stones to sunlight when the solid material was moved. Sample material was taken on geoglyphs' faces from their sides under the surface—original material was shaded by the upper layer. Some examples have an age of tens of thousands of years, and are therefore the natural movements of the alluvial area, while others show an age of change in their last position between [38–40].

1.2. HTW-CTU Project

In 1994, the Maria Reiche Association was established in Dresden, Germany. The main goal was to extend and continue the science work of Dr. Maria Reiche in the Nasca and Palpa regions. To extend Maria Reiche's work, several expeditions to the Nasca region in Peru were done. In 1995, the Nasca project started at the Dresden University of Applied Sciences (HTW Dresden, B. Teichert, and Ch. Richter). The main goals of the Nasca project are to collect information from Nasca cultural heritage site and its neighbourhood and to store and preserve it in a digital form [5,21].

At the CTU in Prague (Czech Technical University), a long-term research for the use of drones started in 2011. The main goals were developing and testing of low-cost photogrammetrical technology for the mapping and monitoring of small areas, such as the documentation of archaeological sites [2,4]. In this project, many types of drones were tested und used. Multicopters are very good at small local areas or objects modelling (like historical constructions) [36]; winged drones seem to be better at the mapping of larger areas [2,4].

In 2004, a German—Czech cooperation on the documentation of historical objects and monuments in Peru was started. After many expeditions, the team has collected a lot of scientific material. VHR images were used as an important basic material in the research and documentation of objects and possible astronomical orientations of geoglyphs and lines in Peru [1,8,11,12]. In 2016, a small German (HTW Dresden)—Czech (CTU in Prague) expedition to the Nasca and Palpa neighbourhood took place. The main project aim was the documentation of aqueducts [5], based on collaboration with the Nasca town hall. In the frame of this project, the measurement of the geoglyph Pista was done as the second project goal using EBee drone (Figure 2).

Some known geoglyphs are not visible in satellite images because of their contrast and width in decimetres only. If there are some known but invisible geoglyphs or lines, there should also be unknown geoglyphs or lines because of their width, place and extent.

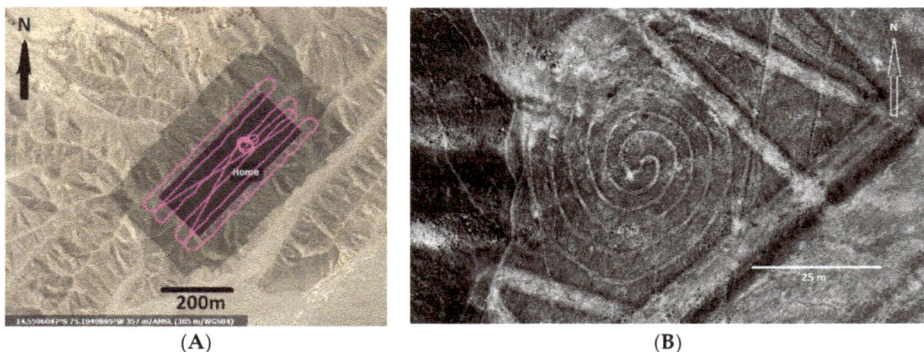

Figure 2. (**A**) Flight over Geoglyph Pista near Palpa and (**B**) a detail of the great spiral near the main geoglyph Pista.

2. Study Area

The additional project of our small expedition in 2016 was a detailed documentation of the selected large area geoglyph. For this purpose, a geoglyph with the local name "Pista" (the Llipata site, Nr. PP01-36) was selected. It is a few kilometers away from Palpa city in a mountainous, dry, stony landscape (Figure 1). The Pista geoglyph is a trapezoid at first glance but after zooming it consists of many other elements (zig-zag lines, spiral, other lines) [26,40–43]. Individual elements overlap, which leads us to believe that the shape of the geoglyph evolved over a longer time period. Its area is really large (0.5 km^2), on the first impression it acts as an artificially flattened mountain peak. Similar geoglyphs are nearby (for example, just above the cemetery of Palpa). As the protection of these geoglyphs is not significant, they are damaged by modern graffiti; on the geoglyphs above the Palpa cemetery there are tracks after *huaqueros*—grave robbers (we think that these graves were probably not a part of the original geoglyph; as ordinary graves are usually not situated on the geoglyphs) [11].

2.1. Investigation of Geoglyphs and Lines near Palpa

Many projects have dealt with archaeological research near Nasca and Palpa using aerial or satellite data [14–16,25,27,28,30,43,44]. It is logical, that more projects were connected to the more famous Pampa de Nazca. The last comprehensive documentation specialized on Palpa geoglyphs and lines using aerial data was conducted in 2006 [26,28,33]. A photo set of black-and-white aerial photographs taken by photogrammetric company in nineties of twentieth century was used. As a partial result, an orthophoto with GSD 28 cm and vector maps of all geoglyphs in Palpa area were created (Figure 3). Even before the era of the drones started, the archaeological settlement near Palpa had been documented using a remote-controlled helicopter in 2004 [3].

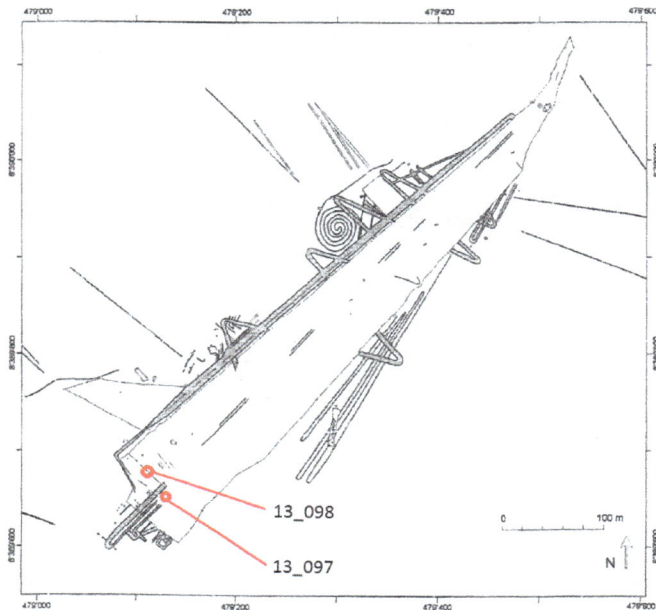

Figure 3. Interpretation of Pista geoglyph based on aerial images (original by K. Lambers, 2006), [26,42]. Red points are samples positions (for ^{14}C analysis, expedition 2012, Table 1), [19].

2.2. Pista Geoglyph Past Survey and Research

Geoglyph Pista (the Llipata site, PP01-36) has been researched several times. Complex research has been carried out by prominent archaeologists, who speak about "Open-air temples" [33–35,43]. The theory is based on the archaeological excavations of the small stone hills on the Pista geoglyph, where ceramic finds and remains of Spondylus shells linked to rituals associated with water were found [40,41,43]. These finds support the opinion that this site was used for ceremonies for a certain period.

However, there are several similar trapezoids in the neighbourhood of Pista geoglyph. From the technical point of view, an appropriate access road or staircase for a large number of people to the ceremonial site should be visible, but nothing similar was found (maybe, it should be a triangular part on the north-west part of geoglyph, but it is questionable). This is the argument for other researchers who claim that the area was not intended for a large number of people. For example, J. Sonnek considers the trapezoids to be a technical work for spinning ropes [11,17,18]. He demonstrates it with practical tests and experiments.

A geophysical (magnetometric) research has been used within archaeological and conservation research. The result has not shown unknown subsurface features like unknown constructions, but older stratigraphic layers of the geoglyph constructional could be detected [38,40,41]. Some places with traces of fire were found [41]. It explains in particular the geology of the substrate. It is also worth mentioning that DDSM (differential digital surface model—the subtraction output of original DSM and filtered DSM) technology based on drone survey detected significant erosion of the Pista geoglyph. Comprehensive documentation of geoglyphs and lines in the Palpa area dealt with older and newer projects that used aerial images and later also satellite imagery [3,27,35]. It can be said that most of the larger geoglyphs and lines have already been discovered; so unknown objects are probably only remaining objects that are under the resolution of satellite images or classic aerial data. The newest technology uses drones [2,5]. The success of drone mapping, survey method and discovering of new objects is documented by this article based on our work in 2016 and other recent findings of geoglyphs near Palpa city using a multicopter in 2018 [3,4,37].

2.3. Dating of Geoglyphs near Palpa

There were some geoglyphs investigated near Palpa city. Logical procedures (younger geoglyphs lie over the older ones), or iconographic similarities can be typically used for the dating. There are also physical dating methods—thermoluminescence and ^{14}C isotope analysis. In the case of the "Pista" geoglyph (the Llipata site, PP01-36, Figure 1), the age was calculated as a time span between 100 BC and 1100 AD [11,38,40,42]. Age variation is widespread; ages of the possible last exposure to Sun-radiation point to multiple cultures (Paracas, Nasca, and Wari culture) [39,42]. An additional dating is based on ^{14}C method and correlated to the period 1360–1410 AD [39,40]. Dating using ^{14}C is an acknowledged method, but the analysed artefacts cannot be 100% matched with the geoglyph. This is evident from the findings when their age variance is considerable. It certainly depends on the analysed sample [11,42]. For other geoglyphs in the neighbourhood (Goddess of fertility and Cerro Carrapo) it was the time span of between 1100 AD and 1560 AD [11,38,40–42]. This would not support the often quoted hypothesis that geoglyphs were created by Nazca cultures (the peak of which had occurred around 400–600 AD) and the older Paracas culture [38], and it would indicate that different types of geoglyphs are fundamentally different in nature, that they originated in different cultures and served therefore for a variety of purposes [11]. Some wooden artifacts could have been brought to the geoglyph later. One cannot eliminate possible material recycling—in a dry climate with a low number of timbers it was certainly possible to use the material several times. It cannot be assumed that organic residues (wood, corn residues) would reach the geoglyph by wind or water (geoglyph is a sort of high plateau). Dispersion in dating can mean that the area was used diversely for many centuries.

Other research [38,39] based on thermoluminiscence also shows that some geoglyphs near Palpa (San Ignatio and Sacramento) can be from the period 450–650 AD, which corresponds with pottery found [35,37,40], late Nazca culture and with frequent opinion of some scientists [38,42,43]. However, the thermoluminescence method has an error margin of up to several hundred years. The problem of finding the age of the geoglyphs is still not resolved. However, most scientists attribute them to the Nazca culture (100 BC to 800 AD) and Paracas culture (1200–800 BC to 100 BC). Dating through [14]C [11,42] is also uncertain in this case and inconclusive if organic (wooden) evidence is not found directly in the geoglyphs as a part of them. Dating was done successfully only in the "Mandala" geoglyph, which is after analysing the wooden parts (evidently used by geoglyph construction) using ^{14}C method, detected as evidently new [11,45]). This could serve not only as a guideline for dating but also for defining the use of some geoglyphs (Table 1).

Some wooden artefacts found directly on the "Pista" geoglyph by the Czech expedition to the Nasca region in 2012 could be probably a part of the geoglyph or original technical equipment, but it is not possible to prove it [11,19].

Table 1. An example of samples dating. During expedition in 2012, several samples from Nasca and Palpa locality were collected. [14]C age is derived from samples gathered on pampa Palpa, Peru (Pista), in 2012, and the same samples were processed later by better ^{14}C AMS technology (Accelerator Mass Spectrometry) in 2015. This analysis shows that e.g., sample Nr.13-098 is from the pre-Columbian age, but it is not evident if this sample has a relationship with "the Pista" geoglyph [19]. We cannot prove it. However, it is interesting why a 500-year-old piece of wood (in the desert area) lays on the geoglyph.

Sample	Description	Conventional Radiocarbon Age (BP)	Calibrated Age (AD)	P (%) Probability
13_097	Palpa, 1, geoglyph Pista	76 ± 88	1666–1783 1796–1949 1666–1949	39(2σ) 61(2σ) 96(2σ)
13_098	Palpa, 2, geoglyph Pista	434 ± 74	1440–1511 1572–1622 1415–1643	58 (σ) 35(σ) 95(2σ)
13_098 (AMS)	Palpa, 2, geoglyph Pista	473 ± 32	1420–1499	92

If the situation permits so, the documentation and subsequent research of geoglyphs can be done from the surface, from the air and from Earth's orbit. The results can be captured cartographically and can be analysed in a suitable environment, e.g., GIS.

3. Documentation Technologies

For the basic documentation of geoglyphs and selected lines (there are some lines several kilometres long), a contactless areal documentation technology is very suitable. It can provide a general overview of the investigated territory and also help us with the searching and detecting of new objects. This can be advantageously used for areas with low levels of vegetation in locations with archaeological sites like in Peru near cities of Nasca and Palpa.

3.1. Aerial and Satellite Imaging

The best overview and analyse of the whole investigated area is provided by aerial and satellite imagery. The old, available black and white aerial photographs from Palpa are not of sufficient quality for the today's point of view, as they cover only part of the territory and access to them is not easy, or they do not exist anymore [26,33]. Selected parts were even military areas at a certain time [9–11]. In some cases, satellite images do not have the sufficient geometric resolution (several lines have a width of only 10–30 cm), although this disadvantage has recently been receding with better geometric resolution of modern VHR (very high resolution) satellites like WorldView-3 with GSD 31 cm or WorldView-1, 2, GeoEye-1, Pleiades, and others with GSD 50 cm [25,27,44,46]. The easiest way to get appropriate local and detailed image data is to get it with the help of smaller airlines that operate from

Nasca airport. It is also possible to use the above mentioned modern drone technology; but this can be very problematic in terms of official authorization and a nearby airport.

Of course, existing satellite or aerial data is a perfect tool, but they cannot be considered completely perfect for all purposes. A final geometric resolution is still too small for fine details or it is too expensive. It is necessary to combine property with terrain exploration, if possible (in many locations or far countries it is not possible; for example, Pampa de Nasca with most known geoglyphs can be visited personally only with a special permission, with appropriate footwear and by accompaniment of a specialist from INC). Documentation or mapping is the only one part of the complex research. However, information can be obtained only by a personal visit of the site, samples collection, study of historical sources and their analysis and other scientific methods.

3.2. RPAS (Drones)

The newly-used abbreviation for drones, RPAS, explains accurately the nature of the device. It is a remotely piloted instrument equipped with various sensors (in particular, it is important to note that there is a person—the pilot, responsible for the equipment which is necessary by law). However, RPAS is often called UAV (unmanned aircraft vehicle) or UAS (unmanned aircraft system, it means UAV with terrestrial control system) or simple in slang "drone" [2,4,5]. These popular instruments serve for the non-contact mapping and monitoring of small areas. It is important to say, that RPAS technology combines close range photogrammetry with aerial photogrammetry and remote sensing and it can produce very accurate outputs in comparison with classical aerial photogrammetry, of course, on limited areas.

Based on the miniaturization sensors and control parts, RPAS provide typical photographic data, but also other data types such as multispectral, thermal, or laser scanned data can be taken. Due to this, it is possible to supply remotely sensed data for small areas very quickly and at a low cost. In comparison to photogrammetry or remote sensing, RPAS give us image data with GSD in cm. From overlapped images, DSM can be derived with a GSD better than one decimetre. RPAS are nowadays used in a wide field of usage; they can be used well in culture heritage for objects documentation, or in archaeological surveying. In this study, we show an archaeologically focused case project, where the outputs from RPAS are excellent [6,7].

We found a good solution in the eBee drone (model 2015); it is simple to use, and relatively obtainable for our university research and it is easy transportable. The main body and demountable wings consist of styropore foam. The drone contains an electrically powered engine with a pusher propeller, battery pack, inertial navigation system (INS), which includes GNSS and IMU (inertial measurement unit), radio-modem and piloting control system. A laptop or tablet with navigation software can be used for navigation and piloting.

The initial necessary information is pixel size (GSD), image overlapping, flight time, monitored area given by rectangle or polygon dimensions. The typical flight covers one square kilometre from an altitude of 100–200 m. Maximal flight time reaches 40 min. GSD depends on the flight altitude and sensor, and can be set from 2 cm for very low flights (normally 4–5 cm from a flight high 120–130 m).

As an output, a set of overlapped images is generated with an approximately known external orientation of photos from changeable cameras. Based on image correlation, DSM and orthophotos can be automatically generated. EBee weighs less than 1 kg and it is easily transportable and safe to use.

3.3. Measurement

An infrared camera was used due to the mild turbidity of the atmosphere and a haze. Altogether, 94 images using the Canon NIR camera ELPH 110 HS with 16 MPx resolution) were taken (Tables 2 and 3), Figure 2.

Table 2. Absolute Geolocation Variance (Pix4D software) in all coordinates (WGS-84).

	X [m]	Y [m]	Z [m]
Mean	−0.003030	−0.004991	0.122334
σ	0.706982	0.444677	1.748451
RMS Error	0.706988	0.444705	1.752725

Table 3. Flight and processing parameters based on photo set above the Pista geoglyph (Piux4D software).

Number of 3D points	12.563.698
Point density (per m^2)	21.06
DSM and orthophoto resolution (GSD)	5.05 cm/pixel
Output coordinate system	WGS84/UTM zone 18S

The image data was processed in Agisoft Photoscan software to a sparse and dense point cloud [2,27] and in Pix4D software directly to an orthophoto and DSM (Figures 4 and 5). From the dense point cloud, the high resolution DSM and orthophoto were derived with a pixel size (GSD) of 5 cm. The orthophoto was processed in ArcGIS to a very precise orthophoto map based on the GNSS receiver on-board the eBee drone. The GSD reaches 5 cm, but of course, the absolute position is not so high, because there was not an RTK (real-time kinematic—it is a technique of real-time corrections, providing up to centimeter-level accuracy based on measurement of a reference station) system at disposal. From time and technical problems, no GCP (ground control points) were used for the geolocation of all images and a created model. For this reason, the absolute accuracy was calculated in selected software Pix4D and it is approximately 4–5 m. However, the absolute accuracy in position is not so important in this case—the main objective was to analyse the previously invisible structure of the geoglyph.

3.4. DDSM—A Valuable Product

An orthophoto is the most used output from drones, but this type of image data has other capabilities in the creation of DSM [2,4,29]. A typical DSM in raster form and grey scale shows only little details. It is better to use coloured hypsometry, which is a shaded relief of differential DSM (DDSM) for example, which enhances small terrain details. There is not a common procedure how the DDSM can be computed—it depends on the scale of details, GSD, terrain and searched features. After some experiments, two different DDSM were created in ArcGIS software. Both were computed as the difference between the original DSM and the filtered DSM (with filter window 20 pixels and 50 pixels). Final DDSM and DSM have a GSD of 10 cm after filtration and resampling (Figures 4 and 5).

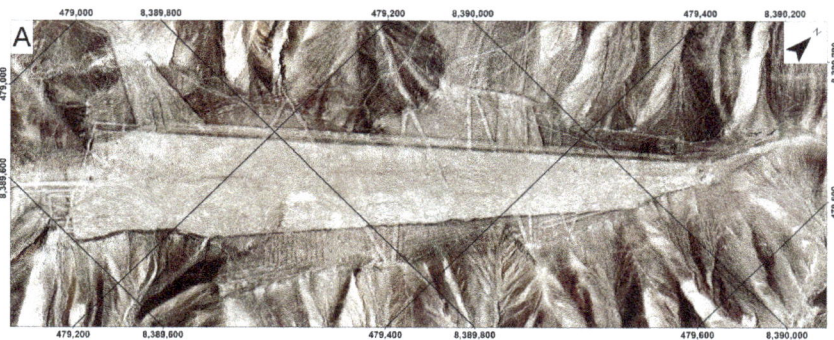

Figure 4. *Cont.*

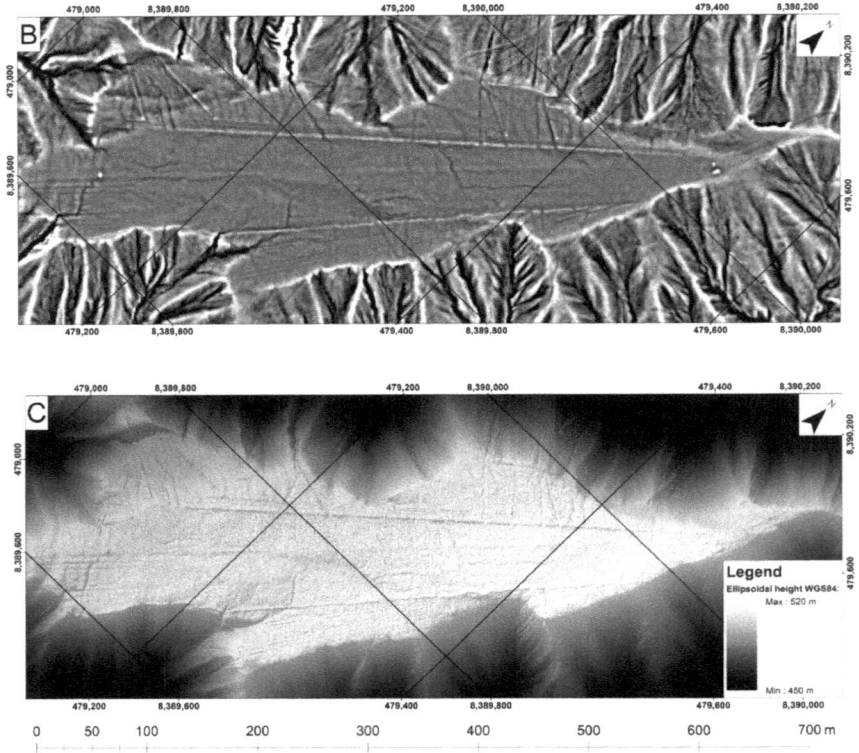

Figure 4. (**A**) Map of the "Pista" geoglyph in false colors, pixel size 5 cm, flight height approx. 120 m; (**B**) differential DSM (digital surface model) geoglyph "Pista", pixel size 10 cm, with visible erosion disturbances due to enormous precipitation in the episode El-Niño; and (**C**) original DSM geoglyph "Pista", pixel size 10 cm (software ArcGIS).

Figure 5. (**A**) Detail of the resulting colored orthophoto with the great spiral and (**B**) differential DSM shows otherwise invisible details—there are visible typical erosion rills (erosion rills are the result of torrential rainfall that is common during El-Niño's episodes; drainage water removes the surface material; in the next episode, the rills are further deepened. There is clearly visible damage of the geoglyph by this phenomenon).

3.5. RPAS Orthophoto

A new very high resolution orthophoto was computed from the RPAS photo set. Currently, these orthophoto and DSM are the most detailed cartographic outputs of the Pista geoglyph. The results have an important documentation value, showing the gradual destruction of the geoglyph and its destruction by the attacks of tourists and vandals who create their own drawings on the geoglyph or in its neighborhood. A detailed orthophoto (with a GSD of 5 cm) has been created, which revealed many previously unknown details. Unexpected new geoglyphs inside the main geoglyph were found).

3.5.1. Orthophoto Interpretation

The geoglyph Pista has good visibility in VHR satellite images after 2000. For example, the QuicBird-2 satellite supplied data with the GSD of 60 cm in panchromatic range from 2001 to 2014; QuickBird-2 was followed by other VHR satellites like OrbView, Pleiades, GeoEye, and others. This geometrical resolution allowed the creation of orthophotos during that time with an unprecedented very high resolution. Many archaeological sites were thus researched remotely based on satellite imagery [1,11,21,25,27,29,31,32,47]. However, there are features, which cannot be observed by satellites. The reason is their very small thickness, especially in their linear structures. Using satellite data, a vector map of geoglyph Pista was created, with visible linear and flat structure. The resolution of this output is limited by linear structures of approximately 40–50 cm [1,11], (see Figure 3).

Based on our documentation and research in 2016, an orthophoto with super resolution (GSD 5 cm) was computed. Canon NIR camera ELPH 110 was used for taking 94 images from a flight altitude of 130 m; the NIR camera was used, because it had better resolution compared to a second RGB camera, which was at our disposal, but at this time it was disturbing because of internal pollution by fine dust during past measurements and in a hazed atmosphere it gave better image quality (infrared rays pass more easily through the atmosphere). However, on the created orthophoto at high magnification it can be seen that the cheap camera that was used does not have the quality of a professional device—the image is blurred in colour and it is generally fuzzy The documented area with the Pista geoglyph and near neighbourhood was 0.6 km^2.

After the vectorising of visible features, a vector map was created which shows more detail than older outputs made from satellite or old aerial data (Figure 6).

3.5.2. New Discovered Features

In the created precise orthophoto, especially yet unknown geoglyph of a flying bird was discovered (dimension 30 × 30 m, thickness of the line 10–15 cm only, Figure 7), which is very similar to the well-known geoglyph [11] "El Condor" or "Hummingbird" from Pampa Nasca (Figure 8A, B). Discovering new geoglyphs was possible just thanks to a very detailed survey with RPAS. In a detailed analysis of a created orthophoto, a number of other geoglyphs can be found, and unfortunately a lot of modern works as well, particularly inscriptions or graffiti that destroy this remarkable monument. Unfortunately, this area is not under UNESCO and it is protected only by a local law.

Another interesting newly found small geoglyph apparently shows a guinea pig (Kecuan "cui", Figure 9). The guinea pigs were reared in the Andes as a source of meat already in pre-Columbian times and later they spread around the world. For some new findings, it is uncertain whether they are original geoglyphs or modern creations. The solitary stone at the beginning of the geoglyph Pista, containing many petroglyphs is definitely interesting; close to this stone a small geoglyph can be found, it was documented from the surface as the only one in 2012 during the inexpensive Czech university expedition. According to the technique and typology, one can say, that some newly found geoglyphs are original, some cannot be decided unambiguously and some are obviously modern (inscriptions, graffiti).

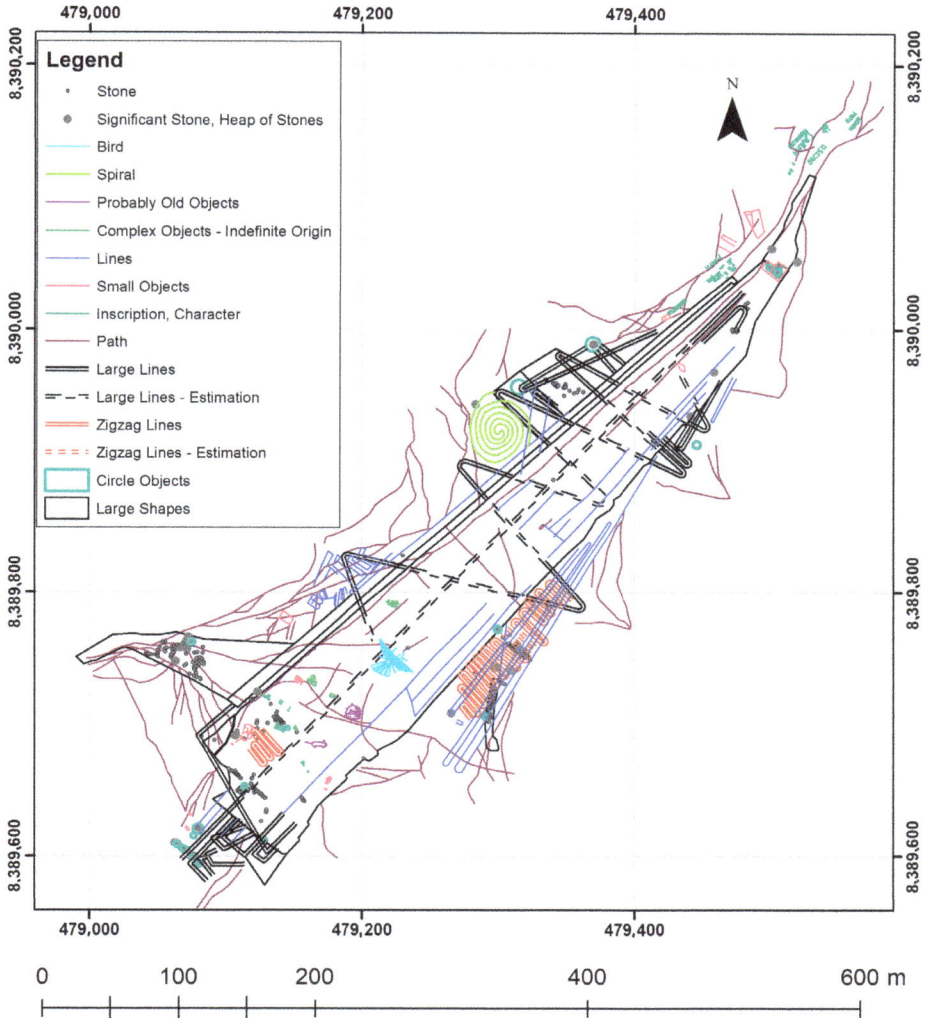

Figure 6. New vector interpretation of Pista geoglyph, based on a remotely piloted aircraft system (RPAS) orthophoto, shows other details unknown before.

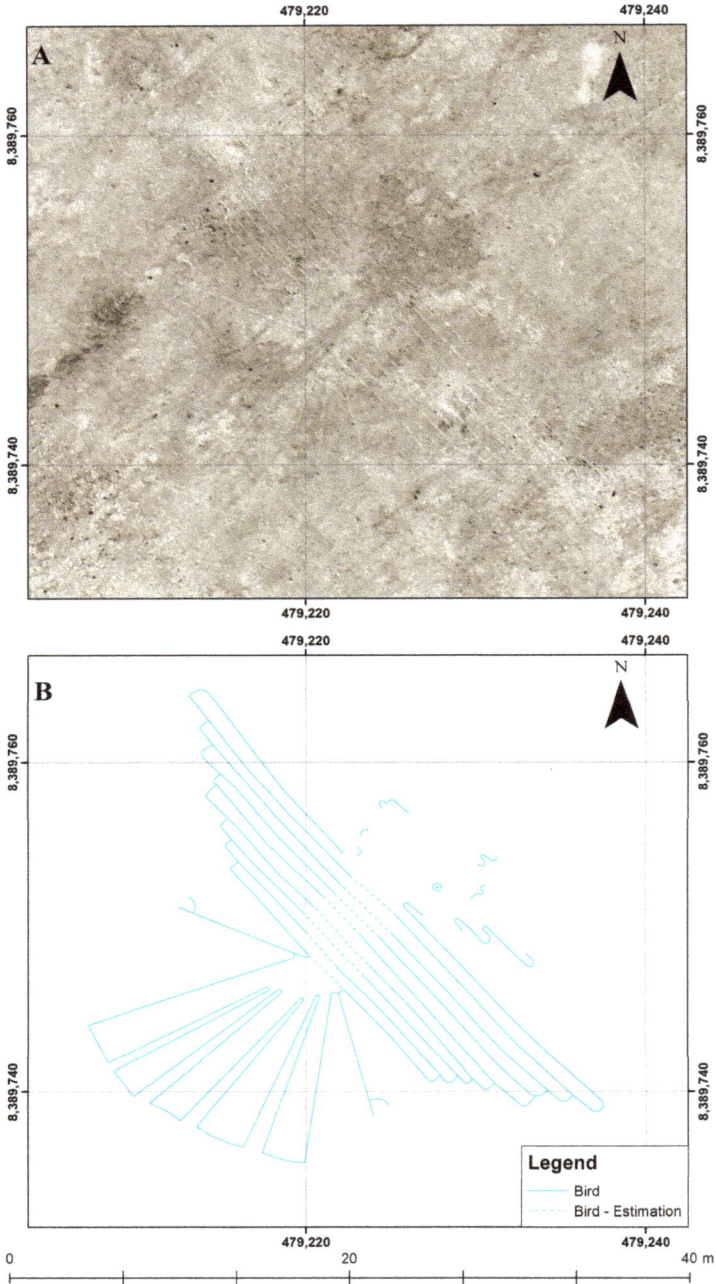

Figure 7. (**A**) Newly found geoglyph inside the well-known geoglyph Pista, similar to the "El Condor" or the "Hummingbird" geoglyph on the Pampa Nasca; line thickness 10–15 cm only (a detailed part of the original orthophotomap, created from aerial images taken by the eBee drone; false color) and (**B**) vector interpretation of the newly found geoglyph.

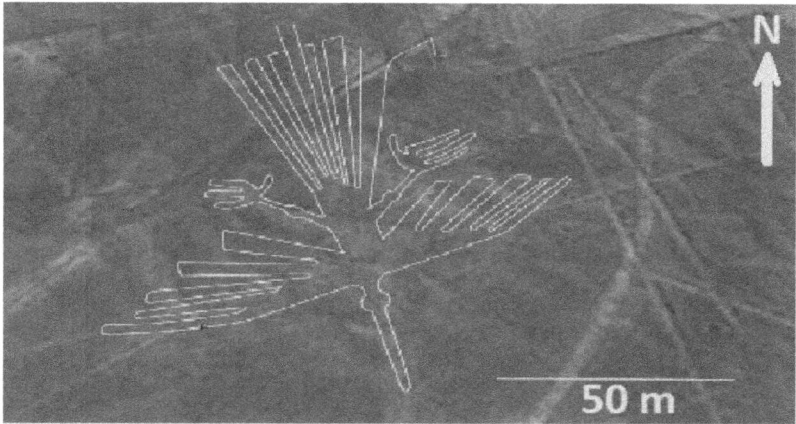

(A)

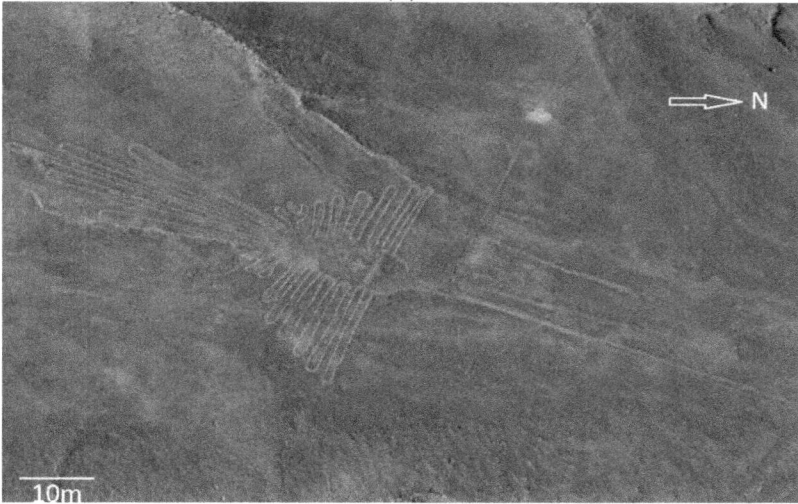

(B)

Figure 8. (**A**) Schema of the well-known geoglyph "El Condor" (Google Earth); (**B**) the "Hummingbird" geoglyph on the Pampa Nasca, hand held photo taken by K. Pavelka with approximate scale and orientation; 2004.

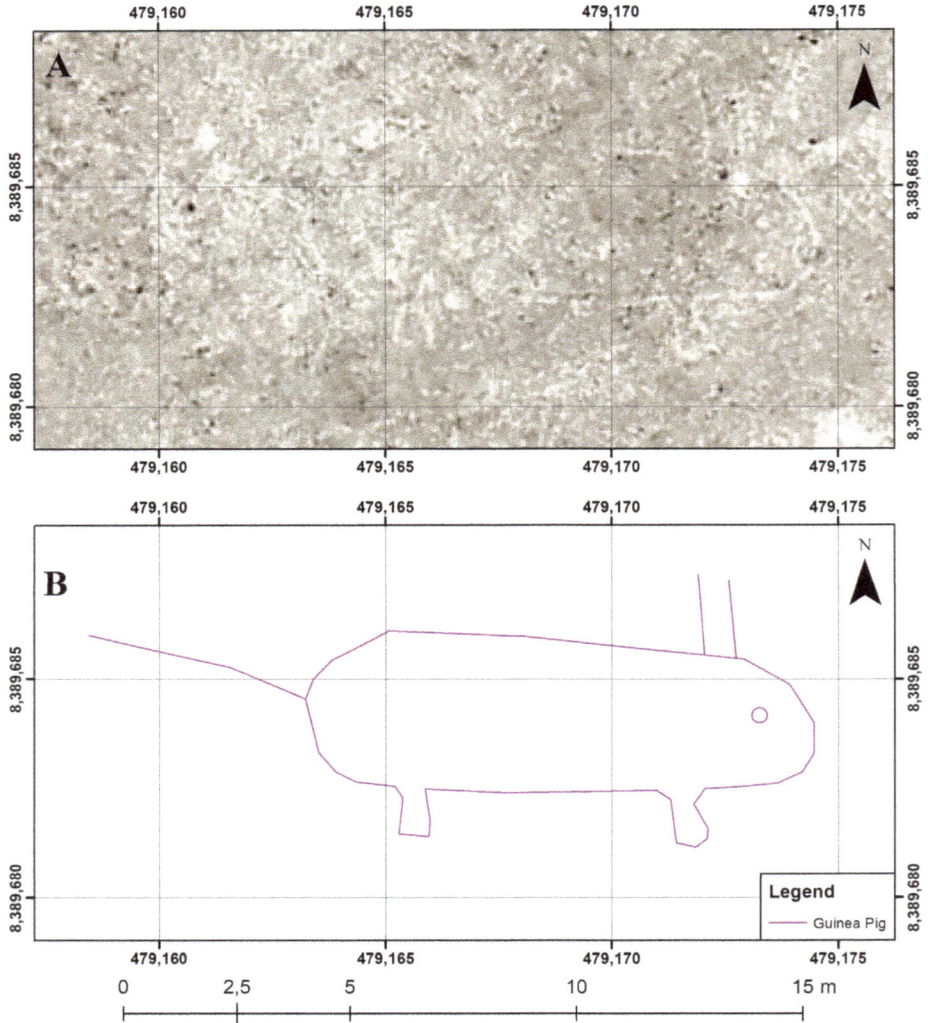

Figure 9. (**A**) Newly found geoglyph inside the well-known geoglyph Pista (may be Guinea pig): a detailed part of the original orthophotomap, created from aerial images taken by the eBee drone) and (**B**) a part of the orthophoto in false color (left) and vector interpretation of the geoglyph (right).

4. Final Remarks

Previous measurements that used satellite imagery or older aerial photography had limited geometric resolution that ends currently at approximately 0.3–0.5 m (Figure 3). There was no information about geoglyphs that exist as a part of previously documented and described ones. Overlapping lines and geoglyphs generally exist mainly in the Pampa Nasca and Palpa area. Linear shape objects with a thickness less than 10–15 cm are not visible on this data [1,11].

New documentation and research measurements using RPAS give practically better detailed results in an order of magnitude. In particular, it bought more detailed DSM and orthophoto. The created orthophoto has a radiometrically inferior quality, which is given by the type of ordinary cheap camera used in the RPAS and also by the flight plan and atmospheric haze (the results would be better for overflight in two perpendicular directions, but it was not possible due to time and weather conditions) [48].

Interpretation of new findings is not easy; some newly found geoglyphs are similar to the known geoglyphs and also can have similar meaning (the bird, the guinea pig, and several lines). As described above, it is very difficult to date geoglyphs. However, it is possible to use the recognized iconography. For example, there is a significant likeness of the newly found bird geoglyph with the well-known geoglyph "El condor" on Pampa de Nasca (Figure 8A) or with other geoglyph, which depict a bird (for example by M. Reiche called "Kosock's bird" [8,9] or with the "Hummingbird" geoglyph, Figure 8B) [9,21].

Newly found geoglyphs must be younger than the original Pista. They are from the time when Pista had not already fulfilled its original (may be religious) purpose. It can be assumed that the religious site could not be covered with graffiti. They show for example bird and guinea-pig, which are the typical observed objects; the guinea-pig is still kept in Peru as a domestic animal. A very detailed orthophoto from RPAS shows not only new geoglyphs, but also geoglyph destruction (modern graffiti and erosion). Some objects are difficult to decide whether they are new or historical. According to the technique and typology, one can say, that some newly found geoglyphs can be original (the Bird, the Guinea pig, and some lines), some cannot be decided unambiguously (small figures and remnants of drawings) and some are obviously modern (inscriptions, graffiti).

Some works refer to the occurrence of stone-houses, towers or simple stone-structures on a geoglyph [40,41]. According to the description, they were stone-houses, they would rather be shelters for few people and this fact would support Sonnek's theory of the technical use of the geoglyph [17,18]. Whoever created a megalithic geoglyph for religious purpose, would probably not build simple shelters on it, but rather altars, which is the most cited theory today [34,35,49]. These stone structures were discovered and investigated on the surrounding geoglyphs (trapezoids) [33,35]. It is also possible that these structures may be younger; dating with ^{14}C may bind to recycled (collected and found) material. However, todays recognized theory views these structures really as altars [34,35].

When using drones, many other discoveries in this locality can be expected. It is certain, that dating methods are improving in the similar way as documentation imaging and geophysical technologies, which can also bring about a change in our view of historical objects.

5. Conclusions

Satellite images can be used for the basic mapping of historical or archaeological sites around the world without personally participating in the site. Data from satellites can be processed to an orthophoto (today with GSD 50 cm) or to a digital relief or digital surface model (DRM and DSM), but based on experiences, it is evident that the resolution of satellite DSM (typically reaches GSD 1 m or worse) is not usually sufficient for archaeological purposes for fine structure detection. For this reason, RPAS mapping and documentation capability was tested and successfully used for the creation of a very detailed orthophoto and DSM. DDSM derived from DSM seems to be very helpful because it shows more details than ordinary DSM. In the above mentioned project, it was normally hardly visible interesting erosion furrows that were uncovered. In the case of documentation of the Pista geoglyph near Palpa in Peru, unexpectedly unknown geoglyphs inside the well-known Pista geoglyph were found. RPAS technology is reported to be very successful in archaeology (for small areas), because it is low cost, mobile, variable and gets very detailed DSM and orthophotos in an order of magnitude more accurate than typical VHR satellite images. This is very important, because some features are small or they are very thin. A great development of this technology in the future and improvements of sensors can be predicted; however, there is a big problem with the safety of RPAS and legislation, which can be totally varied in different countries of the world. It is also worth noting that there is competition to keep improving satellite and standard aerial data. It is obvious, that modern, more detailed research based in this case on RPAS technology still brings new insights. New geoglyphs found from very high resolution RPAS data may change existing interpretations of documented objects.

Author Contributions: Conceptualization, K.P. and J.Š.; methodology, K.P.; software, K.P. and J.Š.; validation, K.P., J.Š., and E.M.; formal analysis, K.P.; writing—original draft preparation, K.P.; writing—review and editing, E.M.; visualization, J.Š.; supervision, K.P.; project administration, E.M.

Funding: This research received no external funding.

Acknowledgments: This preparation of this paper was partly supported by an internal grant of the CTU in Prague, SGS18/056/OHK1/1T/11. We would like to acknowledge the support of the Instituto Nacional de Cultura del Perú (INC) and town hall of Nasca city (mayor Alfonso Canales Velarde, 2016).

Conflicts of Interest: The authors declare no conflict of interest.

References

1. Klokočník, J.; Kostelecký, J.; Pavelka, K. Google earth: Inspiration and instrument for the study of ancient civilizations. *Geoinform. FCE CTU* **2011**, *6*, 193–211. [CrossRef]
2. Raeva, P.; Šedina, J.; Dlesk, A. Monitoring of crop fields using multispectral and thermal imagery from UAV. *Eur. J. Remote Sens.* **2018**. [CrossRef]
3. Eisenbeiss, H.; Lambers, K.; Sauerbier, M.; Zhang, L. Photogrammetric documentation of an archaeological site (Palpa, Peru) using an autonomous model helicopter. In Proceedings of the International Archives of Photogrammetry, Remote Sensing and Spatial Information Sciences, Riva del Garda, Italy, 23–25 June 2014; Volume XXXIV-5/C34, pp. 238–243.
4. Pavelka, K.; Šedina, J.; Matoušková, E.; Faltýnová, M.; Hlaváčová, I. Using remote sensing and RPAS for archaeology and monitoring in Western Greenland. In Proceedings of the International Archives of the Photogrammetry, Remote Sensing and Spatial Information Sciences, 2016 XXIII ISPRS Congress, Prague, Czech Republic, 12–19 July 2016; Copernicus Publications: Göttingen, Germany, 2016; pp. 979–983.
5. Šedina, J.; Hůlková, M.; Pavelka, K.; Pavelka, K., Jr. RPAS for documentation of Nazca aqueducts. *Eur. J. Remote Sens.* **2018**. [CrossRef]
6. Cowley, D.C.; Moriarty, C.; Geddes, G.; Brown, G.L.; Wade, T.; Nichol, C.J. UAVs in Context: Archaeological Airborne Recording in a National Body of Survey and Record. *Drones* **2018**, *2*, 2. [CrossRef]
7. Campana, S. Drones in Archaeology. State-of-the-art and Future Perspectives. *Archaeol. Prospect.* **2017**, *24*, 4. [CrossRef]
8. Kosok, P. *Life, Land and Water in Ancient Peru*; Long Island University Press: Brooklyn, NY, USA, 1965; 256p.
9. Reiche, M. *Mystery on the Desert. Nazca (Peru)*; Offizindruck AG: Stuttgart, Germany, 1968; p. 92.
10. Schulze, D.; Zetzsche, V. *Bilderbuch der Wüste. Maria Reiche und die Bodenzeichnung von Nasca (Picture Book of the Desert. Maria Reiche and the Geoglyph of Nasca)*; Mitteldeutscher Verlag: Halle, Germany, September 2005; ISBN 3-89812-98-0. (In German)
11. Pavelka, K.; Klokočník, J.; Kostelecký, J. *Astronomical-Historical Question Marks of Mesoamerican and Peru*; Czech Technical University in Prague: Prague, Czech Republic, 2013; p. 235, ISBN 978-80-01-05219-8. (In Czech)
12. Teichert, B. Astronomical investigations of the Nasca Lines. In Proceedings of the Nasca Symposium, Bielefeld, Germany, 21–23 June 2006; pp. 87–101.
13. Johnson, D.W.; Proulx, D.A.; Mabee, S.B. The correlation between geoglyphs and subterranean water resources in the Rio Grande de Nasca drainage. In *Andean Archaeology II: Art, Landscape, and Society*; Silvermann, H., Isbell, W.H., Eds.; Kluwer Academic: New York, NY, USA, 2002; pp. 307–332.
14. Masini, N.; Orefici, G.; Danese, M.; Pecci, A.; Scavone, M.; Lasaponara, R. Cahuachi and Pampa de Atarco: Towards Greater Comprehension of Nasca Geoglyphs. In *The Ancient Nasca World New Insights from Science and Archaeology*; Lasaponara, R., Masini, N., Orefici, G., Eds.; Springer International Publishing: Cham, Switzerland, 2016; pp. 239–278.
15. Aveni, A.F. *Between the Lines: The Mystery of the Giant Ground Drawings of Ancient Nasca, Peru*; University of Texas Press: Austin, TX, USA, 2000; p. 88, ISBN 978-0-292-70496-1.
16. Aveni, A.F. (Ed.) *The Lines of Nazca*; American Philosophical Society: Philadelphia, PA, USA, 1990; ISBN 0-87169-183-3.

17. Sonnek, J. *From the Buttonhole to Unravel the Nazca Mystery*; EAN: Hlučín, Czech Republic, 2015; 148p, ISBN 978-80-260-8105-0. (In Czech)
18. Sonnek, J. The Nasca Plains without Mysteries. *Živá Archeologie (Living Archaeol.)* **2011**, 63–67. Available online: https://anzdoc.com/obrazy-planiny-nazca-jii-sonnek-bez-zahad.html (accessed on 26 October 2018).
19. Klokočník, J.; Sonnek, J.; Hanzalová, K.; Pavelka, K. Hypotheses about geoglyphs at nasca, Peru: New discoveries. *Geoinform. FCE CTU* **2016**, *15*, 85. [CrossRef]
20. Däniken, E.V. *Arrival of the Gods: Revealing the Alien Landing Sites of Nazca*; Tantor eBook: Old Saybrook, CT, USA, 1998; ISBN 978-1-86204-353-4.
21. Richter, C. Nasca GIS—An Application for Cultural Heritage Conservation. In Proceedings of the Nasca Symposium, Bielefeld, Germany, 21–23 June 2006; pp. 115–126.
22. Hanzalová, K.; Pavelka, K. Archaeo-astronomical investigation of historical objects in Peru. In Volume I Informatics, Geoinformatics and Remote Sensing. In Proceedings of the 13th International Multidisciplinary Scientific Geoconference, Bulgaria, Albena, 16–22 June 2013; pp. 321–328, ISBN 978-954-91818-9-0. [CrossRef]
23. Gullberg, S.R. Inca Solar Orientation in Southeastern Peru. *J. Cosmol.* **2010**, *9*, 2078–2091.
24. Hanzalová, K.; Pavelka, K. Documentation and virtual reconstruction of historical objects in Peru damaged by an earthquake and climatic events. *Adv. Geosci.* **2013**, *35*, 67–71. [CrossRef]
25. Pavelka, K.; Matoušková, E.; Hanzalová, K. New generation of satellite data and GoogleEarth: Instrument for the study of ancient civilization in Peru. EARSeL Workshop. In Proceedings of the 4th Workshop on Cultural and Natural Heritage, Matera, Italy, 6–7 June 2013.
26. Lambers, K. *The Geoglyphs of Palpa, Peru: Documentation, Analysis, and Interpretation*; Lindensoft Verlag: Aichwald, Germany, 2006; ISBN 3-929290-32-4.
27. Matoušková, E.; Hanzalová, K. Documentation of geoglyphs on Nazca Plain, Peru using remotely sensed data. In Proceedings of the 23rd CIPA Symposium, Prague, Czech Republic, 12–16 September 2011; Faculty of Civil Engineering, Czech Technical University in Prague: Prague, Czech Republic, 2011; pp. 358–365, ISBN 978-80-01-04885-6.
28. Sauerbier, M.; Lambers, K. A 3D model of the Nasca Lines at Palpa (Peru). In Proceedings of the International Archives of Photogrammetry, Remote Sensing and Spatial Information Sciences, Visualization and Animation of Reality-based 3D Models, Tarasp-Vulpera, Engadin, Switzerland, 24–28 February 2003. XXXIV-5/W10 (CD-ROM).
29. Sauerbier, M.; Schrotter, G.; Eisenbeiss, H.; Lambers, K. Multi-resolution image-based visualization of archaeological landscapes in Palpa (Peru). From Space to Place: BAR International Series 1568. In Proceedings of the 2nd International Conference on Remote Sensing in Archaeology, CNR, Rome, Italy, 2–4 December 2006; Oxford Archaeopress: Oxford, UK, 2006; pp. 353–359.
30. Pavelka, K.; Hanzalová, K.; Hlaváčová, I. Using radar data in archaeological sites. In Proceedings of the SGEM 2014, International Multidisciplinary Scientific GeoConference, Albena, Bulgaria, 17–26 June 2014; pp. 383–390.
31. Cigna, F.; Tapete, D. Tracking Human-Induced Landscape Disturbance at the Nasca Lines UNESCO World Heritage Site in Peru with COSMO-SkyMed InSAR. *Remote Sens.* **2018**, *10*, 572. [CrossRef]
32. Comer, D.C.; Chapman, B.D.; Comer, J.A. Detecting Landscape Disturbance at the Nasca Lines Using SAR Data Collected from Airborne and Satellite Platforms. *Geosciences* **2017**, *7*, 106. [CrossRef]
33. Reindel, M.; Gruen, A. The Nasca-Palpa Project: A cooperative approach of archaeology, archaeometery and photogrammetry. In *Recording, Modeling and Visualization of Cultural Heritage*; Baltsavias, E., Gruen, A., van Gool, L., Pateraki, M., Eds.; Balkema: Rotterdam, The Netherlands, 2006; pp. 21–32.
34. Conlee, C. The Mystery of the Nazca Geoglyphs. 2010. Available online: https://www.ceskatelevize.cz/porady/10314142270-tajemstvi-obrazcu-nazca/ (accessed on 10 October 2018).
35. Isla, J.; Reindel, M. New Studies on the Settlements and Geoglyphs in Palpa, Peru. *Andean Past* **2005**, *7*, 8. Available online: https://digitalcommons.library.umaine.edu/andean_past/vol7/iss1/8 (accessed on 26 October 2018).
36. Greshko, M. Exclusive: Massive Ancient Drawings Found in Peruvian Desert. *National Geographic*. 5 April 2018. Available online: https://news.nationalgeographic.com/2018/04/new-nasca-nazca-lines-discovery-peru-archaeology/ (accessed on 26 October 2018).

37. Silverman, H. Paracas in nazca: New data on the early horizon occupation of the rio grande de nazca drainage, peru. *Lat. Am. Antiq.* **1994**, *5*, 359–382. [CrossRef]
38. Greilich, S.; Wagner, G.A. Light Thrown on History—The Dating of Stone Surfaces at the Geoglyphs of Palpa Using Optically Stimulated Luminescence. In *New Technologies for Archaeology. Natural Science in Archaeology*; Reindel, M., Wagner, G.A., Eds.; Springer: Berlin/Heidelberg, Germany, 2009; pp. 271–283.
39. Rink, W.J.; Bartoll, J. Dating the geometric nasca lines in the peruvian desert. *Antiquity* **2005**, *79*, 390–401. [CrossRef]
40. Fassbinder, J.; Unkel, I.; Lambers, K. *Nasca: A Tiny Tool for a Large Line. Magnetometry and dating of Nasca Lines in Palpa, Southern Peru; Small Samples, Big Objects*; Eu-ARTECH Seminar: Munich, Germany, 2007; pp. 27–39.
41. Gorka, T.; Fassbinder, J. Geophysical survey with caesium magnetometry on the geoglyphs of Palpa/Nasca, southern Peru. In *Metalla*; Hauptmann, A., Stege, H., Eds.; Deutsches Bergbau-Museum Bochum: Bochum, Germany, 2009; pp. 45–47.
42. Unkel, I. AMS—^{14}C Analyzes on the Reconstruction of Landscape and Cultural History in the Palpa Region (Peru). Original in German Language: AMS—^{14}C Analysen zur Rekonstruktion der Landschaft-und Kulturgeschichte in der Region Palpa (S-Peru). Ph.D. Thesis, Ruprecht-Karls Universität Heidelberg, Heidelberg, Germany, 2006; p. 212.
43. Reindel, M.; Wagner, G.A. (Eds.) *New Technologies for Archaeology: Multidisciplinary Investigations in Nasca and Palpa, Peru*; Springer: Heidelberg/Berlin, Germany, 2009.
44. Hanzalová, K.; Pavelka, K. Map of Nazca geoglyphs. In Volume XL-5/W2 ISPRS Archives. In Proceedings of the TC V XXIV International CIPA Symposium, Strasbourg, France, 2–6 September 2013; Copernicus GmbH: Göttingen, Germany, 2013; Volume 2013, pp. 309–311. [CrossRef]
45. Pavelka, K.; Matoušková, E.; Faltýnová, M. Dating of Artefacts from Nasca Region and Falsification of "Mandala" Geoglyph. In Proceedings of the 16th International Multidisciplinary Scientific GeoConference SGEM 2016, Albena, Bulgaria, 30 June–6 July 2016; Volume 1, pp. 165–172, ISBN 978-619-7105-58-2.
46. Richter, C.; Teichert, B.; Hanzalová, K.; Pavelka, K. The Nasca Project—A German-Czech cooperation. In Proceedings of the 23rd CIPA Symposium, Prague, Czech Republic, 12–16 September 2011; ISBN 978-80-01-04885-6.
47. Nickell, J. *Unsolved History: Investigating Mysteries of the Past*; The University Press of Kentucky: Lexington, KY, USA, 2005; pp. 9–17, ISBN 978-0-8131-9137-9.
48. Marčiš, M. Quality of 3D models generated by SFM technology. *Slovak J. Civ. Eng.* **2014**, *21*, 13–24. [CrossRef]
49. Reindel, M.; Isla, J.; Lambers, K. Altares en el desierto: Las estructuras de piedra sobre los geoglifos Nasca en Palpa. *Arqueología y Sociedad* **2006**, *17*, 179–222. Available online: http://kops.uni-konstanz.de/handle/123456789/20392 (accessed on 26 October 2018).

geosciences

MDPI

Article

Locating Cave Entrances Using Lidar-Derived Local Relief Modeling

Holley Moyes [1],* and Shane Montgomery [2]

[1]   Department of Anthropology and Heritage Studies, University of California, Merced, CA 95343, USA
[2]   Cornerstone Environmental Consulting, Flagstaff, AZ 86001, USA; montgomery.shane.m@gmail.com
*   Correspondence: hmoyes@ucmerced.edu; Tel.: +1-520-820-6748

Received: 12 December 2018; Accepted: 6 February 2019; Published: 20 February 2019

Abstract: Lidar (Light detection and ranging) scanning has revolutionized our ability to locate geographic features on the earth's surface, but there have been few studies that have addressed discovering caves using this technology. Almost all attempts to find caves using lidar imagery have focused on locating sinkholes that lead to underground cave systems. As archaeologists, our work in the Chiquibul Forest Reserve, a heavily forested area in western Belize, focuses on locating potential caves for investigation. Caves are an important part of Maya cultural heritage utilized by the ancient Maya people as ritual spaces. These sites contain large numbers of artifacts, architecture, and human remains, but are being looted at a rapid rate; therefore, our goal is to locate and investigate as many sites as possible during our field seasons. While some caves are entered via sinkholes, most are accessed via vertical cliff faces or are entered by dropping into small shafts. Using lidar-derived data, our goal was to locate and investigate not only sinkholes but other types of cave entrances using point cloud modeling. In this article, we describe our method for locating potential cave openings using local relief models that require only a working knowledge of relief visualization techniques. By using two pedestrian survey techniques, we confirmed a high rate of accuracy in locating cave entrances that varied in both size and morphology. Although 100% pedestrian survey coverage delivered the highest rate accuracy in cave detection, lidar image analyses proved to be expedient for meeting project goals when considering time and resource constraints.

Keywords: Lidar; GIS; Mesoamerica; Archaeology; Caves; Landscape; Ritual; Visualization; Maya; Belize; Sacred

---

1. Introduction

The ancient Maya civilization (250 AD–950 AD) of Central America developed and adapted to a tropical jungle environment that is the second largest continuous expanse of tropical forest in the world next to the Amazon Basin [1]. For archaeologists, the heavily forested terrain and thick understory has long plagued pedestrian surveys designed to locate and identify anthropogenic features and settlement traces across the landscape, rendering such studies time consuming and costly. Because of these challenges, Mayanists have long sought to use high-altitude imaging to map archaeological sites and settlements, beginning with aerial surveys by Charles Lindbergh [2] and advancing to satellite imagery [3,4]. Satellite imagery revolutionized landscape archaeology in areas that were not heavily forested revealing structures, roads, and extensive landscape modifications, such as raised fields, terracing, and buried features, as well as natural large-scale remnant features that were difficult to detect with pedestrian survey [5,6]. This has not only afforded discovery, but allowed archaeologists to monitor changes over time including heritage destruction [7,8]. However, these surveys had less impact on archaeologists working in tree-covered areas because the satellite or aerial imagery could not penetrate the forest canopies to the earth's surface. It is only with the advent of airborne light detection and range (lidar) technology that we are now able to look "through" the trees and see the ground below at high resolution [9–16].

Lidar scans are carried out with instruments fitted to aircraft that emit pulses of light. In heavily forested terrain, some light will bounce off the canopy and vegetation, and some will rebound from the earth's surface (ground returns), such that physical models of the ground surface may be derived from those points. The returned data are then classified and displayed as 3D point clouds and can be further manipulated to create relief models of the earth's surface known as bare earth models. This enables researchers to locate both anthropogenic and natural landscape features over large areas that can be scanned in only a few days. Originally employed on a small scale in Europe [17], lidar-derived models are quickly becoming one of archaeology's most important tools.

While most archaeologists concern themselves with both unmodified landscapes and anthropogenic features, such as buildings, terraces, and roadways, cave archaeologists are most interested in subterranean spaces. Caves are archaeologically important because they are often sheltered or protected from wind, water, or other erosional or taphonomic processes, and contain a wealth of information due to their excellent preservation and deep stratigraphic deposits [18–20]. For archaeologists working in the Americas, cave sites are critical to understanding religious practices [21,22]. Among many ancient Amerindians, caves were thought of as sacred features of the landscape and the homes of powerful deities, an essential feature of ancient cosmologies. As such, they were sanctified places to conduct rituals. Among the ancient Maya, ritual cave use dates to as early as 1200 BC [23], and sites continue to be used today by traditional Maya peoples [24,25]. These underground spaces are not only of interest to archaeologists, but research in caves is germane to other fields such as climatology, hydrology, terrain analyses, and biological studies.

What exactly is a cave? Definitions are typically dependent on human interaction. For instance, in the Encyclopedia of Caves, they are described as "a natural opening in the Earth, large enough to admit a human being [26]. Similarly, in the Encyclopedia of Cave and Karst Science, the term is applied to "natural openings, usually in rocks, that are large enough for human entry." [27]. This non-specific term has come to mean any cavity in the earth and ontologically they are holes. There are three basic types of holes [28]: superficial hollows dependent on surfaces (that can include rockshelters, shallow, and deep caves with single entrances), perforating tunnels through which a string can pass (caves with multiple entrances), and internal cavities like holes in Swiss cheese, which are dependent on three-dimensional objects and have no contact with the outside environment (that could include caves that are closed off either naturally or anthropogenically).

In remotely sensed data, it is not possible to locate caves, but rather researchers seek to locate cave openings. This creates a challenge because the openings come in many shapes and sizes, can be horizontal or vertically aligned, and may be occluded by rock outcroppings [29]. Attempts to locate these sites have employed airborne infrared thermal scanners to identify thermal variation around potential openings [30,31], while other endeavors used multispectral scans and electromagnetic sensors to distinguish possible anomalies associated with karstic features [32,33]. These techniques were hindered by coarse resolution or issues arising from variables such as vegetation cover or atmospheric processes.

Lidar imagery has proven more successful, but most research has focused on locating sinkholes using automated techniques [34–40]. Sinkholes, located in karstic landscapes, are closed depressions in the earth's surface with internal drainage caused by subsurface dissolution of soluble bedrock [36]. Because they can be quite large, they are more easily viewed using remote sensing techniques. The impetus for most research in locating sinks is that these features may collapse as water tables drop, creating hazardous conditions and property damage. Few researchers have attempted to locate caves associated with sinkholes but there are some notable exceptions. For instance, Weishampel and his colleagues [38], investigating the karstic features in western Belize, were able to detect sinkholes and vertical shafts with diameters of over 5 m. However, this study omitted smaller shafts, and inflow and exsurgence systems, as well as horizontally accessed caves.

Yet, the caves of interest for archaeologists are often those with horizontal entrances that could be accessed without specialized equipment. While ancient people could build ladders or supply other means for entering vertical shafts, archaeological studies suggest that these were not the preferred venues for ritual activities [29]. Sinkholes can be quite deep and difficult to access, so typically, ancient people rarely used deep sinks; therefore, for archaeologists these locations are not necessarily desired targets. Sinkhole shafts are not only difficult to access, but they frequently contain "bad air" (high levels of CO_2), causing difficulty in breathing and rendering them potentially deadly. Additionally, for our purposes, many caves in Mesoamerica were host to large groups of ritual participants, and public access would have been a necessity in the ceremonial planning process, favoring horizontal entrances. While caves with smaller narrower openings were likely to have hosted rites that were smaller or more private with fewer participants, our research suggests that they would still have favored horizontal entrances with easier access.

Furthermore, for archaeologists, caves with smaller entrances are desirable targets because they are less likely to have been discovered by looters sustaining less damage to archaeological deposits and retaining a more complete artifact record. Typically, archaeological discovery of cave sites has depended on pedestrian survey or taking a "gumshoe" approach in which local people escort archaeologists to known sites. Once sites are discovered, they are by nature of their discovery disturbed simply by entry, whether by local people or researchers. In Mesoamerica, votive offerings left by ancient people in caves most often manifest as surface deposits that are particularly vulnerable to theft, looting, and vandalism. It is extremely rare for researchers to be the first to visit these sites. Therefore, artifact theft and movement, feature damage, and the disturbance of subsurface deposits is assumed in most cave contexts. Of the 82 caves in Belize investigated by the Belize Cave Research Project (BCRP), few had been spared by looters. There are only three known intact caves in western Belize: Actun Tunichil Muknal (Cave of the Crystal Sepulcher), discovered by geologist Tom Miller in 1989 and investigated by the Western Belize Regional Cave project in the mid-1990s [41,42]; Chechem Ha Cave (Poisonwood Water), a cave with a blocked entrance discovered by a local family who curated the site for tourist development [43]; and Ch'en Pi'x (Cave of the Awakening), a cave with a difficult entrance drop discovered by geologist Phillip Reeder and his crew [44].

There are a few ways that archaeologists can counteract this sort of heritage destruction. Public education is vital, but is a long-term process, whereas looting is occurring at a rapid rate as populations move into forested areas. Due to their remote locations, it would be difficult to protect even gated caves that could be easily broken into. Another complication is that many cave entrances are too large to be gated. Archaeological projects strive to record sites in their current conditions, but most caves must be considered "salvage" operations and we must acknowledge the incompleteness of the archaeological record [23]. Locating caves using lidar-derived models will potentially mitigate this problem by presenting an effective method for the discovery of sites so that they can be recorded, and research undertaken prior to looting.

In this paper, we present a methodology to locate potential caves sites by employing aerial lidar-derived point clouds. To analyze the lidar data, we use local relief models that require only a cursory knowledge of relief visualization techniques and no specialized skills in computer programming. We take a hybrid approach engaging both automation and manual evaluation, which has proved effective in discovering promising potentialities [29]. In our systematic study, we have been able to effectively predict locations of sink and horizontal cave entrances with fissure openings as small as 1 m in height. We begin by describing the area of our study's concentration, proceed to illustrate typological categories of cave openings, illustrate our methodologies and testing protocols, and discuss our results.

Setting: Geology and Karstic Background

Our survey area surrounds the site of Las Cuevas, a small to medium-sized ancient Maya pilgrimage center [45], located in Belize, Central America (Figure 1). The site has been under investigation by the Las Cuevas Archaeological Reconnaissance (LCAR) project, directed by Holley Moyes, since 2011. The central site core consists of 26 structures organized on an east/west axis facing two plazas (Figure 2). The architectural features surround a dry sinkhole that leads to the entrance of an extensive cave system running beneath the buildings. Archaeologists know little about settlement surrounding pilgrimage centers in the archeological record or their underlying economies; therefore, the current phase of the LCAR project seeks to better understand settlement patterns and ritual landscapes surrounding the site core [29]. Among the ancient Maya, ritual landscapes can include caves, mountains, and water features, but our study focuses on ritual cave sites and their surrounds, employing both lidar imagery and pedestrian survey to identity natural and cultural features. Although many caves were used by the ancient Maya, not all caves show evidence of human activity. Our research on ritual landscapes investigates why some caves were used and others were not by studying cave morphology, spatial patterning over the landscape, and proximity to surface sites. To do this, we look at both utilized and vacant cave sites; therefore, the goal of our survey work is to find all caves across the landscape, not just the ones that show evidence of ancient Maya ritual use.

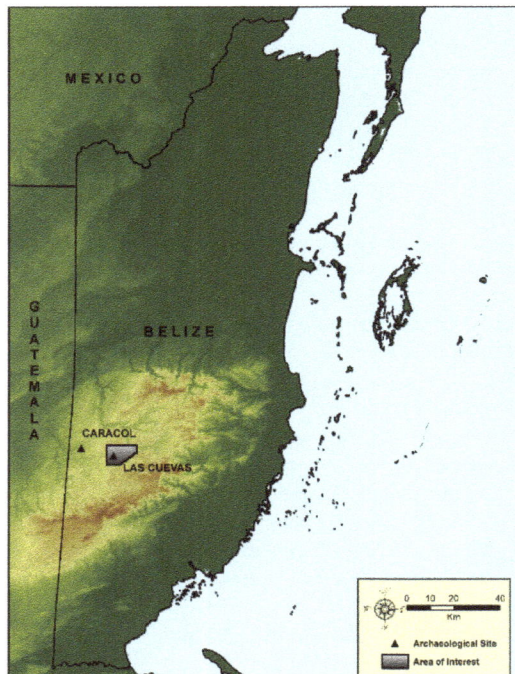

Figure 1. Belize map illustrates location of project area within the Chiquibul Forest Reserve.

The Las Cuevas site resides within the present-day Chiquibul Forest Reserve in western Belize near the southeastern extent of the Vaca Plateau (Figure 3), an area of protected land covered by tropical evergreen broadleaf forest [46] (Figure 4). Much of the Vaca Plateau falls within a geologic zone of Cretaceous limestone, with older Paleozoic layers found further west of the site and metamorphic and granitic material located within the Maya Mountains, 20 km to the north [46,47]. Covering more than 1000 km^2 [48], the Vaca Plateau extends east from Guatemala to the Maya Mountains, north to the boundary fault of the Belize River Valley, and south to the gorge of the Chiquibul River. Speleogenesis (cave origin and development) within the Vaca Plateau is associated with structural weaknesses in carbonate breccia derived from the Upper Cretaceous Campur Formation [49].

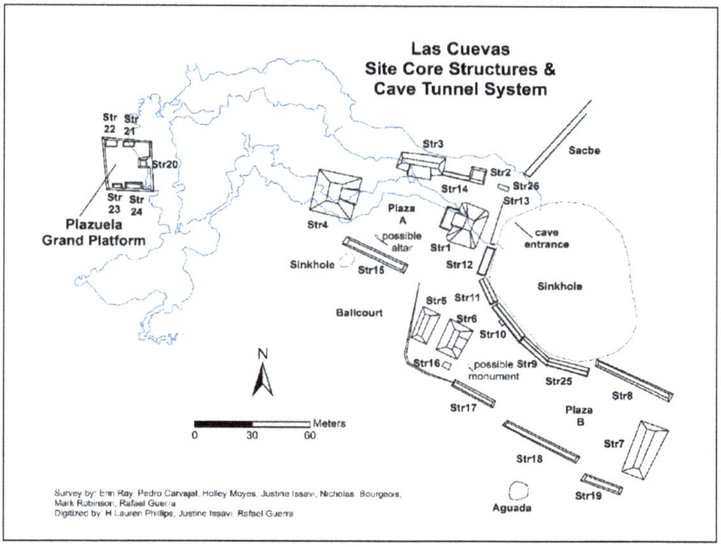

Figure 2. Map of Las Cuevas site illustrating cave system running beneath surface architecture (image courtesy of LCAR).

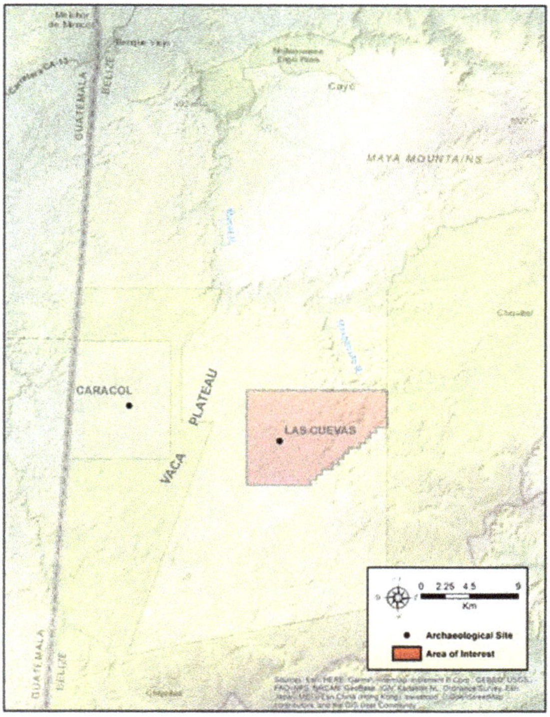

Figure 3. Regional map of Vaca Plateau illustrating location of Las Cuevas site and lidar coverage.

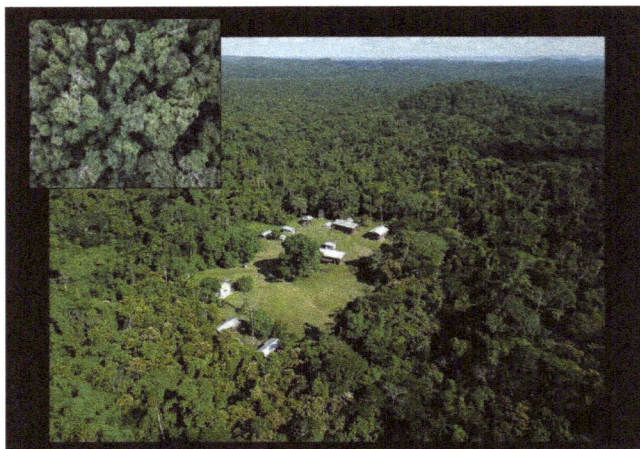

Figure 4. Aerial view of Las Cuevas site and research station in Chiquibul Forest Reserve. Inset shows view from above the tree canopy (images courtesy of LCAR).

In the vicinity of Las Cuevas, prevalent karst features include expansive sinkholes (dolines), restricted shafts and fissures, inflow and exsurgence (outflow) caves, rock shelters, isolated phreatic caves formed by dissolution, and solution scarps (escarpments created by corrosional undercutting). Through-flowing trunk conduits produced by surface stream penetration are rare compared to the extreme south of the Vaca Plateau, where the Chiquibul System forms the largest known hydrologically linked cave network in Central America [50]. The lack of conduits and the less dissected nature of the Las Cuevas area relate to the severe incising of the Macal and Raspaculo rivers to the east, which cut 200–300 m into the limestone, dropping the water table and preventing the formation of significant surface drainages on the plateau above [48].

Previous research indicates that many of the cave entrances detected within the Vaca Plateau are vertical, situated on sinkhole bottoms, hilltops, and the sides of residual hillslopes [51,52]. Shelter caves and shallow rock shelters are most common in association with solution scarps or at the bases of residual hill slopes. Likewise, inflow and exsurgence caves appear at the margin between hill slopes and valley bottoms, where stream flow has penetrated the limestone. Some morphologically complex systems exhibit a combination of entrance types, illustrating the evolution of the formation processes.

The goal of our project is to locate not only sinkholes and vertical drops, but horizontal entrances as well. Caves with exclusively horizontal access occur mainly on side slopes at elevations of 50–150 m above existing dry valley bottoms. For our purposes, we classified cave entrances by size and morphology [29]. Horizontal entrances are large (>5 m in width and over 2 m in height), medium (1–5 m in width and 1–2 m in height), small (<1 m in width and >1 m in height), and fissures (<1 m in height with varying width) (Figure 5). Vertical entrances necessitate down climbs or technical drops. They come in a variety of forms (cylindrical, conical, bowl, or pan-shaped) and can be quite small or may measure hundreds of meters across and tens of meters in depth [53,54], may contain water or be filled in with sediment. Sinkholes and shafts were classified by the maximum width of the diameter of the opening as small manhole-like entrances (<1–5 m), medium (<5–10 m), and large (<10 m). Sinkholes containing horizontal cave entrances at their base were noted.

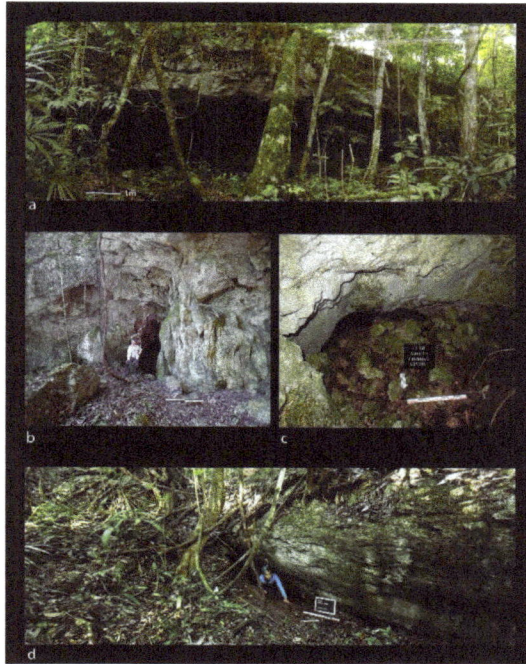

Figure 5. Horizontal cave entrances: (**a**) large entrance, Bird Tower Cave; (**b**) medium-sized entrance, Eduardo Quiroz Cave; (**c**) small entrance, Actun Uo (Uo Frog Cave); and (**d**) fissure entrance, Actun Z'uhuy Ch'en (Untouched Cave). (Photos courtesy of LCAR).

2. Methods

The Las Cuevas regional dataset was obtained in spring 2013 through the Western Belize lidar Consortium initiative, representing a cooperative effort between multiple archaeologists working within the country. The collection area covered approximately 1057 km^2 in western Belize, including much of the Belize River Valley and Vaca Plateau. Lidar acquisition was performed by the National Center for Airborne Laser Mapping (NCALM) over fourteen successive flights. An Optech Gemini Airborne Laser Terrain Mapper (ALTM, manufacturer, city, state abbreviation if in US or Canada, country) capable of 5 to 10 cm (2 to 4 in) vertical accuracy was mounted on a Cessna 337 Skymaster aircraft and flown at an altitude of 600 m above ground level with a ground speed of 60 m/s. The 325 survey lines were 137 m apart, resulting in a 300 percent swath overlap. The laser had a pulse rate of 125 kHz, a scan frequency of 55 Hz, and a scan angle of 18 degrees, resulting in a density of 15 points per square meter averaged throughout the entire area of interest [11,12]. Within the Las Cuevas lidar set, ground point returns ranged from 0.1 to 10 points per square meter based on associated canopy density and topography (Figure 6). Higher ground point returns were correlated with steep slopes, ridge tops, and modern infrastructure, while lower returns were found in low-lying bajo areas with thicker middle-level vegetation.

Point clouds were created from arrays of reflected light pulses emitted from the ALTM. In addition to flat or gently undulating terrain, vertical or steeply sloped terrain was captured within certain limits due to the scan angle (Figure 7). The entire point cloud represented the totality of pulse returns from upper canopy, understory, objects, and ground floor. The ground returns were classified by NCALM with automated methods utilizing TerraSolid Software and rendered to produce bare earth digital elevation models (DEMs) for further analysis of the natural and anthropogenic features.

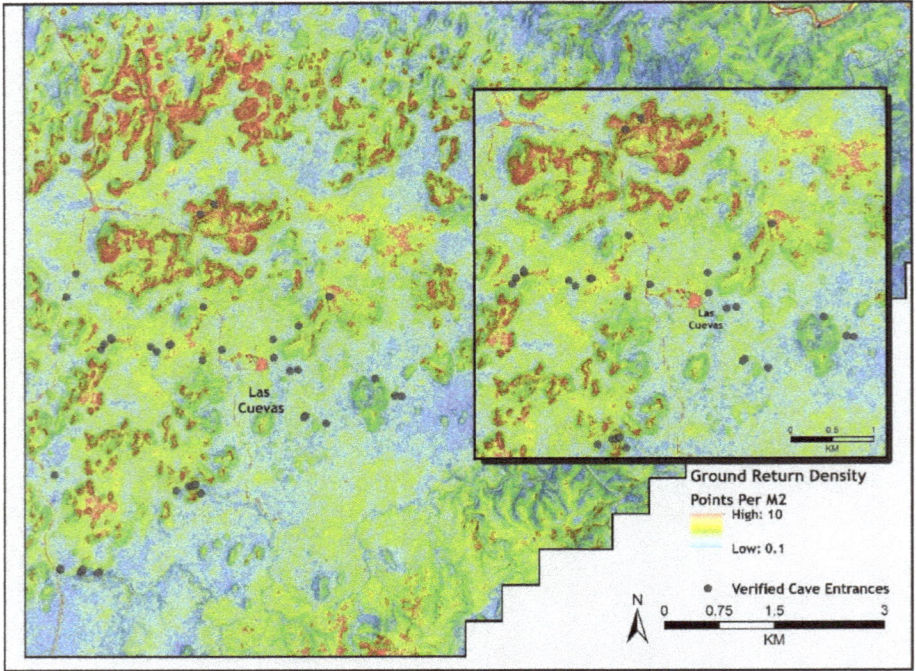

Figure 6. Map illustrating the density of LiDAR ground point returns correlated with topography.

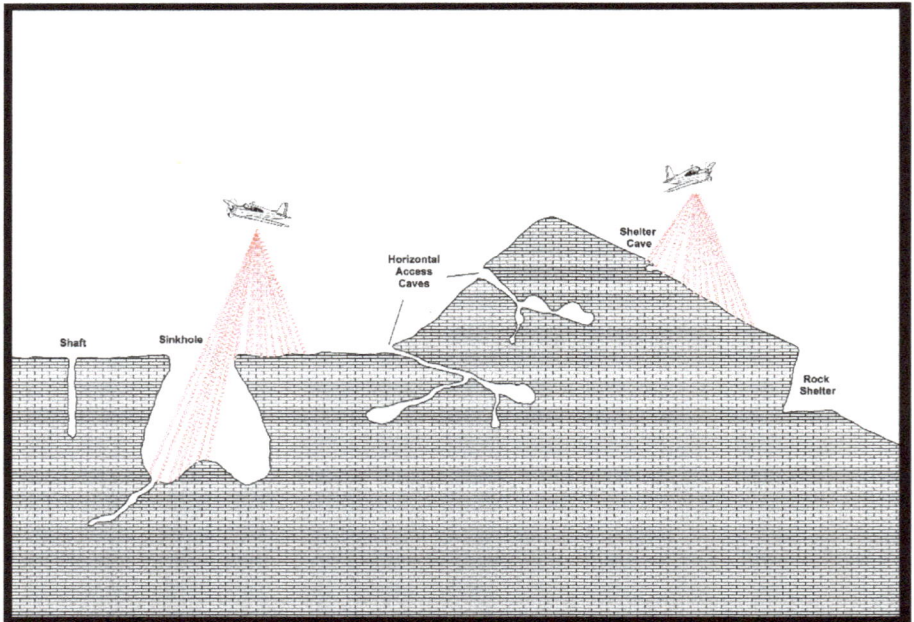

Figure 7. Schematic of lidar pulse angles (image by Shane Montgomery).

Data derived from lidar scanning, on the most basic level, contains X-, Y-, and Z-coordinates. Lidar data can be stored in a number of different formats, such as ASCII (text), binary, and LASer (LAS) format. LAS format has been endorsed by the ASPRS (American Society for Photogrammetry and Remote Sensing) [55] and was utilized for the Western Belize Lidar Consortium data. In addition to the spatial coordinates, the LAS files contained point classification, pulse intensity, RGB value, return number, scan angle, and overlap points. Once the lidar-derived DEM was created from this information, other relief visualizations and surface models were generated focusing on the detection of potential cave features.

Numerous raster data visualization techniques exist for the illumination of small-scale positive and negative relief features [56], including hillshade [57], sky-view factor [58], openness [59], local dominance [60], and local relief modeling (LRM). Each technique possesses strengths and weaknesses when applied to the task of remote cave detection. Hillshading has been widely utilized by archaeologists and other researchers for landscape recreations across diverse topographic settings. However, one disadvantage inherent within hillshading relates to its single illumination source, which can obscure individual features in shadow. Furthermore, hillshading is produced through cosine values based on the illumination and angle of the relief surface; the technique does not classify topographic change based on immediate variation and often oversaturates areas facing directly towards or away from the illumination source. Detection of these subtle elevation differences represents a crucial step in the documentation of all potential cave features across a given landscape.

After experimentation with multiple techniques, the project selected LRM for its ability to highlight both positive and negative anomalies in relation to the generalized landscape [56]. The method, when properly utilized, extracts human modifications and potential cave features on the immediate level and provides effective understanding of the associations between prehistoric constructions and ritual cave spaces. Although automated algorithms have been used to streamline sinkhole detection [37,61], manual methods are best employed in local relief modeling when attempting identification of the range of karstic features across a given landscape. Manual detection of potential cave features offers the user the ability to discern illusive entrance types (i.e., horizontal and inflow/exsurgence caves) not readily captured by automated methods. Additionally, natural landforms displaying concave characteristics, such as restricted valley bottoms, appear as negative relief and would require purging before additional analyses [62].

Detection of potential cave entrances was conducted primarily through ArcMap and LP360. Raster creation and editing was performed within ArcMap version 10.3, while LP360 was used visualize point clouds in profile and 3D views [29] An initial local relief model was created within ArcMap using the LRM Toolbox [63]; the LRM Toolbox ran the lidar-derived DEM through eight processing steps designed to extract minor variations in elevation based on overall trends within the larger surrounding landscape (Figure 8). A circular neighborhood radius of 25 m was established for the kernel standard size based on the presumed maximum extent of possible sinkhole openings within the Vaca Plateau [48]; this kernel size has been utilized in previous landscape modeling research on similar gradually sloping terrain with positive results [56,62].

Once initial processing was complete, the 222 km^2 Las Cuevas LRM was draped over a slope model produced from the original lidar-derived DEM and divided into 500 m × 500 m tiles to correspond with the LAS point cloud grid size previously established by NCALM. Alphanumeric designations were given to each grid tile to provide a unique identification number to each remotely detected cave entrance or karstic feature. Negative topographic anomalies were identified within each grid through LRM index value analysis and then compared to the existing hillshade layer provided by NCALM. Spatial and morphological characteristics of potential cave entrances were explored through a combination of manual visualization of point cloud data within LP360 and semi-automated LRM analysis within ArcMap. Features of interest were examined in profile (Figure 9) and 3D views of all points classified as ground returns in order to generate data on entrance width/height and maximum depth. Identified negative anomalies were classified based on morphology (horizontally or vertically accessed) and topographic setting (i.e., slope, valley, hill top). These potential cave entrances were then plotted onto the existing slope model for further analysis and uploaded to handheld GPS devices for field verification.

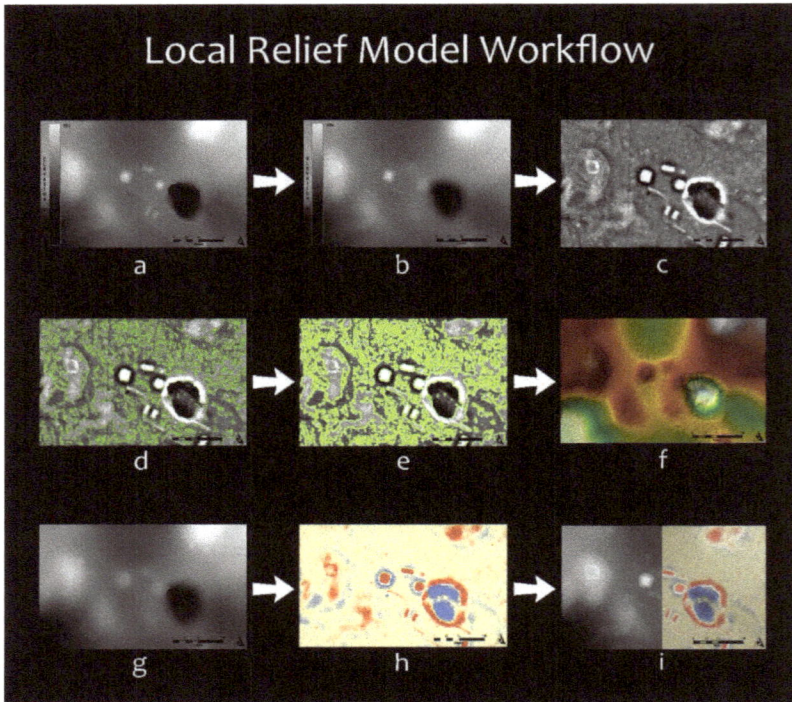

Figure 8. LRM Workflow Chart: (**a**) original 1-m DEM of Las Cuevas, (**b**) split image of original DEM and smoothed DEM, (**c**) difference of smoothed and original DEM, (**d**) creation of contour lines, (**e**) contours converted to elevation points, (**f**) Triangulated irregular network(TIN) created from elevation points, (**g**) Digital terrain model (DTM) generated from TIN, (**h**) LRM created from subtraction of DTM from original DEM, and (**i**) split image of original DEM and LRM.

Analysis of Remotely Detected Cave Entrances

Research conducted on our lidar coverage [29] identified 377 remotely sensed cave entrances within the 222 km^2 zone encapsulating a significant portion of the southeastern Vaca Plateau. Due to the size of the data set, we chose to narrow the project's focus to a 95.25 km^2 region representing the maximum extent of the prehistoric settlement zone of Las Cuevas. The updated study area contained approximately 42% (n = 157) of all formerly identified remotely sensed cave entrances (Figure 10).

Analyses of the lidar data suggested that closed depression sinkholes were the most commonly encountered cave entrance type within the sample, accounting for 66% (n = 104) of all remotely detected features. The morphology of these sinkholes was easily visualized in cross-section, with multiple returns located below the presumed ground level. Vertical shafts with openings less than one meter were rare throughout the study area (n = 3; 3%) and over half of the lidar-detected sinkholes displayed small openings between 1.1–5 m (n = 55; 53%) (Figure 11a). Medium-sized sinkholes with openings between 5.1–10 m comprised 31.7% of the closed depressions (n = 33). Sinkholes with diameters exceeding 10 m were infrequently encountered within the Las Cuevas study area, consisting of 12.5% (n = 13) of the doline features. Sinkhole depths ranged from 1.3–30 m, with an average negative relief of 5.7 m. The Las Cuevas central site core is built around a large sinkhole that measures 85 m with a depth ranging from 13–15 m. The cave entrance is 9 m in height and can be observed in profiles view in the lidar imagery (Figure 11b). The second largest sinkhole in the Las Cuevas region, Actun P'ook, measures 60 m across, and along with Las Cuevas is considered an outlier in this portion of the Vaca Plateau (Figure 12).

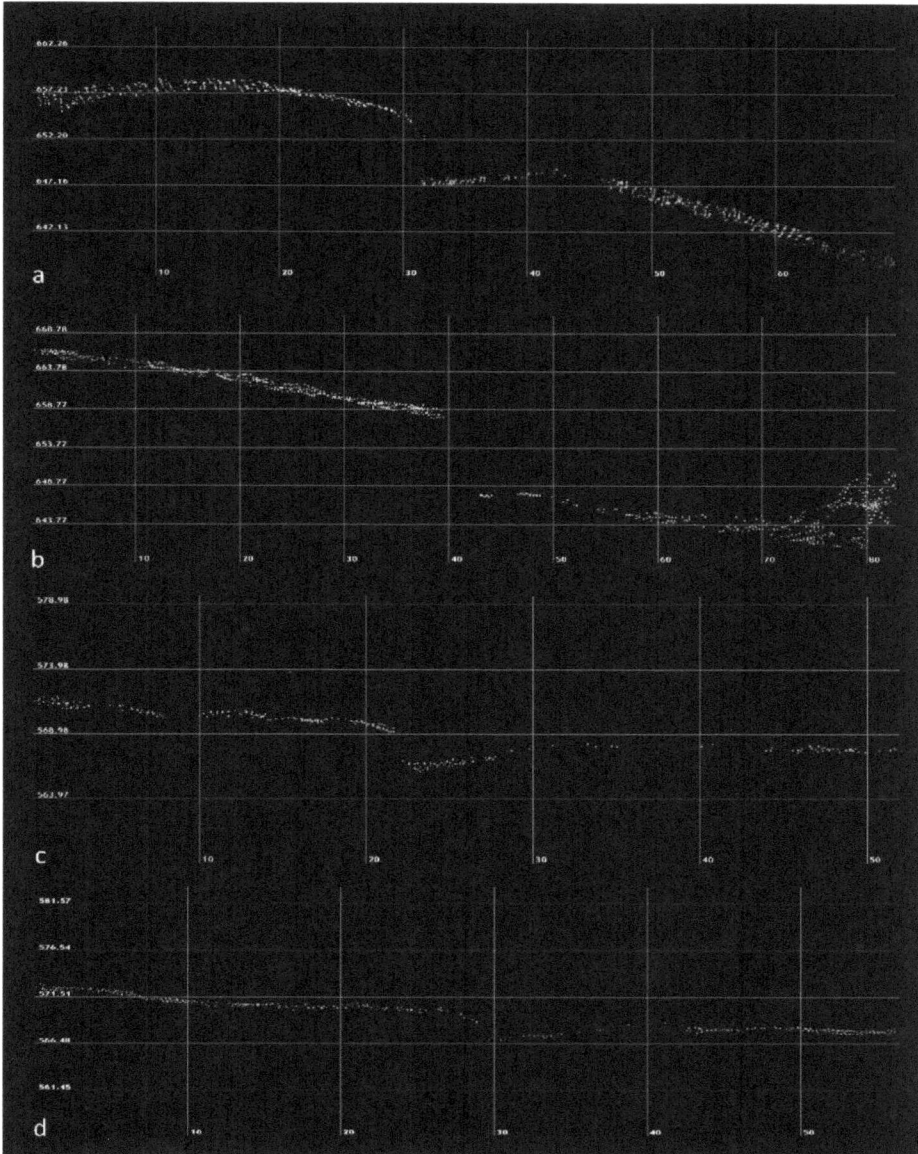

Figure 9. Point cloud profile images of four horizonal cave entrances rendered in LP360: (**a**) large entrance, Bird Tower Cave; (**b**) medium-sized entrance, Eduardo Quiroz Cave; (**c**) small entrance, Actun Uo (Uo Frog Cave); and (**d**) fissure entrance, Actun Z'uhuy Ch'en (Untouched Cave).

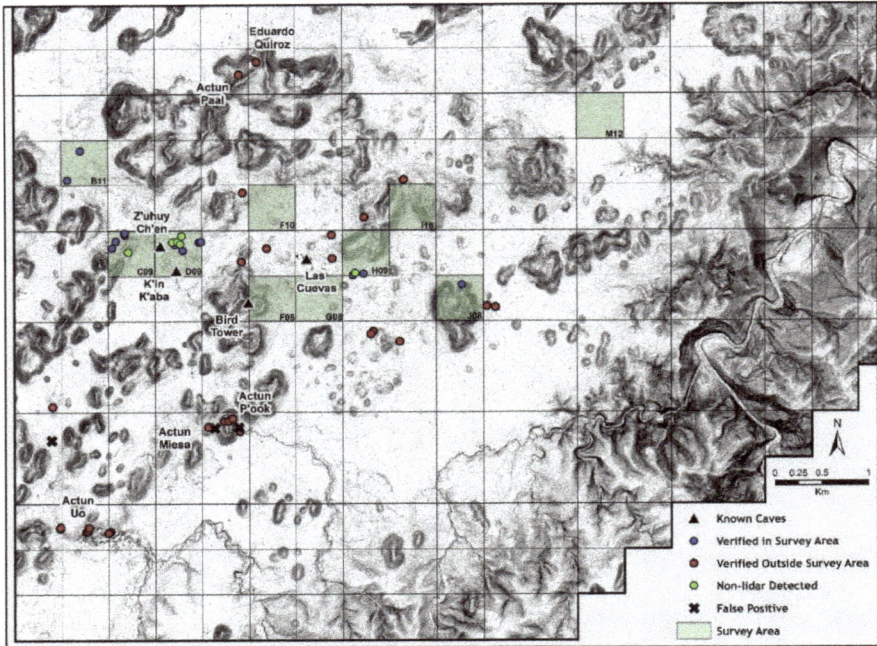

Figure 10. Map of pedestrian survey area for 2017/2018 field seasons. Green tiles indicate area systematically surveyed.

Horizontally accessed caves accounted for 28% (n = 44) of the remotely detected features. Caves of this type were less obvious when viewed in cross-section compared to the more common sinkhole features, and care was taken to differentiate between cliff faces and genuine horizontal entrances. In profile, sheer cliffs retained ground points and were associated with slopes angling away from the exposure. By contrast, potential horizontally accessed caves lacked ground returns in the void between the entrance ceiling and base. Furthermore, many examples displayed sloped areas angling towards the entrance, indicating movement of colluvial soils into the cave opening. Estimated entrance heights ranged from 1.8 to 16.6 m, with an average clearance of 4.7 m. Excluding one previously known cave (Z'uhuy Ch'en or Pure Cave) with a fissure entrance less than 1 m in height, no other horizontally accessed fissure entrances were detected within the study area. Fourteen caves classified with horizontal access (31.8%) display estimated entrance heights of less than 3 m; an additional 15 horizontally accessed caves (34.1%) featured clearances between 3.3 and 5 m. The remaining 15 entrances within the classification measure between 5.3–16.6 m in height. Within the sample, horizontal width ranged between 1.2 and 17.5 m.

The remaining remotely sensed features consisted of caves associated with drainage inflows and exsurgences (Figure 13). Eight inflow caves were identified within the Las Cuevas study area, all displaying horizontally accessed entrances. A majority of these entrances (n = 6; 75%) were small to medium in height and opening, with only two examples exceeding 5 m tall and 4.5 m in width. A single exsurgence cave was detected on an unnamed tributary of the Monkey Tail Branch, displaying a 5.5-m-wide opening and an estimated entrance of 12.6 m. No through-flowing trunk conduits were identified within the study area.

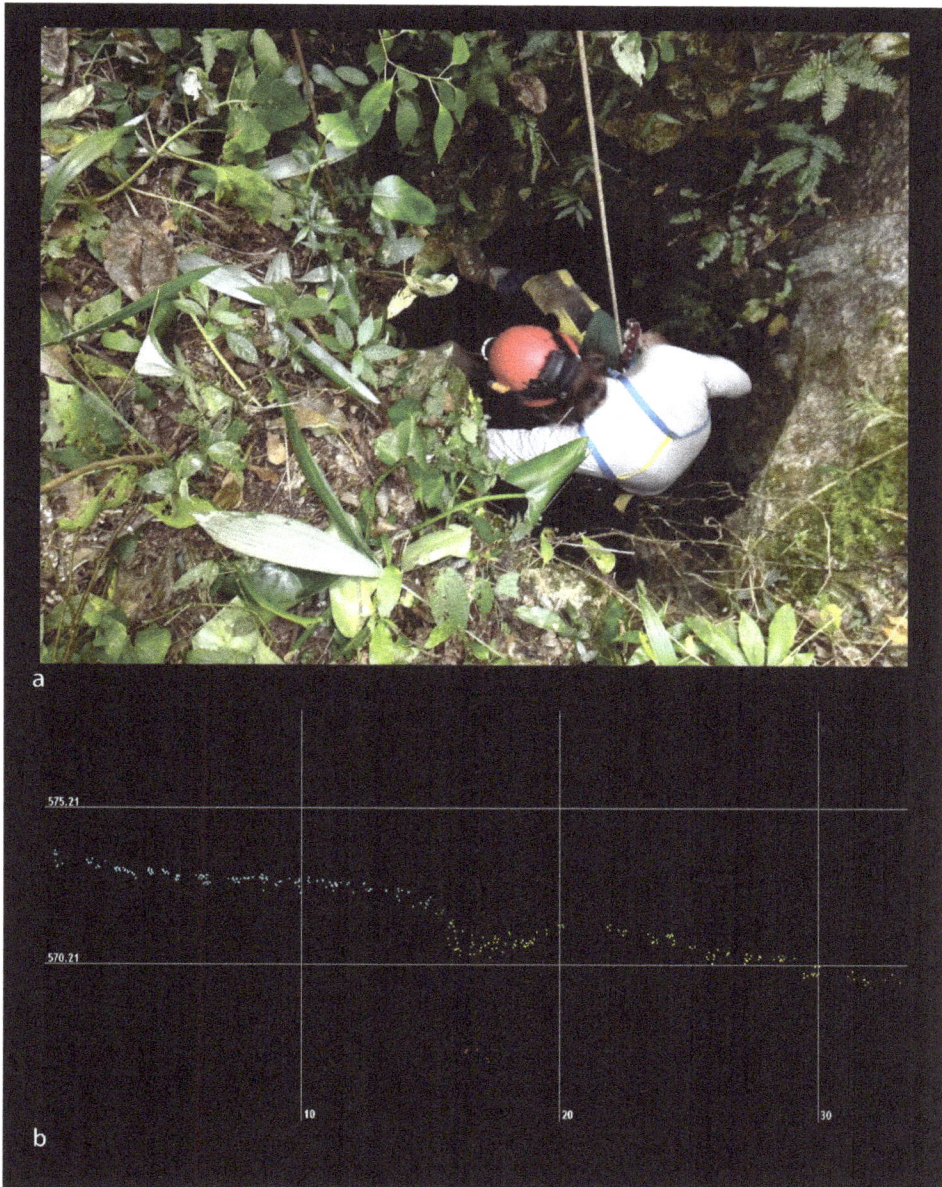

Figure 11. Photo of small shaft opening (**a**). (Matt Oliphant pictured in opening (top), Photo courtesy of LCAR), (**b**) and lidar point cloud in profile using LP360.

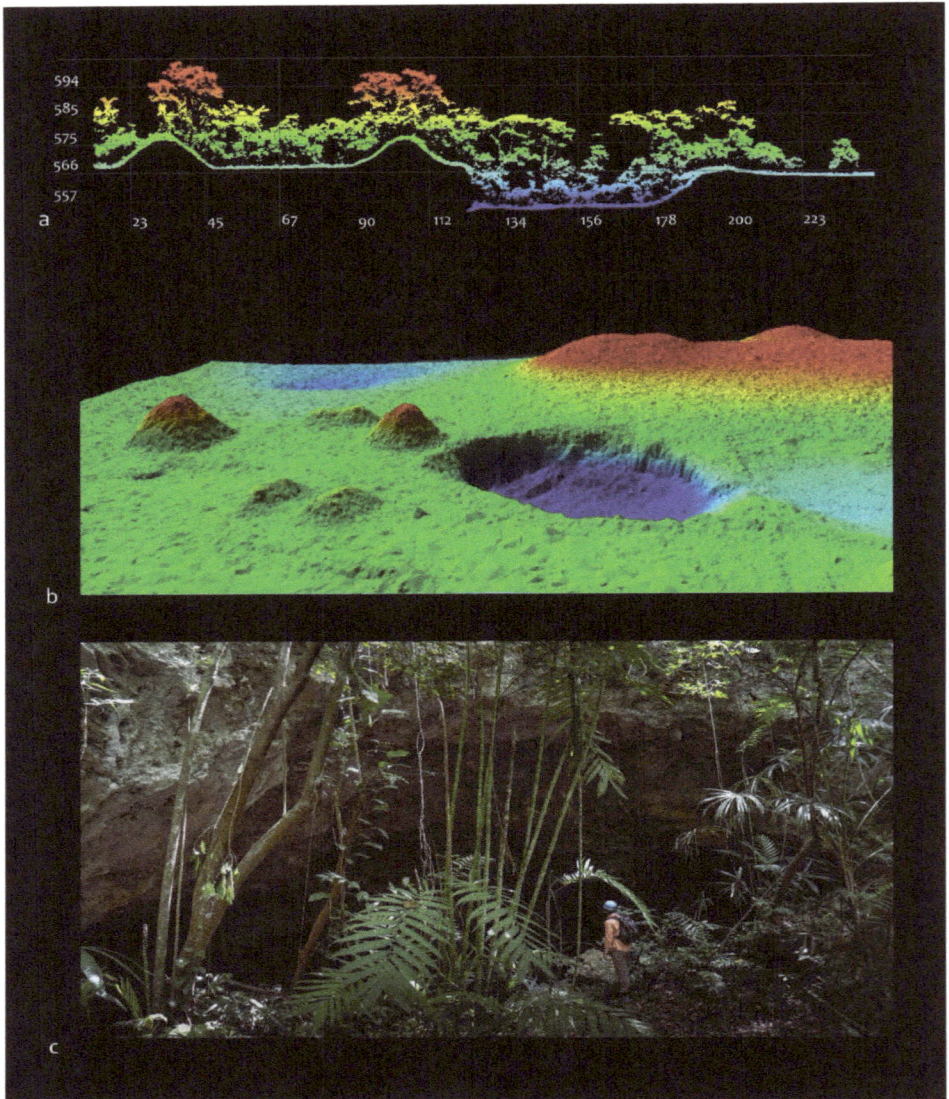

Figure 12. Las Cuevas surface site and sink hole: (**a**) point cloud of cave and sinkhole viewed from the north using LP360, (**b**) 3D image of sinkhole and surface structures (LP360), (**c**) photo of large Las Cuevas cave entrance (courtesy of LCAR).

Analysis of the topographic setting of all remotely detected cave entrances within the study area indicated that a majority of all features occur on hill slopes (n = 57, 42.7%), with a lower percentage (n = 55, 35%) situated within flat or gently undulating valleys. Potential cave entrances were less common along slope bases (n = 21, 13.4%) and within larger shallow depressions (n = 6, 3.8%). Seven remotely detected cave entrances (4.5%), exclusively minor sinkhole features, had been incorporated into ancient Maya agricultural terracing. Only one karstic feature, Actun P'ook, was located at the apex of a ridge. The distribution of karstic features within the Las Cuevas region illustrates the diverse formation of cave systems within this portion of the Vaca Plateau.

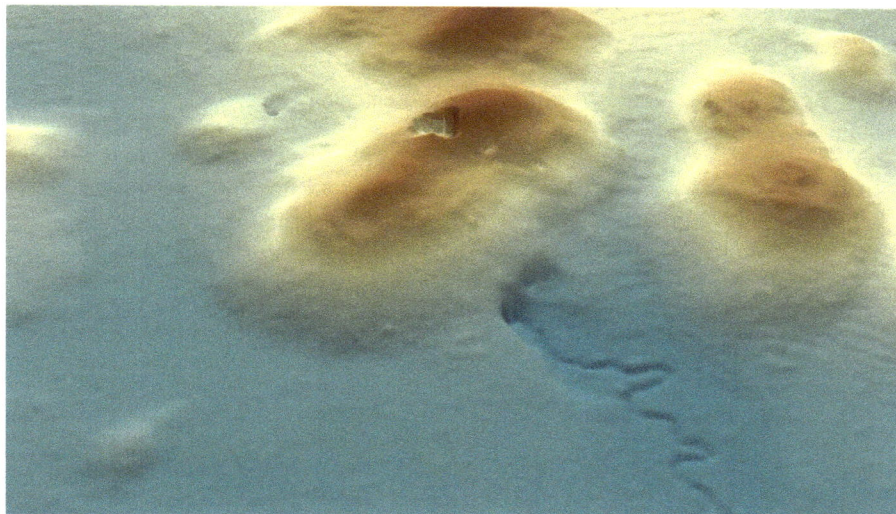

Figure 13. Lidar image showing drainage inflow indicating the presence of a cave entrance (Actun P'ook or Sombrero Cave).

Before our lidar survey, the project was aware of four known caves in the 95.25 km^2 survey area: the cave at Las Cuevas (large sink with horizontal entrance), Bird Tower Cave (large sink with horizontal entrance), K'in Kaba (large horizontal entrance with two openings), Zuhuy Ch'en (fissure entrance). The lidar images suggested that there were many more potential cave sites and possibly large cave systems in the project area. However, despite advanced technology, the accuracy and validity of remotely detected data remained in question, particularly regarding sub-surface features. We expected to find two types of errors: false positives (anomalies in the lidar images that resembled cave openings), and cave openings that went undetected in the lidar visualizations.

To test our findings, we used two methods of survey, each designed to uncover different types of errors and that would help to correlate features in the imagery and those on the ground. First, we directed an opportunistic survey (Survey 1) to verify select caves identified on the lidar. In this method, we directly visited potential cave openings based our the lidar imagery. This method would help to ferret out false positives and help us to calibrate our interpretations with real-world data. In Survey 2 we conducted a systematic pedestrian survey within 10 dispersed grids around the Las Cuevas site center. The grids were located in various terrains including bajos (low-lying swamps and low hills) (B11, F10, G08, M12), karstic ridges, hills, and valleys (C09, D09, F08, H09), and steep slopes (J08, I10). The aim of this method was to obtain total coverage of specified target areas to account for undetected openings, as well as false positives within the data set. This method was much more laborious and time consuming.

The systematic survey was completed over two six-week field seasons in the summers of 2017 and 2018 using two to three crews of three people each. The 95.25 km^2 survey area was divided into 381 contiguous 500 m × 500 m tiles. Each potential cave entrance previously detected through remote analysis was provided an identification number based on an alphanumeric grid system (i.e., A03_C01). Only caves with moderate to heavy ancient Maya usage were given names by our Maya in-country partners. Ten tiles, a combined 2.5 km^2 were selected for systematic pedestrian survey comprising 2.6% of the target area. The objective in tile selection was to survey a variety of micro-landscapes and topographies (such as hilly or flat, boggy vs. dry) to determine the active factors influencing lidar detection within the project area. By selecting a subset of the data, it was then possible to cover 100% of the terrain within the tile, assuring that no karstic features escaped observation.

The logistical goal of the Las Cuevas survey was to inspect targeted areas in a methodical fashion to document the totality of natural and anthropogenic features in a dense, dynamic sub-tropical forest environment. Traditional survey methods often employed in more open landscapes proved non-viable in the thick rainforests of the Vaca Plateau. Instead, a method more suitable to the environment was used. Within ArcMap, a fishnet grid was created for each tile consisting of 50-m lattices aligned to the cardinal directions. The fishnet, along with the remotely detected cave point data, were uploaded to Garmin 64s GPS units for each survey crew. Survey crews cut through vegetation to create paths along the 50-m spaced transect, stopping at each vertex and radiating outward counterclockwise in a 25-m arc, providing total coverage between each point.

3. Results

Of the 152 cave openings revealed in our lidar imagery, 29 (19%) were visited opportunistically in Survey 1 (Table 1). Of these, 26 (86%) proved to be cave openings or sinks, though some represented multiple entrances to the same cave system. For instance, Actun Uo (Uo Frog Cave) had five entrances, all of which were detected on the lidar image. Survey 1 located three false positives as well, which included one sheer limestone face that contained no rock shelter area or shelter cave (E05_C05), one depressed interior drain with dirt slopes (A05_C03), and one area of dead trees and thick vegetative growth incorrectly identified as a small horizontally accessed cave (E05_C07). Out of ten predicted horizontally accessed sites, we located eight actual entrances on pedestrian survey. Therefore, 70% of the predicted horizontal entrances were confirmed.

Table 1. Caves located in Survey 1 and Survey 2, and four previously known caves.

Survey 1—Cave Openings Suveyed Opportunistically								
Cave Name	Cave_ID	Cave_Type	Size	Max Width of Opening (m)	Max Sink Depth/Hor Ent Ht (m)	Topography	Verified	Artifacts Present
Actun Uo	B03_C02	Inflow_Drain	Small	0.7	1	Valley	Yes	No
Actun Uo	B03_C01	Inflow_Drain	Small	1.2	2	Valley	Yes	No
Actun Uo Ent 1	A03_C01	Hor Ent	Large	5	2.9	Depression	Yes	Yes
Actun Uo Ent 2	A03_C02	Hor Ent	Medium	2.8	2	Depression	Yes	Yes
Actun Uo Ent 3	B03_C03	Hor Ent	Large	7.8	7.4	Solution Scarp	Yes	Yes
Actun P'ook Ent 1	E05_C03	Hor Ent	Large	16	6.5	Side Slope	Yes	Yes
Actun P'ook Ent 2	E05_C02	Sink	Large	60	30	Hill Top	Yes	Yes
Actun P'ook Ent 3	E05_C04	Hor Ent	Medium	4.4	8	Side Slope	Yes	Yes
Actun P'ook Ent 4	E05_C06	Hor Inflow Drain	Large	14.2	11.5	Slope Base	Yes	Few
Actun Miesba	E05_C01	Sink-Hor Ent	Large	11.9	10	Side Slope	Yes	Yes
Actun Paal	E13_C01	Hor Ent	Medium	4.3	4	Side Slope	Yes	Yes
Eduardo Quiroz	F13_C01	Hor Ent	Large	6	7.4	Side Slope	Yes	Yes
Dinoaur Egg Cave	A06_C01	Inflow_Drain	Large	17	8.7	Valley	Yes	Yes
NA	I07_C01	Inflow_Drain	Medium	2.8	2.2	Slope Base	Yes	No
NA	K08_C02	Inflow_Drain	Small	1	0.5	Valley	Yes	No
NA	B03_C04	Sink	Large	11	2.5	Side Slope	Yes	No
NA	C03_C01	Sink	Large	17.5	7.2	Side Slope	Yes	No
NA	G09_C02	Sink	Small	1.5	3.2	Side Slope	Yes	Few
NA	H10_C01	Sink	Medium	6	5.8	Side Slope	Yes	No
NA	E09_C02	Sink	Medium	7.5	3	Valley	Yes	No
NA	E10_C02	Sink	Medium	5.2	7.7	Valley	Yes	No
NA	F09_C01	Sink	Small	4	3.3	Valley	Yes	No
NA	G09_C03	Sink	Medium	5.1	3.7	Depression	Yes	No
NA	H07_C01	Sink	Large	11.1	4.5	Valley	Yes	Few
NA	H07_C02	Sink	Large	23	4.3	Valley	Yes	No
NA	I11_C01	Sink	Small	2.9	2.9	Valley	Yes	No
NA	K08_C01	Hor Ent	Medium	3.3	3.5	Slope Base	Yes	No
NA	E05_C05	Hor Ent	Small	4.1	6.4	False Positive	Yes	No
NA	E05_C07	Hor ent	Small	3.6	2.5	False Positive	Yes	No
NA	A05_C03	Sink	Large	25	5.2	False Positive	Yes	No

Table 1. Cont.

Cave Name	Cave_ID	Cave_Type	Size	Max Width of Opening (m)	Max Sink Depth Hor Ent Ht (m)	Topography	Verified	Artifacts Present
Survey 2—Systematic Coverage of Survey Tiles								
NA	B11_C03	Shaft	Small	0.5	5.5	Valley	Yes	No
NA	J08_C01	Shaft	Small	1	2.3	Slope Base	Yes	No
NA	C09_C03	Sink	Small	2.4	3.9	Slope Base	Yes	No
NA	C09_C04	Sink	Small	2.5	2.2	Valley	Yes	No
NA	D09_C02	Sink	Medium	5.5	7.9	Valley	Yes	No
NA	D09_C04	Sink	Medium	6.8	5.4	Valley	Yes	No
NA	C09_C01	Sink	Medium	7.9	5.3	Valley	Yes	Few
NA	D09_C03	Sink	Medium	9.1	3.5	Valley	Yes	No
NA	B11_C01	Sink-Hor Ent	Large	10.5	10.5	Valley	Yes	Few
NA	H09_C02	Sink	Large	11.5	2	Valley	Yes	No
NA	H09_C01	Sink	Large	12	8	Valley	Yes	No
NA	D09_C05	Sink	Large	13	6.3	Valley	Yes	No
NA	C09_C02	Sink-Hor Ent	Large	15.2	12.5	Valley	Yes	Yes
Cave openings not detected by lidar								
NA	H09_T02	Sink	Small	1	6	Valley	Yes	No
NA	D09_T09.1	Sink	Small	2.1	3.3	Valley	Yes	Yes
NA	D09_T08	Sink	Small	2.7	0.9	Valley	Yes	No
NA	D09_T10	Sink	Small	4.4	0.5	Valley	Yes	No
NA	D09_T09.2	Sink	Medium	5	2	Valley	Yes	No
Actun Chichan	C09_T06	Horizontal Ent		Blocked Ent		Side Slope	Yes	Yes
Known caves prior to survey								
Las Cuevas	NA	Sink-Hor Ent	Large	85	14	Valley	Known	Yes
Bird Tower	NA	Horizontal Ent	Large	14.8	7.3	Side Slope	Known	Yes
K'in K'aba Ent 1	NA	Horizontal Ent	Large	17.3	4.5	Slope Base	Known	Yes
K'in K'aba Ent 2	NA	Horizontal Ent	Medium	5.1	3.5	Side Slope	Known	Yes
Z'uhuy Ch'en	NA	Horizontal Ent	Fissure	0.6	0.7	Slope Base	Known	Yes

Within our systematic Survey 2, 13 potential cave openings were found in 5 of the 10 tiles surveyed (B11, C09, D09, H09, J08). These consisted of medium to large-sized sinkholes (n = 11) and small constricted shafts (n = 2). Six caves within the tiles were not identified through remote sensing methods and were detected exclusively via pedestrian survey. These included three shallow sinkholes, one constricted shaft, one horizontally accessed cave, and one cave with a blocked entrance. Within the local relief model, the shallow sinks were largely indistinguishable from other minimal relief depressions across the project area, such as household reservoirs or chultuns (small underground storage areas carved into limestone bedrock). None of the three sinks exceeded two meters in depth. Similarly, the undetected shaft was situated within a slight depression ubiquitous throughout this portion of the Vaca Plateau. The actual shaft was obscured by a collection of palm fronds, further concealing the minor feature. The single horizontally accessed cave occurred in a clearing where massive hardwood trees had previously toppled, allowing sunlight into the lower near ground levels and leading to the growth of thick vines and grasses. Due to this thick low vegetation, few lidar points were able to penetrate to ground level, resulting in poor coverage. Finally, the blocked entrance, a small horizontally accessed cave, Actun Chichan (Small Cave), was almost completely infilled with limestone blocks and small boulders, giving no indication on the surface that would be remotely detectable. It is likely that the entrance was closed by the ancient Maya, though there were a few artifacts inside.

In areas in which our survey coverage was 100%, we found cave openings not detected in the lidar. This is the most common error reported by archaeologists when using lidar imagery to detect small low-lying structures (<2 m in height) in areas of dense low vegetation [29,64–67]. In other areas of Belize, notably the Belize River Valley, invasive Guinea grass and secondary-growth vegetation has limited the effectiveness of lidar in the identification of minor archaeological features [68]. In our survey area consisting primarily of high forest canopy, we were likely to encounter this error as a result of tree fall and rarely due to low-growth vegetation covering cave openings. False positives were few (n = 3) and were the result of steep slope angles or tree fall, but not correlated with vegetation types as reported in other archaeological studies [66,68–70]. Likewise, density of ground returns did not affect the overall detection of karstic features, even when potential sinkhole openings were denoted by a single anomalous return point. These findings indicate that within our survey area, only a small percentage of cave entrances will escape detection due to human-constructed blockages or vegetative cover (see Figure 6 which illustrates the relationship between detected cave entrances and point cloud densities).

4. Discussion

After two years of field research, 39 newly identified, remotely sensed cave openings were confirmed and an additional six caves were encountered through pedestrian survey, greatly increasing both the number of known sites in the greater Las Cuevas area. Prior to the 2017 field season, the project located and investigated only four caves in the immediate area: the cave at Las Cuevas, Z'uhuy Ch'en (Pure Cave), K'in Kaba (Birthday Cave), and Bird Tower. These were either known caves or had been found by pedestrian survey. All four of these sites were visible within the local relief models, including the fissure entrance associated with Z'uhuy Ch'en (see Figure 9). Working with the lidar imagery was instrumental in increasing the number of known sites from 4 to 45.

Combined, 25% of the 157 potential cave openings within the projected area were verified over two summer field sessions. The 39 confirmed cave entrances led to the identification of 15 fully or partially developed cave systems (37.5%). Actun Uo (Uo Frog Cave), Actun P'ook (Sombrero Cave), Eduardo Quiroz, Actun Paal (Child's Cave), Actun Miesa (Nothing Cave), and one unnamed cave (C09-C02) had moderate to heavy prehistoric use ($n = 6$), and four others showed minimal use. Three of the remaining systems remain underexplored due to technical requirements (i.e., climbing gear) or bad air. Prehistoric use by the ancient Maya within smaller sinks and shafts was scant, often denoted by low quantities of ceramic material. The morphology of the features rendered them difficult to access without the aid of ropes or ladders, and organic matter and the lack of secondary entrances impeded ventilation in many of these systems. High levels of carbon dioxide were registered in chambers associated with these openings, which would have proven dangerous and potentially fatal to past populations attempting any prolonged ritual activities.

Although horizontally accessed caves were more difficult to detect, local relief modeling and ground-truthing revealed a strong correlation between remotely-sensed entrances and developed cave systems. Two caves in particular proved heavily utilized in prehistory. Actun Uo (Uo Frog Cave), located four kilometers southwest of Las Cuevas, featured three horizontal entrances, numerous chambers and passages, and a through-flowing subterranean stream. The cave was architecturally modified by the ancient Maya, who constructed multiple platforms, partitions, standing stone-altar complexes, and enclosed masonry rooms throughout the system. Eduardo Quiroz Cave, a previously identified system excavated more than half a century ago by A.H. Anderson and David Pendergast [71], was originally thought to be a new complex due to its incorrect plotting by the original researchers. We later identified the site using extant archaeological features.

Inflow and exsurgence caves occur rarely throughout the greater Las Cuevas area. Besides the previously mentioned inflow associated with Actun P'ook, four other cave entrances of this classification were authenticated. Nearly all were inaccessible due to water flow, collapse, or obstructions of organic material. A single example (Dinosaur Egg Cave), developed into a complex system of large chambers and tunnels following a series of initial constrictions.

5. Conclusions

Most research to date in cave detection utilizes algorithms that search for depressions in local landscape topography, but our hybrid methodology employs both automated and non-automated techniques to find not only sinks, but horizontal cave entrances as well. Using local relief modeling from lidar-derived images, our project was able to identify hundreds of potential cave openings in the Chiquibul Forest Reserve from large sinks to small fissures. We located a wide variety of cave entrances, both vertical and horizontal, ranging from small features that measured less than 1 m in height or diameter to much larger and more easily discovered openings.

In the heavily forested area of the Chiqubul Reserve in western Belize, pedestrian survey is inevitably slow and requires cutting trails and often negotiating rough karstic terrain. Guided by lidar imagery, our search for cave openings during two six-week summer field seasons, increased our knowledge of confirmed caves in the project area from 4 to 14 sites that were used by the ancient Maya, and an additional 26 unused sites detected in the lidar imagery. We evaluated the accuracy of our lidar detection using two pedestrian survey methods.

Based on other surveys in areas with thick ground cover, the results of our systematic survey were as expected. Using this survey technique, we discovered six karstic features within the domain that were not detected in the lidar image analysis. This is a reminder that that pedestrian survey remains the most reliable method of obtaining the best survey coverage in forested areas. However, it is not the most efficient or cost-effective. Recall that of the 381 survey tiles created for the Las Cuevas surrounds, we were only able to complete 100% pedestrian survey coverage of 10 tiles during two summer field seasons. At this rate, it would take us 74 years to complete the survey!

The results of our opportunistic survey in which we targeted and visited caves detected on the lidar imagery, met with an 86% success rate with only 3 false positives, verifying 26 cave openings. This method proved to be expedient in meeting the project goals of locating and investigating unknown cave sites. We recognize that there will always be cave openings that lidar cannot detect due to intentional or unintentional entrance blockages or thick forest cover, but in terms of heritage preservation, if we are to locate unlooted or relatively intact cave sites in a timely manner, lidar imagery provides the most cost-effective and directly applicable method for detection, far superior to pedestrian survey alone.

Author Contributions: Conceptualization, H.M.; methodology, H.M. and S.M.; investigation H.M. and S.M.; formal analysis, S.M.; resources H.M.; writing—original draft preparation, H.M. and S.M.; writing—review and editing, H.M. and S.M.; visualization, H.M. and S.M.; supervision S.M.; project administration, H.M.; funding acquisition, H.M.

Funding: Funding for the lidar survey was made available from the Alphawood Foundation to the Western Belize Lidar Consortium. The 2017/2018 field seasons were supported by grants to Moyes from the National Geographic Society (#9544-14, GR-000000589) and the Alphawood Foundation.

Acknowledgments: Thanks to the Belize Institute of Archaeology and particularly to John Morris for permitting our research. The survey crew deserves special acknowledgement for their untiring efforts, particularly Nicholas Bourgeois, and our in-country partners Antonio Mai, Javier Mai, and Israel Canto.

Conflicts of Interest: The authors declare no conflict of interest.

References

1. Minty, C.; Bridgewater, S. Introduction. In *A Natural History of Belize: Inside the Maya Forest*; University of Texas Press: Austin, TX, USA, 2012; ISBN 9780292726710.
2. Kidder, A. Five Days over the Maya Country. *Sci. Mon.* **1930**, *30*, 193–205.
3. Saturno, W.; Sever, T.L.; Irwin, D.E.; Howell, B.F.; Garrison, T.G.; Wiseman, J.R.; El-Baz, F. Putting Us on the Map: Remote Sensing Investigation of the Ancient Maya Landscape. In *Remote Sensing in Archaeology*; Wiseman, J., El-Baz, F., Eds.; Springer: New York, NY, USA, 2007; pp. 137–160, ISBN-10: 038744615X.
4. Sever, T.L.; Irwin, D.E. Landscape archaeology: Remote-sensing investigation of the ancient Maya in the Peten rainforest of northern Guatemala. *Anc. Mesoam.* **2003**, *14*, 113–122. [CrossRef]
5. Blumberg, D.H.; Neta, T.; Margalit, N.; Lazar, M.; Freilikher, V. Mapping exposed and buried drainage systems using remote sensing in the Negev Desert, Israel. *Geomorphology* **2004**, *61*, 239–250. [CrossRef]
6. Breeze, P.S.; Drake, N.A.; Groucutt, H.S.; Parton, A.; Jennings, R.P.; White, T.S.; Clark-Balzan, L.; Shipton, C.; Scerri, E.M.L.; Stimpson, C.M.; et al. Remote sensing and GIS techniques for reconstructing Arabian palaeohydrology and identifying archaeological sites. *Quart. Int.* **2015**, *382*, 98–119. [CrossRef]
7. Hritz, C. Contribution of GIS and Satellite-Based Remote Sensing to Landscape Archaeology in the Middle East. *J. Archaeol. Res.* **2014**, *22*, 229–276. [CrossRef]
8. Parcak, S. Satellite Remote Sensing Methods for Monitoring Archaeological Tells in the Middle East. *J. Field Archaeol.* **2007**, *32*, 65–81. [CrossRef]
9. Chase, A.F.; Chase, D.Z.; Weishampel, J.F. Lasers in the Jungle: Airborne sensors reveal a vast Maya landscape. *Archaeology* **2010**, *63*, 27–29.

10. Chase, A.F.; Chase, D.Z.; Fisher, C.T.; Leisz, S.; Weishampel, J.F. Geospatial Revolution and Remote Sensing Lidar in Mesoamerican Archaeology. *Proc. Natl. Acad. Sci. USA* **2012**, *109*, 12916–12921. [CrossRef]

11. Chase, A.F.; Chase, D.Z.; Awe, J.J.; Weishampel, J.F.; Iannone, G.; Moyes, H.; Jaeger, J.; Brown, K.M. The Use of LiDAR in Understanding the Ancient Maya Landscape. *Adv. Archaeol. Pract. A J. Soc. Am. Archaeol.* **2014**, *2*, 208–221. [CrossRef]

12. Chase, A.F.; Chase, D.Z.; Awe, J.J.; Weishampel, J.F.; Iannone, G.; Moyes, H.; Jaeger, J.; Brown, K.M.; Shrestha, R.; Carter, W.; et al. Ancient Maya Regional Settlement and Inter-Site Analysis: The 2013 West-Central Belize Lidar Survey. *Remote Sens. New Perspect. Remote Sens. Archaeol.* **2014**, *6*, 8671–8695. [CrossRef]

13. Gallagher, J.M.; Josephs, R.L. Using Lidar to detect cultural resources in a forested environment: An example from Isle Royale National Park, Michigan, USA. *Archaeol. Prospect.* **2008**, *15*, 187–206. [CrossRef]

14. Hofton, M.A.; Rocchio, L.E.; Blair, J.B.; Dubayah, R. Validation of vegetation canopy lidar sub-canopy topography measurements for a dense tropical forest. *J. Geodyn.* **2002**, *34*, 491–502. [CrossRef]

15. Weishampel, J.F.; Blair, J.B.; Dubayah, R.; Clark, D.B.; Knox, R.G. Canopy topography of an old-growth tropical rainforest landscape. *Selbyana* **2000**, *21*, 79–87.

16. Weishampel, J.F.; Blair, J.B.; Knox, R.G.; Dubayah, R.; Clark, D.B. Volumetric lidar return patterns from an old-growth tropical rainforest canopy. *Int. J. Remote Sens.* **2000**, *21*, 409–415. [CrossRef]

17. Sittler, B. Revealing Historical Landscapes by Using Airborne Laser Scanning: A 3-D Model of Ridge and Furrow in Forests near Rastatt, Germany. In Proceedings of the Natscan, Laser-Scanners for Forest and Landscape Assessment–Instruments, Processing Methods and Applications, Freiburg im Breisgau, Germany, 3–6 October 2004; Thies, M., Koch, B., Spiecker, H., Weinacker, H., Eds.; International Archives of Photogrammetry and Remote Sensing: Freiburg im Breisgau, Germany, 2004; Volume XXXVI, Part 8/W2. pp. 258–261.

18. Sherwood, S.C.; Goldberg, P. A Geoarchaeological Framework for the Study of Karstic Cave Sites in the Eastern Woodlands. *Midcont. J. Archaeol.* **2001**, *26*, 145–168.

19. Straus, L.G. Underground Archaeology: Perspectives on Caves and Rockshelters. In *Archaeological Method and Theory*; Schiffer, M.B., Ed.; University of Arizona Press: Tucson, AZ, USA, 1990; Volume 2, pp. 255–304, ISBN 9780120031023.

20. Woodward, J.C.; Goldberg, P. The Sedimentary Records in Mediterranean Rockshelters and Caves: Archives of Environmental Change. *Geoarchaeology* **2001**, *16*, 327–354. [CrossRef]

21. Moyes, H.; Brady, J.E. Caves as Sacred Space in Mesoamerica. In *Sacred Darkness: A Global Perspective on the Ritual Use of Caves*; Moyes, H., Ed.; University Press of Colorado: Boulder, CO, USA, 2012; pp. 151–170, ISBN 9781607323600.

22. Nicolay, S. Footsteps in the Dark Zone: Ritual Cave Use in Southwest Prehistory. In *Sacred Darkness: A Global Perspective on the Ritual Use of Caves*; Moyes, H., Ed.; University Press of Colorado: Boulder, CO, USA, 2012; pp. 171–184, ISBN 9781607323600.

23. Moyes, H.; Kosakowsky, L.; Ray, E.; Awe, J.J. *The Chronology of Ancient Maya Cave Use in Belize. Research Reports in Belizean Archaeology*; Institute of Archaeology, NICH: Belmopan, Belize, 2017; pp. 327–338.

24. Brown, L. Planting the Bones: Hunting Ceremonialism at Contemporary and Nineteenth-Century Shrines in the Guatemalan Highlands. *Lat. Am. Antiq.* **2005**, *16*, 131–146. [CrossRef]

25. Scott, A.M. Communicating with the Sacred Earthscape: An Ethnoarchaeological Investigation of Kaqchikel Maya Ceremonies in Highland Guatemala. Ph.D. Dissertation, Department of Anthropology, University of Texas, Austin, TX, USA, 2009.

26. White, W.B.; Culver, D.C. Definition of Cave. In *Encyclopedia of Caves*; White, W.B., Culver, D.C., Eds.; Academic Press: Burlington, MA, USA, 2012; pp. 103–107, ISBN 978-0123838322.

27. Klimchouk, A. Caves. In *Encyclopedia of Caves and Karst Science*; Gunn, J., Ed.; Fitzroy Dearborn: New York, NY, USA, 2004; pp. 417–421, ISBN 9781579583996.

28. Casati, R.; Varzi, A.C. *Holes and Other Superficialities*; MIT Press: Cambridge, MA, USA, 1994; ISBN 9780262032117.

29. Moyes, H.; Montgomery, S. Mapping ritual landscapes using lidar: Cave detection through local relief modeling. *Adv. Archaeol. Pract.* **2016**, *4*, 249–267. [CrossRef]

30. Rinker, J.N. Airborne infrared thermal detection of caves and crevasses. *Photogramm. Eng. Remote Sens.* **1975**, *44*, 1391–1400. Available online: https://www.asprs.org/wp-content/uploads/pers/1975journal/nov/1975_nov_1391-1400.pdf (accessed on 25 November 2018).

31. Wynne, J.J.; Titus, T.N.; Diaz, D.C. On developing thermal cave detection techniques for earth, the moon and mars. *Earth Planet. Sci. Lett.* **2008**, *272*, 240–250. [CrossRef]

32. Cooper, A.H. Airborne multispectral scanning of subsidence caused by Permian gypsum dissolution at Ripon, North Yorkshire. *Q. J. Eng. Geol. Hydrogeol.* **1989**, *22*, 219–229. [CrossRef]

33. Beard, L.P.; Nyquist, J.E.; Carpenter, P.J. Detection of karst structures using airborne EM and VLF. In *SEG Technical Program Expanded Abstracts 1994*; Society of Exploration Geophysicists: Tulsa, OK, USA, 1994; pp. 555–558. [CrossRef]

34. Gutierrez, F.; Cooper, A.H.; Johnson, K.S. Identification, Prediction, and Mitigation of Sinkhole Hazards in Evaporite Karst Areas. *Environ. Geol.* **2008**, *53*, 1007–1022. [CrossRef]

35. Kobal, M.; Bertoncelj, I.; Pirotti, F.; Dakskobler, I.; Kutnar, L. Using Lidar Data to Analyse Sinkhole Characteristics Relevant for Understory Vegetation under Forest Cover-Case Study of a High Karst Area in the Dinaric Mountains. *PLoS ONE* **2015**, *10*, e0122070. [CrossRef] [PubMed]

36. Miao, X.; Qiu, X.; Wu, S.S.; Luo, J.; Gouzie, D.R.; Xie, H. Developing efficient procedures for automated sinkhole extraction from lidar DEMs. *Photogramm. Eng. Remote Sens.* **2013**, *79*, 545–554. [CrossRef]

37. Wu, Q.; Deng, C.; Chen, Z. Automated delineation of karst sinkholes from LiDAR-derived digital elevation models. *Geomorphology* **2016**, *266*, 1–10. [CrossRef]

38. Weishampel, J.F.; Hightower, J.N.; Chase, A.F.; Chase, D.Z.; Patrick, R.A. Lidar Detection and Characterization of Karst Depressions. *J. Cave Karst Stud.* **2011**, *3*, 187–196. [CrossRef]

39. Weishampel, J.F.; Hightower, J.N.; Chase, A.F.; Chase, D.Z. Remote sensing of below-canopy land use features from the Maya polity of Caracol. In *Understanding Landscapes: From Discovery through Land Their Spatial Organization*; Caracol Archaeological Project, Deptartment of Anthropology, University of Las Vegas: Las Vegas, NV, USA, 2013; pp. 131–136. Available online: http://www.caracol.org/wp-content/uploads/2016/05/WeishampelEtAl2013.pdf (accessed on 12 November 2018).

40. Zhu, J.; Taylor, T.P.; Currens, J.C.; Crawford, M.M. Improved Karst Sinkhole Mapping in Kentucky using LiDAR Techniques: A Pilot Study in Floyds Fork Watershed. *J. Cave Karst Stud.* **2014**, *76*, 207–216. [CrossRef]

41. Awe, J.J. The Western Belize Regional Cave Project: Objectives, Context, and Problem Orientation. In *The Western Belize Regional Cave Project: A Report of the 1997 Field Season*; Awe, J.J., Ed.; University of New Hampshire, Department of Anthropology Occasional: Durham, NH, USA, 1998; pp. 1–22.

42. Moyes, H. Cluster Concentrations, Boundary Markers, and Ritual Pathways: A GIS Analysis of Artifact Cluster Patterns at Actun Tunichil Muknal, Belize. In *The Maw of the Earth Monster: Mesoamerican Ritual Cave Use*; Brady, J.E., Prufer, K.M., Eds.; University of Colorado Press: Boulder, CO, USA, 2005; pp. 269–300, ISBN-13: 9780292705869.

43. 43 Moyes, H.; Awe, J.J.; Brook, G.; Webster, J. The Ancient Maya Drought Cult: Late Classic Cave Use in Belize. *Lat. Am. Antiq.* **2009**, *20*, 175–206. [CrossRef]

44. Colas, P.R.; Reeder, P.; Webster, J. The Ritual Use of a Cave on the Northern Vaca Plateau, Belize, Central America. *J. Cave Karst Stud.* **2000**, *62*, 6–10.

45. Moyes, H.; Robinson, M.; Voorhies, B.; Kosakowsky, L.; Arksey, M.; Ray, E.; Hernandez, S. *Dreams at Las Cuevas: A Location of High Devotional Expression of the Late Classic Maya*; Research Reports in Belizean Archaeology; Institute of Archaeology, NICH: Belmopan, Belize, 2015; Volume 12, pp. 239–249.

46. Bridgewater, S. *A Natural History of Belize: Inside the Maya Forest*; University of Texas Press: Austin, TX, USA, 2012; pp. 96–99, ISBN 9780292726710.

47. Bateson, J.H.; Hall, I.H.S. *The Geology of the Maya Mountains, Belize*; HM Stationery Office, London.: London, UK, 1977; Volume 3, ISBN 978-0118807654.

48. Miller, T.E. Geologic and hydrologic controls on karst and cave development in Belize. *J. Cave Karst Stud.* **1996**, *58*, 100–120.

49. Vinson, G.L. Upper Cretaceous and Tertiary Stratigraphy in Guatemala. *Bull. Am. Assoc. Petr. Geol.* **1962**, *46*, 425–465.

50. Miller, T.E. Inside Chiquibul. *Natl. Geogr.* **2000**, *197*, 54–71.

51. Feld William, A. The Caves of Caracol: Initial Impressions. In *Studies in the Archaeology of Caracol, Belize*; Chase, D.Z., Chase, A.F., Eds.; Pre-Columbian Art Research Institute Monograph 7; PARI: San Francisco, CA, USA, 1994; pp. 76–82.

52. Reeder, P.; Brinkmann, R.; Alt, E. Karstification on the northern Vaca plateau, Belize. *J. Cave Karst Stud.* **1996**, *58*, 121–130.

53. Benson, R.C.; Yuhr, L.B. *Site Characterization in Karst and Pseudokarst Terraines: Practical Strategies and Technology for Practicing Engineers, Hydrologists and Geologists*; Springer: New York, NY, USA, 2015; pp. 16–25, ISBN 9789401799232.

54. Williams, P.; Gunn, J. Dolines. In *The Encyclopedia of Caves and Karst Science*; Gunn, J., Ed.; Fitzroy Dearborn: New York, NY, USA, 2003; pp. 304–310, ISBN 9781579583996.

55. Gatziolis, D.; Andersen, H.E. *A Guide to LIDAR Data Acquisition and Processing for the Forests of the Pacific Northwest*; Gen. Tech. Rep. PNW-GTR-768; US Department of Agriculture, Forest Service, Pacific Northwest Research Station: Portland, OR, USA, 2008; p. 768, ISBN 9781508770954. Available online: https://www.fs.fed.us/pnw/pubs/pnw_gtr768.pdf (accessed on 19 November 2018).

56. Kokalj, Z.; Hesse, R. *Airborne Laser Scanning Raster Data Visualization—A Guide to Good Practice*; Založba ZRC: Ljubljana, Slovenia, 2017; ISBN 9789612549848.

57. Brassel, K.E.; Little, J.; Peucker, T.K. Automated Relief Representation. *Ann. Assoc. Am. Geogr.* **1974**, *64*, 610–611. Available online: https://www.jstor.org/stable/2569517 (accessed on 19 November 2018).

58. Zakšek, K.; Oštir, K.; Kokalj, Ž. Sky-View Factor as a Relief Visualization Technique. *Remote Sens.* **2011**, *3*, 398–415. [CrossRef]

59. Yokoyama, R.; Shirasawa, M.; Pike, R.J. Visualizing topography by openness: A new application of image processing to digital elevation models. *Photogramm. Eng. Remote Sens.* 2002, 68, pp. 257–266. Available online: https://pdfs.semanticscholar.org/c3d9/a561fdb9e8c34a2b79152aea72b46090bb2e.pdf (accessed on 19 November 2018).

60. Hesse, R. Visualisierung Hochauflösender Digitaler Geländemodelle Mit LiVT. In *Computeranwendungen Und Quantitative Methoden in Der Archäologie. 4*; Workshop Der AG CAA 2013; Lieberwirth, U., Herzog, I., Eds.; Berlin Studies of the Ancient World: Berlin, Germany, 2016; pp. 109–128.

61. Doctor, D.H.; Young, J.A. An Evaluation of Automated GIS Tools for Delineating Karst Sinkholes and Closed Depressions from 1-m LiDAR Derived Digital Elevation Data. In Proceedings of the 13th Multidisciplinary Conference on Sinkholes and the Engineering and Environmental Impacts of Karst, NCKRI SYMPOSIUM 2, Carlsbad, NM, USA, 6–10 May 2013; Land, L., Doctor, D.H., Stephenson, J.B., Eds.; National Cave and Karst Research Institute: Carlsbad, NM, USA, 2013; pp. 449–458.

62. Hesse, R. LiDAR-derived Local Relief Models—A new tool for archaeological prospection. *Archaeol. Prospect.* **2010**, *17*, 67–72. [CrossRef]

63. Novák, D. Local Relief Model (LRM) Toolbox for ArcGIS. Electronic Document. 2014. Available online: http://www.academia.edu/5618967/Local_Relief_Model_LRM_Toolbox_for_ArcGIS_UPDATE_2014-10-7 (accessed on 12 November 2018).

64. Hare, T.; Masson, M.; Russell, B. High-Density LiDAR Mapping of the Ancient City of Mayapan. *Remote Sens.* **2014**, *6*, 9064–9085. [CrossRef]

65. Hutson, S.R. Adapting LiDAR Data for Regional Variation in the tropics: A case study from the Northern Maya Lowlands. *J. Archaeol. Sci.* **2015**, *4*, 252–263. [CrossRef]

66. Prufer, K.M.; Thompson, A.E.; Kennett, D.J. Evaluating Airborne LiDAR for Detecting Settlements and Modified Landscapes in Disturbed Tropical Environments at Uxbenka. Belize. *J. Archaeol. Sci.* **2015**, *57*, 1–13. [CrossRef]

67. Reese-Taylor, K.; Hernández, A.A.; Esquivel, F.C.A.F.; Monteleone, K.; Uriarte, A.; Carr, C.; Acuña, H.G.; Fernandez-Diaz, J.C.; Peuramaki-Brown, M.; Dunning, N. Boots on the Ground at Yaxnohcah: Ground-Truthing Lidar in a Complex Tropical Landscape. *Adv. Archaeol. Pract.* **2016**, *4*, 314–338. [CrossRef]

68. Yaeger, J.; Brown, M.J.; Cap, B. Locating and Dating Sites Using Lidar Survey in a Mosaic Landscape in Western Belize. *Adv. Archaeol. Pract.* **2016**, *4*, 339–356. [CrossRef]

69. Rosenswig, R.M.; López-Torrijos, R.; Antonelli, C.E.; Mendelsohn, R.R. Lidar Mapping and Surface Survey of the Izapa State on the Tropical Piedmont of Chiapas, Mexico. *J. Archaeol. Sci.* **2013**, *40*, 1493–1507. [CrossRef]

70. Crow, P.; Benham, S.; Devereux, B.J.; Amable, G.S. Woodland Vegetation and Its Implications for Archaeological Survey Using LiDAR. *Forestry* **1989**, *80*, 241–252. [CrossRef]

71. Pendergast, D.M. *Excavations at Eduardo Quiroz Cave, British Honduras (Belize)*; Toronto, Royal Ontario Museum, Art and Archaeology Occasional: Toronto, ON, Canada, 1971.

geosciences

MDPI

Article

Automated Detection of Field Monuments in Digital Terrain Models of Westphalia Using OBIA

M. Fabian Meyer [1,*], Ingo Pfeffer [2] and Carsten Jürgens [1]

[1]   Department of Geography, Ruhr-University Bochum, Workgroup of Geomatics, 44801 Bochum, Germany; carsten.juergens@rub.de

[2]   LWL-Archaeology for Westphalia, 48157 Münster, Germany; Ingo.Pfeffer@lwl.org

*   Correspondence: matthias.meyer@rub.de; Tel.: +49-234-32-24791

Received: 15 January 2019; Accepted: 24 February 2019; Published: 28 February 2019

Abstract: While Light Detection and Ranging (LiDAR) revolutionized archaeological prospection and different visualizations were developed, an automated detection of cultural heritage still poses a significant challenge. Therefore, geographers and archaeologists from Westphalia, Germany are developing automated workflows for classifying field monuments from special terrain models. For this project, a combination of GIS, Python, and Object-Based Image Analysis (OBIA) is used. It focuses on three common types of monuments: Ridge and Furrow areas, Burial Mounds, and Motte-and-Bailey castles. The latter two are not classified binary, but in multiple classes, depending on their degree of erosion. This simplifies interpretation by highlighting the most interesting structures without losing the others. The results confirm that OBIA is suitable for detecting field monuments with hit rates of ~90%. A drawback is its dependency on the use of special terrain models like the Difference Map. Further limitations arise in complex terrain situations.

Keywords: automated detection; OBIA; LiDAR; Difference Map; field monument; Burial Mound; Motte-and-Bailey castle; Ridge and Furrow

1. Introduction

The use of airborne laser scanning in archaeology started almost 20 years ago. Since then, LiDAR revolutionized archaeological prospection and new visualizations were invented to ease interpretation of field monuments. In parallel, the need for automated workflows and classifications became evident to handle the growing number of datasets.

One of the first archaeological applications of LiDAR in Germany was performed by Sittler in Rastatt [1]. He successfully searched for Ridge and Furrow structures, digitized some of them manually, and carefully predicted that algorithms would be able to detect structures automatically. Later, Heinzel and Sittler [2] presented an automated approach using Pattern Recognition for the detection and delineation of single Ridge and Furrow structures. The evaluation with reference data produced accuracies up to 84%, depending on the complexity of the terrain.

In parallel, de Boer experimented with Template Matching to detect Burial Mounds in the Netherlands [3]. This technique slides a tiny digital terrain model (DTM, the template or window) over the DTM of a study area and calculates the correlation to the part that is currently covered by the template. The result is a raster with correlations at each position written to the corresponding pixel. The highest pixel values of this raster are considered as hits. De Boer used multiple templates of different sizes and was able to detect most of the reference mounds. In the end, the research was not continued due to too many false positives [4].

Another project to detect field monuments using Template Matching was undertaken by Trier and a Norwegian research team. In multiple publications, they demonstrated how to detect pits and similar structures like Burial Mounds or Charcoal kiln sites [5–7]. An example from Germany is

Schneider et al., presenting a workflow for detecting Charcoal kiln sites using Template Matching as well [8].

Most recently, Freeland et al. provide an example of an automated detection of mounds in the Kingdom of Tonga [9] and Cerrillo-Cuenza presents a workflow for detecting mounds in Spain [10]. In 2018 and 2019, Davis et al. developed and compared workflows for the detection of mounds in South Carolina [11,12].

Up to now, however, not all field monuments are covered and furthermore no attempt was made to adapt the variety of methods to the province of Westphalia, Germany, which has a rich archaeological record including monuments from the Stone Age until World War II (Figure 1). Therefore, the project presented here aims at a provincewide detection of frequently appearing field monuments like Burial Mounds and medieval fields, improving knowledge of Westphalia's historic landscapes and supporting preservation; e.g., medieval fields are strong indicators for lost settlement that often are completely invisible in the terrain and help to reconstruct spatial distribution of settlements in different epochs.

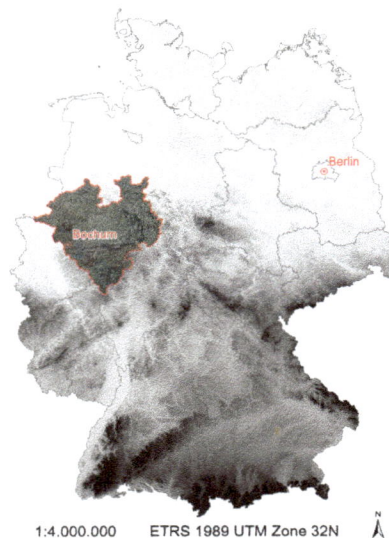

1:4.000.000   ETRS 1989 UTM Zone 32N

Figure 1. Location of Westphalia (red) in Germany along with the cities of Bochum and Berlin for reference. Data source: SRTM & [13].

Westphalian archaeologists are limited regarding funds to spend on projects like this. Therefore, as a secondary objective, this project examines the potential of the currently only available LiDAR data source for automated classifications using OBIA. The quality is far away from today's standard but it is for free, covers the whole province and is presumably precise enough for detecting the desired structures.

This article presents the initial steps of the project, in which workflows are developed and tested. It will be demonstrated that the combination of OBIA and a simple terrain visualization is suitable for the detection of different types of field monuments. These workflows are supposed to be finally used for a provincewide detection of field monuments.

Three types of field monuments are considered. In the beginning, Motte-and-Bailey castles were used to investigate the potential of OBIA for the classification of complex structures. A Motte-and-Bailey castle is a medieval fortification that consists of a mound and a surrounding bailey. The former usually carried a fortified building and the latter was the place for other buildings of varying purposes (Figure 2) [14]. Today, in most instances, only the motte and its surrounding ditch are preserved in varying shapes and conditions (Figure 3). Other common field monuments are areas with Ridge and Furrow structures (Figure 4). A single ridge is the remnant of an early medieval field that was just a few meters wide but hundreds of meters long [15]. The third type of field monument is the Burial Mound. These mounds were common in the Bronze Age and the Iron Age and are preserved quite often.

Figure 2. Motte-and-Bailey castle [14].

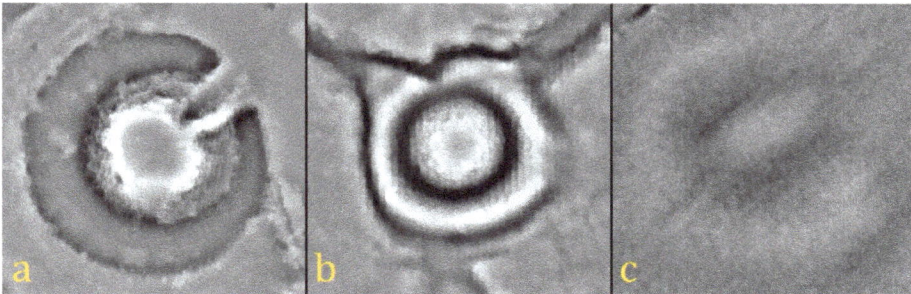

Figure 3. Overview of the different castles from the reference data: (**a**) Borken, (**b**) Paderborn, and (**c**) Warburg [13,16].

Figure 4. Ridge and Furrow area in Dülmen.

Although new monuments are searched systematically, there are still many field monuments hidden in the terrain. Looking at the size of Westphalia, the need for automated approaches becomes obvious. It has an area of about 21,000 km^2, of which 3/4 are unsealed and therefore suitable for archaeological prospection. In northern Westphalia, these areas mostly consist of agricultural land, whereas in the southern mountainous region forests are dominating the landscape. Pastures and some minor land use classes are present all over Westphalia.

2. Materials and Methods

Archaeologists in North Rhine-Westphalia (NRW) benefit from the Open Geodata principle of the provincial government, which provides up-to-date and continuously updated spatial data for free [17]. Table 1 provides an overview of the datasets that were used in this project so far.

Table 1. Overview of the LiDAR datasets.

Field Monument	Motte-and-Bailey Castles	Ridge and Furrow Areas	Burial Mounds
provided as		point data	
date of acquisition	2008–2010		2016
preprocessing	filtered, last pulse		filtered, last pulse, interpolated
point spacing and distribution	irregular, 1–4 pt/m^2		regular grid, 1 pt/m^2

The LiDAR datasets for the detection of Motte-and-Bailey castles and Ridge and Furrow areas were acquired between 2008 and 2010 and provided as filtered point data in an irregular distribution with a point density of 1–4 pt/m^2. The data for the detection of mounds were acquired in 2016 and provided online [17] as filtered point data in an interpolated regular grid with 1 pt/m^2—these point data can be converted to DTM without interpolation.

Additionally, a digital land use model (DLM) containing information about the current land use as point, line and polygon features is used [11,17]. From this dataset, polygons representing unsealed areas were extracted once and merged into a 'positive layer', to which the search is limited. This avoids misclassifications in areas where no monuments can be preserved, e.g., in settlements or mining areas (Figure 5).

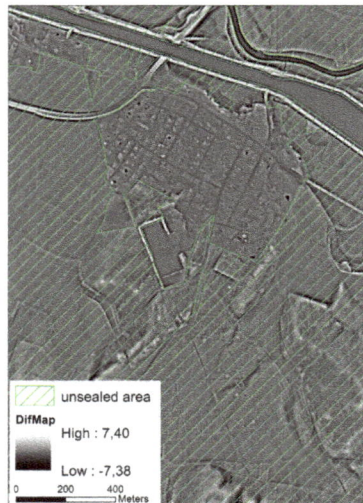

Figure 5. Difference Map of a part of the investigation area in Haltern. The search is limited to the areas that are unsealed (green). The others, primarily settlements, are not considered [13].

For the development and evaluation of the classification algorithm, reference data were taken from *FuPuDelos*, the official database of archaeological records, maintained by the Westphalian archaeology [18]. There are three things to consider when working with this dataset:

(1) To ease evaluation of the Burial Mounds, a GIS will be used to flag all results having a reference point feature inside as *true positives*. This requires the points to be very precise, but some are offside up to a few meters due to imprecise recording in the field. They need correction based on visual inspection of the terrain model, which is done manually for all monuments in the investigation area. (2) Some monuments that are not protected by law were destroyed between their record and the acquisition of the terrain model. They are not present anymore and therefore cannot be classified. The reference points corresponding to destroyed monuments were removed from the dataset to avoid their impact on the classification. (3) The dataset tends to be incomplete and contains some records that were not yet confirmed in the field. This must be considered in the evaluation. Therefore, results are also inspected visually in the terrain models and aerial photography.

To evaluate the classification, especially with the Burial Mounds, results are interpreted as *true positives* (TP) if they are present in the reference dataset. For the sake of simplicity, results that are absent are *false positives* (FP), although a few of them might be revealed as new findings. Therefore, all FP were interpreted visually based on different terrain models and aerial photography. Even without inspection in the field, some of them are most likely new discoveries—these are called *new positives* (NP). Finally, evaluation and interpretation are difficult and therefore the workflows are supposed to lead the interpreter's focus to the most promising features. In this context, the authors agree that the final interpretation still cannot be done by a computer (e.g., Sevara et al. [19]).

In this project, two approaches are considered. On the one hand, discrete field monuments are classified by tracing their shape. In doing so, castles and mounds are not only classified binary as *true* or *false*, but furthermore by their degree of erosion in multiple classes, e.g., to what extent they are leveled or stretched by farming activities (Figure 3c). This is a similar result organizing approach to that of Trier et al. [6].

In contrast to binary classifications, it is not necessary to optimize the algorithms towards high completeness or high correctness. Full completeness is guaranteed by the fact that the classes are derived by the worst preserved reference monuments. At the same time, the classes containing well-preserved results show high values of correctness. Their results are more likely to be true and can be interpreted at first. The drawback is that evaluation becomes difficult since calculating reasonable values for an overall correctness and completeness is hardly possible.

On the other hand, tracing the shape of areal monuments, e.g., Ridge and Furrow areas, is difficult if single elements looks like nonarchaeological structures. Therefore, the second approach is not to find every single ridge or furrow but to tag the location of the area by highlighting objects indicating an area of similar structures.

The workflow (Figure 6) starts in a GIS with the calculation of the desired visualization(s) and ends in the same place with the interpretation. Just as important as an accurate classification is an easy access to the visualizations and a user-friendly workflow. Therefore, an *ArcGIS*-tool was written in Python that is designed to work with the mentioned LiDAR data. It is capable of calculating different visualizations, currently a conventional DTM as well as the special visualizations Difference Map (DM) and Local Relief Model (LRM) that were both invented by Hesse [20]. The latter two exclusively show the microrelief, which is essential for a classification using OBIA (see below).

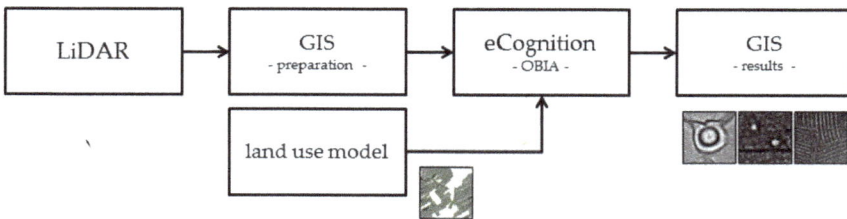

Figure 6. Overview of the general workflow for the detection of field monuments [13].

For the classification, OBIA, implemented in *eCognition Developer* by *Trimble*, was used. This technique does not classify single pixels, but objects representing homogeneous areas within an image. In a DTM, they correspond to areas of the same height. Objects are generated in the initial segmentation step that has three adjustable thresholds: Most importantly, the *scale* parameter defines the final size of the objects. Adjustments are to be done very carefully to represent the real-world object as best as possible. The following two parameters describe the homogeneity of the objects: The *shape* parameter defines how much the algorithm takes the shape of the emerging object or the spectral values of its pixels into account. Finally, the *compactness* parameter defines the smoothness of an object border, which is a relation between length and width of an object and its area. The segmentation starts with converting every pixel into a single object and then iterates over the objects, merging the current one with the most similar neighbor. This is done until the adjusted thresholds are reached [21]. The segmentation is perfect when the object borders match the borders of their corresponding field monuments.

Afterwards, statistical values (*features*) are calculated for every object. Some of which refer directly to the object (e.g., *length* and *width*) and some to its neighbors (e.g., *rel. border to brighter objects*). From these, the user can choose features to describe classes. This is the advantage over pixel-based approaches because objects can be seen in a relation to their neighbors and therefore be discriminated by their location, which is essential for the detection of field monuments. In terms of OBIA, remnants of a Motte-and-Bailey castle can be described as a local maximum (the motte) or as an object completely surrounded by darker (lower) objects, which is in close proximity from a ring-shaped local minimum (a ditch surrounding the motte). Up to now, three *rulesets* where developed that run fully automated and export the classification results to GIS-compatible shapefiles.

Although OBIA does not 'see' an object in the way the human eye does, it nevertheless benefits from special terrain visualization like the DM or the LRM that were originally developed for manual interpretation. In this project, the DM was preferred over other Visualizations because of its excellent ratio of benefit and calculation effort. Calculation is done in two simple steps using a GIS: At first, the terrain model is smoothed using a low-pass filter, which removes the microrelief while preserving hills and valleys. The Difference Map is then calculated by subtracting the smoothed DTM from the initial one, only preserving the microrelief with its small-scale features [20]. Because hills and valleys were removed by the subtraction, the contrast increases significantly but most importantly, all monuments appear in a leveled situation because slopes are removed as well (Figure 7).

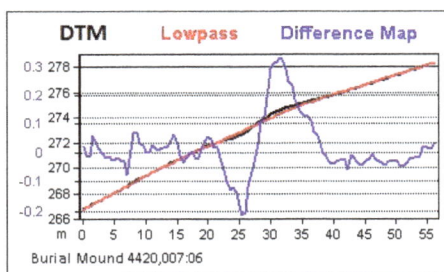

Figure 7. Profile graphs of a Burial Mound in a digital terrain model (DTM) (black), a smoothed DTM (red) and the final Difference Map (blue) [13].

The latter is necessary because object borders in LiDAR datasets follow the contour lines making field monuments invisible to OBIA if they are located on a hillside. Figure 8 demonstrates this issue as the castle on the left side cannot be detected. On the right side the hill is removed and motte and bailey stand out against the surrounding area even though they are almost eroded completely.

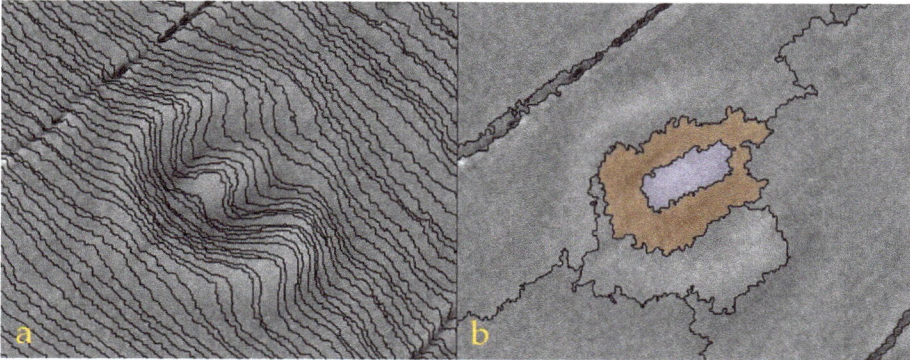

Figure 8. Difference Maps of a highly eroded Motte-and-Bailey castle on a hillside in Warburg with overlaying objects derived from a regular DTM (**a**) and from a Difference Map (**b**) [13,16].

A disadvantage of using a visualization like this is that the calculation will usually generate undesired structures as well, especially in complex situations like slopes (Figure 9). They are a problem for the detection of some field monuments and cannot be handled perfectly.

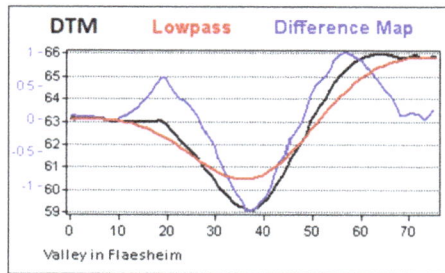

Figure 9. Profile graphs of a valley in a DTM (black), a smoothed DTM (red), and the final Difference Map (blue) [13].

The next chapters will present the classification algorithms (*rulesets*). For the definition of class borders, software-internal statistics (*features*) are used. Based on the *eCognition* reference book [21], details are to be found below.

The following *features* describe the shape or extent of an object.

- *Area* equals number of pixels of an object. It can be converted into different measurements if a unit is available. In this case 1 px equals 1 m^2.
- *Border length/area* describes the ratio of the border length to the area of an object. It is dependent on the shape and the size of object at the same time.
- *Compactness* is calculated by dividing the area of an object by its number of pixels. The ideal compactness equals 1.
- *Density* determines if an object is rather shaped like a square, which is the 'most dense' object, or like a filament. Calculation is done by dividing the number of pixels of an object by its approximate radius.

- *Elliptic fit* is calculated by comparing the object to an ellipse of similar size (area) and proportions (width and length). The overlapping area is then compared to the area of the object outside the ellipse. A value of 0 indicates less than 50% fit, whereas 1 indicates a perfect fit.
- *Length/width*: *eCognition* internally calculates two different length-to-width ratios, one of which using a bounding box. The smaller value is finally assigned to the object and helps discriminating between elongated and well preserved mottes.
- *Max* and *Min pixel value* provide the highest and the lowest pixel value of an object. When using a DM, it is representing the elevation of an object above ground level.
- *Mean* describes the average pixel value of an object.
- *Roundness* is calculated by subtracting the radius of the largest enclosed ellipse from the smallest enclosing ellipse of an object. The lower the value, the more similar the object looks to an ellipse, meaning that the roundness for the detection of mottes and mounds should be as low as possible.
- *Shape index* describes the smoothness of an object border. Low values are desired because they represent smooth borders.
- *Standard deviation* describes the homogeneity of pixel values of an object.
- *Width* is calculated by dividing the number of pixels of an object by its *length/width* ratio.

The second group of *features* describes the relation of an object to its neighbors. They help discriminating objects by their location and relation to others, which is essential for detecting many different types of structures:

- *Distance to [class]* describes the distance in pixels to the closest object of a specified class, measured from center to center.
- *Number of [class] [distance]* defines the number of objects in a specified class within a specified distance.
- *Relative area to [class] [distance]* is similar to *number of*. This value describes the area covered by objects of a specified class relative to the total area around the object within a specified distance.
- *Relative border to [class]* describes the percentage of the border length that an object shares with objects of a specified class relative to the total border length.
- *Relative border to brighter objects [layer]* determines if an object is surrounded by brighter (in terms of LiDAR: higher) objects. If the value equals 0, an object represents a local maximum.

2.1. Detection of Motte-and-Bailey Castles

For the development of the corresponding ruleset, three castles in very different conditions were considered. The first motte in Borken (Figure 3a) still carries a house from the 18th century [22] and is therefore in perfect condition. The second one—the Imbsenburg in Paderborn (b)—was destroyed at some point but is still in very good condition. The last one, located in Warburg, is almost completely leveled off by farming activities over the last centuries (c).

Looking at mottes (a) and (b) in Figure 3 again and comparing the profile graphs of motte (a) in Figure 10, another link between visualization and classification becomes obvious. In the DM, the border of the motte appears higher than its center and the motte becomes a ring (Figure 10, center). This is a result of the low-pass filter that starts flattening the mound from the edges towards the center, which has to be considered in the classification step because the surface of the mound is not homogenous enough to be represented by a single object.

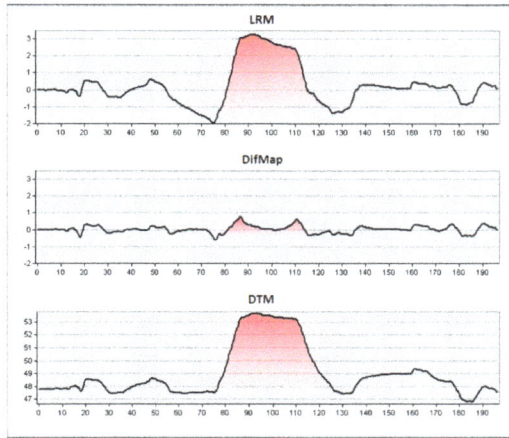

Figure 10. Profile graphs of the motte in Borken in the LRM (**top**), Difference Map (DM) (**center**) and DTM (**bottom**). The motte is highlighted in red [13].

The first task was to find suitable settings for the segmentation that would work in all areas. After several tests, the settings 10/0.001/0.5 (*scale/shape/compactness*) were chosen to generate large objects, which strongly take into account the pixel values (the relative height) and are quite compact, with the following statistical values (Table 2).

Table 2. Overview of the objects representing the real-life mottes and the derived class borders.

Motte-and-Bailey Castles	Borken	Paderborn	Warburg	Class Borders *Possible Mound*
area (px)	1212	1953	1978	>= 1000
border length/area	1.130	0.124	0.169	>= 0.1
compactness	1.222	1.357	1.383	<= 1.6
length/width	1.026	1.019	2.314	<= 2.5
roundness	0.275	0.355	0.489	<= 0.6
shape index	1.134	1.380	1.888	<= 2.0

The classification (Figure 11) starts with the search for big mounds (the mottes, blue). All objects whose shape indicates a motte are collected in a class called *possible mound*. The values of the class borders are set in a way that allows the classification of more eroded structures and to avoid misclassifications as best as possible. In the next step, the software examines if an object is a local maximum and therefore still probably a mound (*rel. border to brighter objects*). Due to the mentioned problem resulting from the low-pass filter, the decision is based on both DM and regular DTM because in the latter a mound with pseudo structures is still intact. An alternative solution is to use a LRM but one wanted to avoid the calculation effort (see Figure 10 top). If an object is a local maximum, it is moved to a class depending on where it appears. These two classes get joined afterward in the class *mound*.

Ditches are classified in almost the same way (red). At first, objects are classified as a *possible ditch*, depending on the shape statistics in Table 3. Secondly, objects that are a local minimum, either in the DM or the DTM, and that are located close to a *mound* object (<= 40 px), are classified in corresponding classes and finally joined in the class *ditch*. Vice versa, only those objects from the *mound* class are classified as a *motte* that are located in close proximity to an object in the *ditch* class (<= 40 px) and share at least 15% percent of their border with a *ditch*.

Table 3. Overview of the objects representing the real-life ditches around the mottes and the derived class borders.

Ditches	Borken	Paderborn	Warburg	Class Borders *Possible Ditch*
density	1.213	1.137	1.877	<= 2.0
length/width	1.268	1.026	1.483	<= 1.5

After the mottes are classified, they are sorted into subclasses by their appearance. Four classes were defined to cover all types that appeared so far as well as possible types in between (Table 4).

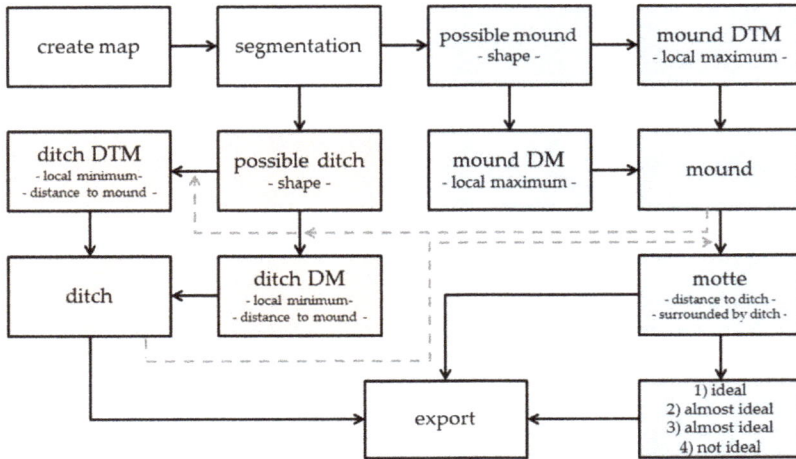

Figure 11. Overview of the workflow in *eCognition* for the detection of Motte-and-Bailey castles.

Table 4. Overview of the classes that represent the mottes by their degree of erosion.

Class	Length/Width	Shape Index	Explanation	Example
(1) ideal	>= 1.0 ... < 1.75	>= 1.0 ... < 1.5	round and high	Borken (a), Paderborn (b)
(2) almost ideal		>= 1.5 ... < 2.0	round and flat	
(3) almost ideal	>= 1.75 ... < 2.5	>= 1.0 ... < 1.5	long and high	
(4) not ideal		>= 1.5 ... < 2.0	long and flat	Warburg (c)

2.2. Detection of Ridge and Furrow Areas

For the development of the workflow two regions with extensive Ridge and Furrow areas were chosen. One of which is located in Dülmen, the other one in Höxter. Both are covered by forests and have an area of 8 km^2.

Ridges and furrows appear in groups and trying to detect every single structure is not promising because the number of false positives would be too high. In addition, the reference dataset only contains the outlines of the areas. Therefore, it is again not possible to calculate reasonable values for an evaluation, which makes visual inspection necessary. The approach is to find single ridges and furrows ('indicating objects') that are surrounded by multiple similar structures, which makes them reveal themselves as a part of a Ridge and Furrow area.

This time, finding segmentation settings turned out to be much more difficult. The problem was that many segmentation settings were unsatisfying because the object borders did not follow the ridges or furrows in the way they were supposed to. After systematically testing, 5/0.001/0.5 (*scale/shape/compactness*) was chosen to work sufficiently. This setting is in a good balance between finding more eroded structures on the one hand and not producing too much useless objects on the other.

The first step after the segmentation is to find long and thin objects (Figure 12). The required statistics for an object to be classified as a *possible ridge or furrow* are listed in Table 5. As with the castles, the second step is to determine if an object in this class is a local maximum or minimum. Sorting in corresponding classes is again done with the *rel. border to brighter objects* criterion, this time using thresholds of > 80% and < 80%. After that, all minima and maxima are collected in *max. or min.* in order to determine which objects are surrounded by at least six objects of the same class in a defined area around the object. Besides, the classifier checks the distance to the next maximum or minimum and how much of the border of an object is shared with an object of the same class (Table 6). If an object meets all requirements, it is classified as a *ridge or furrow* and exported.

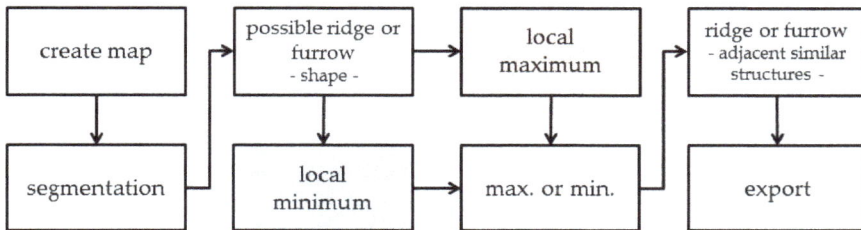

Figure 12. Overview of the workflow in *eCognition* for the detection of Ridge and Furrow areas.

Table 5. Description of the class *possible ridge or furrow*.

Class	Area (px)	Density	Length/Width	Width (px)
possible ridge or furrow	>= 1000	<= 1.4	> 3.0	>= 15 ... <= 55

Table 6. Description of the final class *ridge or furrow*.

Class	Distance to max. or min. (px)	Number of max. or min. (100px)	rel. Border to max. or min.
ridge or furrow	<= 70	>= 6	> 0.4

2.3. Detection of Burial Mounds

The investigation area for the detection of Burial Mounds is located around the Haard, a hilly landscape close to Haltern. It mostly consists of forested areas with a decent amount of Burial Mounds in varying shapes and conditions as well as some other areas to test the robustness towards artificial structures.

Compared to the others, the detection of Burial Mounds is straightforward. In terms of OBIA, the task is to find round local maxima in the DM (Figure 13). Doing this is easy, but the drawback of detecting such a simple structure is a high number of false positives. This is minimized by using the landscape model to reject the results that are located in settlements and are therefore false.

As mentioned above, mounds are not classified binary but in five classes depending on their degree of erosion. After finding segmentation settings (6/0.001/0.1, *scale/shape/compactness*) that sufficiently represent most of the (reference) mounds, class borders were derived by carefully looking at the statistics of the objects representing reference mounds (Table 7). The class borders become more restrictive as they describe more ideal mounds. This is done to separate these mounds from the others. The latter are collected in classes (4) *eroded* and (5) *highly eroded* and are therefore not lost. At first, all possible mounds are classified in class (5). After that, only those objects are classified further that meet the more narrow class borders of class (4). Then, it is the same procedure with classes (3) to (1) and therefore only the very best objects reach class (1).

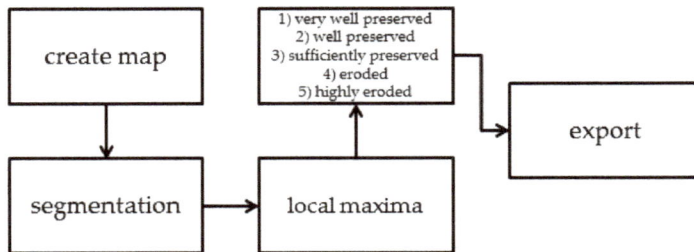

Figure 13. Overview of the workflow in *eCognition* for the detection of Burial Mounds.

Table 7. Overview of the classes that represent the Burial Mounds by their degree of erosion.

Class [1]	(1) Very Well Preserved	(2) Well Preserved	(3) Sufficiently Preserved	(4) Eroded	(5) Highly Eroded
area (px)		>= 85 ... <= 320			>= 20 ... <= 500
compactness		<= 1.5	<= 1.6	<= 1.8	<= 2
elliptic fit		>= 0.86	>= 0.8	>= 0.7	>= 0.6
length/width	<= 1.13	<= 1.25	<= 1.4	<= 1.6	<= 2
roundness		<= 0.25	<= 0.4	<= 0.6	<= 1
shape index		<= 1.25	<= 1.4	<= 1.6	<= 1.85
mean		<= 1.2	... <= 1.35	... <= 1.45	>= 0.2 ... <= 1.55
max (px)		>= 0.6 ... <= 1.65	>= 0.59 ...	>= 0.57 ...	>= 0.55 ... <= 2
min (px)		>= −0.1 ... <= 0.85			>= −0.15 ... <= 1.15
stand. dev.	>= 0.2	>= 0.1 ... <= 0.38			>= 0.05 <= 0.45

[1] If no class border is specified, the border of the next wider class (to the right) is applied.

3. Results

This chapter will present the results that were produced so far. They are structured by the monument that they describe.

3.1. Motte-and-Bailey Castles

Calculating a meaningful correctness and completeness for the detection of Motte-and-Bailey castles (Figure 14) is not possible, at least not at this stage, because the number of samples is too low. All three samples were detected, only in Paderborn a single false positive was classified. Testing some more castles, that did not influence the parameters, revealed some difficulties in the segmentation of distorted mottes and baileys. Other than that, shapes that differ from the three presented above need slight adjustments to be segmented. Additionally, this could be addressed by defining more classes. For further evaluation, a provincewide investigation needed to be carried out, which is not planned so far because the number of unknown castles is presumably not high enough. Nevertheless, the results show that a relatively simple algorithm in OBIA can detect complex multiobject field monuments like Motte-and-Bailey castles in different degrees of erosion, once useful parameters are found.

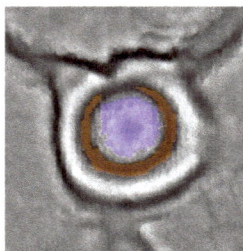

Figure 14. Detected castle in Paderborn [13].

3.2. Ridge and Furrow Areas

The results of the classifier for Ridge and Furrow areas are presented in Figure 15. Red structures represent single ridges or furrows who were left in *max. or min.* They do not have enough neighbors to be definitely part of a Ridge and Furrow area but nevertheless help to identify the blue ones. The latter are the important objects representing single ridges or furrows that have at least six adjacent objects of the same class around and can therefore be addressed as a part of a Ridge and Furrow area (class *Ridge or furrow*). These blue structures were found in every relatively big area. On the other hand, structures like modern drainages, which basically look like a sequence of ridges and furrows, were not detected—probably because their elements is too thin (Figure 15b).

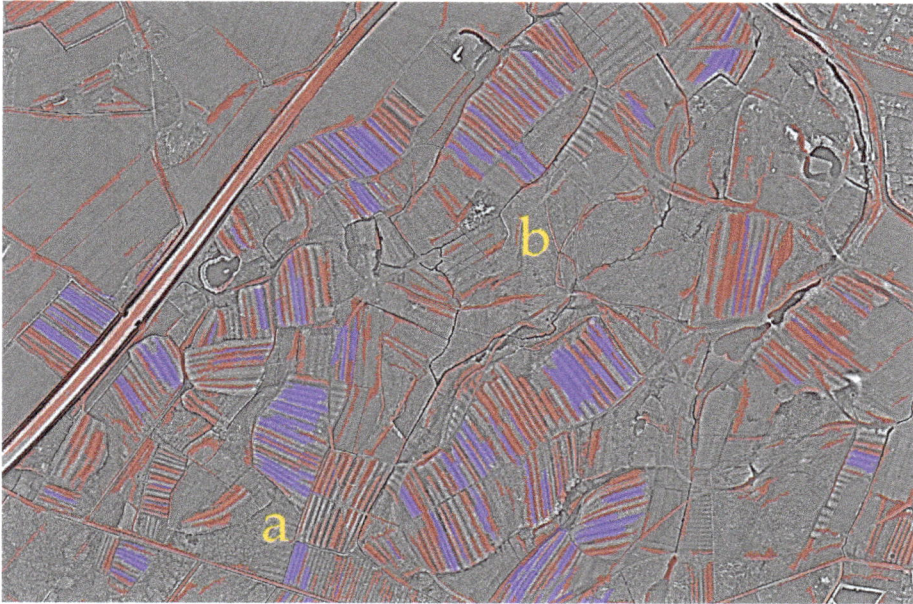

Figure 15. Classification result of the investigation area in Dülmen, overlaying a DM. The blue objects are ridges or furrows that have at least six adjacent objects of the same class and can therefore be addressed as a part of a Ridge and Furrow area. The red structures look like ridges or furrows in terms of shape and are a minimum or maximum, but do not have enough neighbors to be classified in blue [13,16].

According to the approach of finding these blue 'indicating objects', only one object is needed per Ridge and Furrow area. Therefore, having Ridge and Furrow areas being mostly colored red is fine, as long as there is at least one blue structure present. It is possible to identify more of the red objects as blue ones by lowering the threshold, but along with the number of blue structures, the risk to generate misclassifications rises as well.

Problems occur when ridges or furrows are cut (Figure 15a), which is caused by the *scale* parameter in the segmentation algorithm. To solve this problem, the whole process in *eCognition* would to be run again, starting with the time-consuming segmentation.

Other misclassifications occur if multiple paths run parallel to contour lines in hillside areas (Figure 16). The problem is that a sequence of path-slope-path-slope (etc.) looks like ridges and furrows, once the hill is removed in the DM. To address this, searching for slopes in the regular DTM might be a solution but the question arises, what happens to ridges and furrows that are located on a hillside and are oriented along the contour lines.

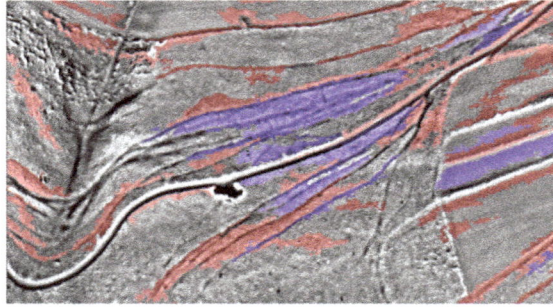

Figure 16. Misclassifications in a hillside situation in Höxter in a DM [13].

3.3. Burial Mounds

Although completeness and correctness are not that meaningful as with binary classifications, Table 8 provides an idea of what is possible under good circumstances. The decreasing correctness is no surprise because the class descriptions get wider in order to find possible eroded mounds as well (Trier et al. 2015 observed the same with their confidence levels [6]). Completeness is 100% by definition because all 172 reference mounds defined the classes. Regarding the values, two things should be noted. On the one hand, some Burial Mounds were segmented insufficiently, meaning that it was impossible to generate objects with borders meeting those of the real-world mounds, and are therefore not included in the dataset with its 172 class defining mounds (see above). The lack of segmentation can be caused by a high degree of erosion, suboptimal segmentation settings or, most likely, a size differing significantly from the average. These mounds are missing in the classification and are therefore not included in the calculation.

On the other hand, the reference dataset is probably not complete. Therefore, some of the *false positives* are actually unconfirmed and might be *true*, once they are interpreted or even excavated (the most promising *false* mounds are called *new positives*). The others are *false positives* until they are definitely interpreted as *true* or *false*.

Despite all these uncertainties, this method is still useful because the interpreter is able to interpret the classes with the most promising results (e.g., classes 1–3) at first and can eventually focus on the others or see the results in their relation to the others (Table 8).

Table 8. Results of the classification of Burial Mounds in the investigation area in Haltern.

Class	TP [1] + NP [2]	FP [3]	Total	Correctness [4] (%)
(1) very well preserved	14	0	14	100
(2) well preserved	20	10	30	67
(3) sufficiently preserved	64 + 9	138	211	35
(4) eroded	52 + 11	404	467	13
(5) highly eroded	22 + 6	905	933	3
total	172 + 26	1457	1655	(12)

[1] true positives, included in the reference dataset, [2] new positives, included in the reference dataset after manual interpretation based on terrain models and aerial photography, [3] not included in the reference dataset, [4] correctness = TP+NP/Total.

Figures 17 and 18 present a subset from the mound detection area in Haltern. The decreasing correctness is displayed by the dominance of orange and red shapes, whereas more ideal shaped mounds are colored yellow and green. It is obvious that the classification works better in flat areas and has problems at edges, because the DM generates pseudo mounds where the terrain is more complex. Nevertheless, the results are useful and simplify the interpretation by highlighting probable monuments. In addition, Figure 18 shows a possible new mound (*new positive*, yellow) that was not included in the reference dataset (missing dot) and looks like its neighbors of the same class.

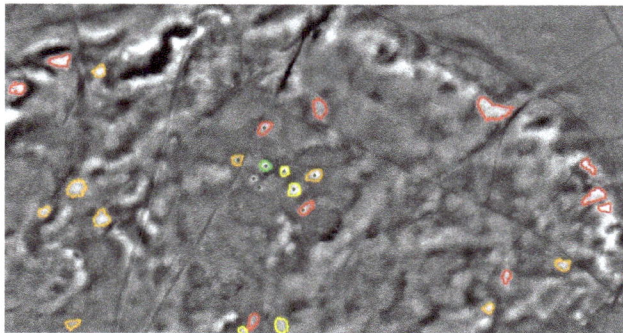

Figure 17. Classification results of Burial Mounds in Haltern. Members from the following classes are presented (Table 8); 3 (yellow), 4 (orange), and 5 (red). Mounds with a dot are included in the reference dataset [13].

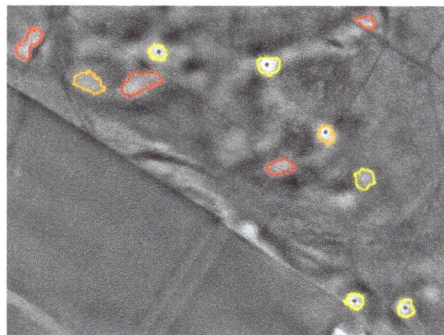

Figure 18. Classification results of Burial Mounds in Haltern. Members from the following classes are presented (Table 8); 3 (yellow), 4 (orange) and 5 (red). Mounds with a dot are included in the reference dataset. The yellow traced mound on the right side without a dot inside is considered as a new discovery (*new positive*) [13].

Although detection of mounds produced useful results, there are some issues to be solved. A major problem is the variety of sizes, making it difficult to find segmentation settings that work for every size of Burial Mound. Multiple segmentation would be a way to address this. Another problem is that some mounds are classified in lower classes (e.g., 4 & 5) although their appearance is almost ideal. This sometimes happens due to only one outlying statistical value (e.g., shape index), making the mound incompatible to the better classes. This could be addressed by reclassifying the statistics of every object to a common scale at first and then by calculating an average value representing the degree of erosion for every mound. This value would finally be the only one to be classified in the classes listed above, which would avoid the strong influence of single outlying statistics.

4. Conclusions & Outlook

This article presented possible ways of classifying typical field monuments in the relatively low-quality Westphalian LiDAR dataset. It was demonstrated that all three presented monuments are basically detectable with automated workflows using OBIA and that it is reasonable to classify monuments by their degree of erosion.

The results reveal the same difficulties that other studies are affected by as well: the more complex the terrain is and the more anthropogenic influence is present, the more difficult it is to discriminate between *true* and *false positives*.

Optimizing the positive layer will be one point to address this issue. Up to now, only unsealed areas where considered. However, there are unsealed, positive areas under strong anthropogenic influence that still need to be rejected. In addition, some structures that cause *false positives* are not recorded as polygons but as points (e.g., foundations of windmills) and lines (e.g., roads and railway tracks) due to their size. These data need further examination to decide which of them can be buffered and rejected as well. This will improve classification results as well as save processing time.

The combination of OBIA and the available LiDAR dataset seems to be suitable for the presented purpose, because especially the calculated quality of the mounds detection looks similar to those from studies with high-quality data (e.g., approximately 10 pt/m^2, [6,7]). The problem with OBIA is its vulnerability to distortions and the implementation in commercial software. Therefore, Template Matching, e.g., as proposed by Davis [12], is investigated and was already implemented for experimental purposes in the *ArcGIS*-tool.

Regarding the latest innovative trend of Machine Learning, e.g., summarized and proposed by Trier et al. [23], the authors are curious to examine its potential compared to 'traditional' approaches like the one presented here. The problem is the large amount of high-quality training data, which may be problematic in regards to archaeological records.

The workflows are supposed to be an addition to the toolbox of archaeological prospection. Hopefully, they can contribute to the provincewide database of archaeological records that was mentioned in the beginning. A precondition is that they produce good results in other areas as well.

Author Contributions: Conceptualization, M.F.M., I.P. and C.J.; methodology, M.F.M.; software, M.F.M.; validation, M.F.M. and I.P.; investigation, M.F.M.; resources, I.P. and C.J.; data curation, M.F.M. and I.P.; writing—original draft preparation, M.F.M.; writing—review and editing, M.F.M. and C.J.; visualization, M.F.M.; supervision, I.P. and C.J.; project administration, C.J.

Funding: This research received no external funding.

Acknowledgments: The corresponding author thanks Ingo Pfeffer and Carsten Jürgens for their support and the motivation to start the Ph.D. project. Last but not least, thanks to the team of the LWL-Archaeology in Münster, especially Michael Rind and Christoph Grünewald for their feedback and supervision of the project.

Conflicts of Interest: The authors declare no conflicts of interest.

References

1. Sittler, B. Revealing historical landscapes by using airborne laser scanning. A 3-D modell of Ridge and furrow in forests near Rastatt (Germany). In *Laser-Scanners for Forests and Landscape Assessment*; Thies, M., Koch, B., Spiecker, H., Weinacker, H., Eds.; ISPRS Archives: Freiburg, Germany, 2004; XXXVI, PART 8/W2; pp. 258–261. ISSN 1682-1750.

2. Heinzel, J.; Sittler, B. LiDAR surveys of ancient landscapes in SW Germany: Assessment of archaeological features under forests and attempts for automatic pattern recognition. In *Space, Time, Place. Third International Conference on Remote Sensing in Archaeology*; Campana, S., Forte, M., Liuzza, C., Eds.; Archaeopress Ltd.: Oxford, UK, 2010; pp. 113–121, ISBN 978 1 4073 0659 9.

3. De Boer, A. Using pattern recognition to search LIDAR data for archeological sites. In *The world is in Your Eyes. CAA2005. Computer Applications and Quantitative Methods in Archaeology*; Figueiredo, A., Velho, G., Eds.; Archaeopress: Tomar, Portugal, 2007; pp. 245–254.

4. Kramer, I.C. An archaeological reaction to the remote sensing data explosion. Ph.D. Thesis, University of Southampton, Southampton, UK, 2015.

5. Trier, Ø.D.; Zortea, M. Automatic Detection of Pit Structures in Airborne Laser Scanning Data. *Archaeol. Prospect.* **2012**, *19*, 103–121. [CrossRef]

6.   Trier, Ø.D.; Zortea, M.; Tonning, C. Automatic detection of mound structures in airborne laser scanning data. *J. Archaeol. Sci.: Rep.* **2015**, *2*, 69–79. [CrossRef]

7.   Trier, Ø.D.; Pilø, L.H.; Johansen, H.M. Semi-automatic mapping of cultural heritage from airborne laser scanning data. *Sémata* **2015**, *27*, 159–186.

8.   Schneider, A.; Takla, M.; Nicolay, A.; Raab, A.; Raab, T. A Template-matching Approach Combining Morphometric Variables for Automated Mapping of Charcoal Kiln Sites. *Archaeol. Prospect.* **2015**, *22*, 45–62. [CrossRef]

9.   Freeland, T.; Heung, B.; Burley, D.V.; Clark, G.; Knudby, A. Automated feature extraction for prospection and analysis of monumental earthworks from aerial LiDAR in the Kingdom of Tonga. *J. Archaeol. Sci.* **2016**, *69*, 64–74. [CrossRef]

10. Cerrillo-Cuenca, E. An approach to the automatic surveying of prehistoric barrows through LiDAR. *Quat. Inter.* **2017**, *435*, 135–145. [CrossRef]

11. Davis, D.; Sanger, M.; Lipo, C.P. Automated mound detection using lidar and object-based image analysis in Beaufort County, South Carolina. *Southeast. Archaeol.* **2018**, 1–15. [CrossRef]

12. Davis, D.; Lipo, C.P.; Sanger, M.C. A comparison of automated object extraction methods for mound and shell-ring identification in coastal South Carolina. *J. Archaeol. Sci.: Rep.* **2019**, *23*, 166–177. [CrossRef]

13. Land NRW (2018), dl-de/by-2-0 (www.govdata.de/dl-de/by-2-0). Available online: https://www.opengeodata.nrw.de/produkte/geobasis (accessed on 20 October 2018).

14. Hinz, H. *Motte und Donjon: Zur Frühgeschichte der mittelalterlichen Adelsburg*; Rheinland-Verlag GmbH: Köln, Germany, 1981; ISBN 3-7929-0433-1.

15. Küster, H. *Geschichte der Landschaft in Mitteleuropa. Von der Eiszeit bis zur Gegenwart*, 4th ed.; C.H. Beck: München, Germany, 2010; ISBN 978-3-406-64438-2.

16. Meyer, M.F. Die automatische Suche nach Bodendenkmälern im Laserscan. In *Archäologie in Westfalen-Lippe 2015*; Rind, M.M., Dickers, A., Eds.; Beier & Beran: Langenweißbach, Germany, 2016; pp. 250–254, ISBN 978-3-95741-052-8.

17. Open Data—Digitale Geobasisdaten NRW. Available online: https://www.bezreg-koeln.nrw.de/brk_internet/geobasis/opendata/index.html (accessed on 20 October 2018).

18. FuPuDelos. Available online: https://www.lwl.org/delos/ (accessed on 20 October 2018).

19. Sevara, C.; Pregesbauer, M.; Doneus, M.; Verhoeven, G.; Trinks, I. Pixel versus object—A comparison of strategies for the semi-automated mapping of archaeological features using airborne laser scanning data. *J. Archaeol. Sci.* **2016**, *5*, 485–498. [CrossRef]

20. Hesse, R. LiDAR-derived Local Relief Models—A new tool for archaeological prospection. *Archaeol. Prospect.* **2010**, *17*, 67–72. [CrossRef]

21. Trimble (Ed.) *eCognition Developer. Reference Book*; Trimble Germany GmbH: München, Germany, 2014.

22. Haus Döring. Available online: https://www.borken.de/tourismus/sehenswertes/haus-doering.html (accessed on 20 October 2018).

23. Trier, Ø.D.; Cowley, D.C.; Waldeland, A.U. Using deep neural networks on airborne laser scanning data: Results from a case study of semi-automatic mapping of archaeological topography on Arran, Scotland. *Archaeol. Prospect.* **2018**, 1–11. [CrossRef]

geosciences

MDPI

Article

GIS and Remote-Sensing Application in Archaeological Site Mapping in the Awsard Area (Morocco)

Ange Felix Nsanziyera [1,*], Hassan Rhinane [1], Aicha Oujaa [2] and Kenneth Mubea [3]

[1] Geosciences Laboratory, University Hassan II, B.P. 3366 Maarif, Casablanca 20100, Morocco;
   h.rhinane@gmail.com
[2] Institut National de Sciences de l'Archéologie et du Patrimoine (INSAP), B.P. 6828, Rabat 10000, Morocco;
   aicha.oujaa@gmail.com
[3] Regional Centre for Mapping of Resources for Development, P.O. Box 632-00618, Nairobi 00100, Kenya;
   kpwmubea@gmail.com
* Correspondence: felixon87@gmail.com; Tel.: +212-606-783-072

Received: 2 April 2018; Accepted: 15 May 2018; Published: 7 June 2018

Abstract: Morocco is famous as one of the archaeologically richest places with many sites. In addition, some of the sites have been listed as UNESCO World Human Heritage sites. In situ observations are used in cultural heritage and archaeological sites mapping. However, this procedure requires periodic observations, which are practically difficult to combine with traditional methods and practices since this is time consuming and expensive. Thus, modern technologies, mainly GIS and remote sensing, are gaining attention as tools for prediction at archaeological sites. The aim of this paper is to assess the application of GIS and remote sensing in order to develop a predictive model, which will be used in locating areas with high potential as archaeological sites in the Awsard area (southern Morocco). The analytic hierarchy process (AHP) as a multi-criteria decision making method, which integrates archaeological data and environmental factors, geospatial analysis and predictive modelling, has been applied to the identification of possible tumuli locations in the study area. The model was developed using a zone of 21 km^2 with 233 known sites. It was later validated using 530 unknown sites within an area of 980 km^2. The acceptable accuracy of 93% was calculated using an estimation of predictive gain, which proves the efficiency of the model's predictive ability.

Keywords: analytic hierarchy process (AHP); archaeology; predictive model; tumuli; remote sensing; multi-criteria; Saharan Morocco

---

1. Introduction

Morocco is full of archaeological remains, from Volubilis and Lixus (ancient roman cities ruins) in the north of the country to funerary mounds, rocks engravings, and pottery across the country especially in the Moroccan Sahara [1–4]. The pre-historic funerary monuments (tumuli) are a primary target for looting, since they are known to contain valuable objects. A large part of these monuments have been vandalised by vandals. A general survey should be undertaken in order to effectively protect and study the archaeological sites in the Awsard area. The study of tumuli in Morocco shows that the major groups are located south of the Atlas and the eastern zones in pre-Saharan zones. These tumuli constitute the cultural and environmental diversity of the landscape characterized by large sandy plains located between the Moroccan Saharan desert, the Atlas mountains and the Atlantic coast [5].

Remote sensing has opened up new horizons and possibilities for archaeology. Satellite application in archaeology is an emerging field that uses high-resolution satellites with thermal and infrared capabilities to pinpoint potential sites of interest on the Earth around a meter or so in depth. Satellite remote-sensed data have become a common tool of investigation, mapping, prediction, and forecast of archaeological sites locations through the development of GIS-based predictive models. Archaeologists have realized the potential of GIS and remote sensing in archaeology. For example, aerial photography can detect phenomena on the surface associated with subsurface relics [6–8], while the use of infrared and thermal electromagnetic radiation can be used to detect underground archaeological remains [9,10]. Some few published studies have shown the importance of using high-resolution satellite images in archaeology and heritage management [11].

Remote sensing in the form of aerial photography as well as ground-based techniques such as ground penetrating radar, soil resistance and magnetic prospection, are well established and widely practiced within archaeological research [12–16]. Such techniques have their merits but also have their disadvantages (e.g., geophysical surveys are time consuming, expensive and tend to be limited in their spatial extent; whilst features in the landscape that reflect beyond the visible spectrum are unable to be detected using conventional aerial photography) [17]. Airborne light detection and ranging (LiDAR) as well as multi- and hyperspectral remote sensing are additional geo-archaeological tools, which have been used sparingly to date within this context [18–22]. Lasaponara and Masini [23] discussed the way the emergence of optical sensors, especially the very high spatial resolution satellite (VHSRS) sensors, initiated a new era in archaeological research. With their multi-spectral capability and very high spatial resolution these sensors are an excellent tool for archaeological investigations. However, SAR (synthetic aperture radar) is encountering more difficulties in realizing its full potential for archaeological applications due to the greater complexity of data processing and interpretation tools. Chen et al. [24] and Tapete and Cigna [25] explored SAR-based approaches for the reconnaissance of archaeological signs and SAR interferometry for the monitoring of cultural heritage sites, and the ways and means to reduce complexity of data processing and interpretation.

Depending on their configuration, spaceborne, airborne and field-based sensors (multi- and hyperspectral) deliver capture and attractive data since those sensors can detect electromagnetic radiation in the visible and beyond into the infrared and the thermal. With technological improvements, the higher spatial and spectral resolutions previously only attainable through airborne remote sensing are increasingly becoming available on spaceborne platforms. Multi-spectral sensors such as Sentinel and Landsat acquire data in broad (~100–200 nm) irregularly spaced spectral bands [26]. This will have a negative effect on distinguish narrow spectral features due to averaging across the spectral sampling range or masking by stronger features [27]. The net effect is the reduction or loss in the information that can potentially be extracted. Hyperspectral sensors, usually made of tens to hundreds of bands, are able to retrieve near-laboratory quality reflectance spectra such that the data associated with each pixel approximate the true spectral signature of the target material [28]. Given the cost implications of acquiring such high spatial and spectral resolution imagery, they are rarely used in archaeology. Consequently, the application of remote sensing is limited to geological units classification based on a multi-spectral Landsat 7 ETM+ images. This geological map will be further used in archaeological modelling [29].

GIS has, to date, been primarily applied to "landscape studies" [30–32]. In the context of GIS, this refers to archaeological analysis undertaken at regional scale as distinct from intra-site scale [33]. GIS has also permitted further advances in the use of classical statistical approaches to archaeological materials by using one sample significant test [34], and recently for the application of geo-statistical approaches to spatial variables [33].

An inductive predictive model aims to predict the archaeological characteristics of places from their non-archaeological, usually environmental, characteristics by using statistical properties of locations where sites occur in order to generate a classification rule that governs the archaeological properties of locations for which archaeological properties are not known [33]. The predictive models are often done for two purposes: to explain the observed distribution of archaeological remains, and hence the behaviour of past communities, or to inform archaeological management strategies.

Many criticisms have been made against inductive predictive models concerning, on the one hand, the fact that those models try to explain the past through asserting the primacy of correlation between behaviour and environmental characteristics. This is taken as de-humanizing the past since the meaningful human actors that we try to understand are reduced to automata who behave according to a rule that connects their behaviour to their environment [34]. On the other hand, correlative predictions of those models can be seen as anti-historical. It assumes that the observed patterns are a product of the immediate surrounding of the communities responsible for them and can, therefore, be explained by the link between the two. The present practice of predictive modelling is still very much environmentally deterministic. Cultural variables are not included in the models, resulting in predictions ultimately based on physical properties of the current landscape [35]. This may be explained by the fact that social and cognitive factors seem to be difficult to model as they are regarded as being too abstract and intangible for use in a predictive mode. Thus, the predictions can only be studied for a very limited range of questions, based on very specialized data sets (mostly relating to the ritual prehistoric landscapes of Wessex in England (e.g., [36,37]). Another issue with predictive modelling is that more effort and funds for archaeological heritage management will be dedicated only to the areas of "high archaeological value". These areas will become better known, whereas the areas that are designated a "low value" on the predictive map will largely be ignored in (commercial) archaeological research [31]. It must be agreed that the predictive models may be telling us something about the behaviour of people in the past. The main advantage of predictive modelling is that it has some pragmatic utility for the management of archaeological resources. In other words: because those models aim at the effective protection of cultural (archaeological) heritage rather than in its understanding [36,37].

In Moroccan archaeology, the use of geospatial technologies is still absent, and some interesting studies have shown the huge role played by geospatial technologies in archaeology and heritage management [38,39]. For a long time, archaeological surveys in Maghreb regions including Morocco have employed conventional methods that rely mainly on a field's inventory where local testimonies play a key role in the localization of archaeological sites in a neighbourhood [39–41]. Unfortunately, this classic method has often shown limitations especially when it comes to work in uninhabited mountainous and limited access areas [39]. This represents a big challenge for Moroccan archaeologists given the vastness of the areas to be surveyed and its limited accessibility as well as the huge financial and human resources required.

In the light of this issue, this paper presents an archaeological predictive model, which is a GIS-based model that suggests the locations where tumuli may be found in the Awsard area, using geospatial techniques. This model constitutes a solution for archaeological heritage management by identifying the areas with high potential for hosting archaeological remains [42]. This research explores the use of GIS, remote sensing and geospatial analysis to locate potential areas of archaeological sites (mainly tumuli). In addition, the research explores the development of an archaeological predictive model.

2. Materials and Methods

2.1. Study Area

The study area is located in the southern Moroccan Sahara near the Morocco-Mauritanian border in the province of Awsard. This zone of 1000 km^2 is delimited by the 14°20′ W and 14°36′ W meridians and by the 22°21′ N and 22°40′ N parallels. The study area has a Saharan desert climate typically dominated by low relief landscapes which are part of the Reguibat uplift, an important geological domain of the West African Craton (Figure 1a). The geological formations are mainly made of crystalline bedrock that consists of various igneous and metamorphic rocks dominated by granites, schist, gneisses and migmatites. These bedrock terrains are very tectonised and almost completely flattened, and covered by Paleozoic formations made mainly of schist, sandstones and quartzite (Figure 1b). The bedrock terrains and its Paleozoic cover deepen westwards under the Meso-Cenozoic formations of the Laayoune-Dakhla-Trafaya basin. Small layers of Quaternary deposits consist of detrital sediments of various types that are often less consolidated: fluvial, river-lake, slope and wind types also exist in the study area. The landscape is characterized by the presence of peneplains which alternate with prominent reliefs made of cuesta, outliers and scattered hills corresponding to inselberg, locally known as *glebs*. Most common quaternary deposits are the plains and river terraces, alluvial fans, glacis and peneplain with a thin mantle of sand and regs. The shapes of the most common quaternary deposits are the plains and river terraces, cones of dejection, glacis and peneplains. Those deposits with thin mantles of sand and regs carry thousands of tumuli and other evidence of prehistoric human occupation such as rock engravings.

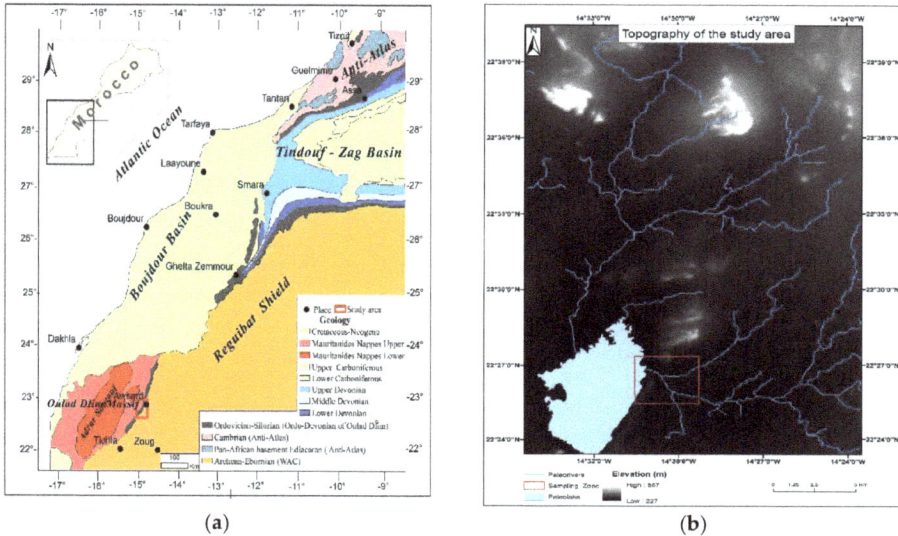

(a)   (b)

Figure 1. Geological maps: (**a**) background after the geological map of Morocco, scale 1:1,000,000; (**b**) topography (digital elevation model, DEM) map of study area.

These tumuli of different types and diverse locations punctuate the landscape of the region around Awsard and clearly reflect those well known in other Moroccan regions [43]. Saharan tumuli are commonly of circular and conical shapes. They measure from 3 to 15 m in diameter and about 1 to 5 m in height (Figure 2a). Other kinds of crescent-moon shaped tumuli are also frequent, with a width of 2 to 6 m and 5 to 15 m length, their height vary from 1 to 5 m and they commonly possess two or rarely three antennas that can reach several tens of meters.

(a)   (b)   (c)

Figure 2. Tumuli in the Awsard area: (**a**) conical tumulus, (**b**,**c**) long stone tumuli.

Rare funerary monuments called *bazinat* are spectacular in shape with 2 or 3 levels and sometimes they have an imposing size of 10 m diameter even more. The last type of Saharan Moroccan tumuli is the "Tumuli with long stone" whose length can reach 5 m (Figure 2b,c).

In order for predictive modelling of archaeological sites to succeed, quantifiable data are needed to create statistical statements about patterns on the locations of the sites [44]. On the other hand, spatial analysis of many environmental and archaeological variables in a GIS environment are required. Satellite imagery processing with a multi-criteria decision making approach based on the analysis of geological, geomorphological, archaeological profiles and GIS spatial analysis including slope, aspect, elevation and hydrologic-surface analysis were carried out. To obtain a wide spectrum of geo-environmental data for analysis, our work was based on topographical analysis of digital elevation models (DEM).

2.2. Data

2.2.1. Geo-Archaeological Database

This survey was conducted in six different field campaigns from 2014 up to 2017. The first field survey was carried between December 2014 and April 2017. Six field surveys were conducted in the Awsard area. The surveys were based on walking on the field with the aim of mapping all visible sites in the area. We worked with small mobile teams of three people equipped with a handheld GPS device, a tape measure and a compass. The task involved taking GPS coordinates, and the size and orientation of tumuli. Ultimately, we were able to cover 100% the area of our zone of interest (around 1000 km^2). In total, the teams recorded 815 structures. The first mission focused on an area zone of 21 km^2 where 233 sites were recorded. The data from this area that we called the "sampling zone" were used to develop the predictive model. After model development, we carried out five field surveys in the whole area, not only in high potential zones but also in low potential zones. A detailed geo-archaeological database was built in the ArcGIS (Esri, Redlands, CA, USA) environment with many attributes (name of site, geographic coordinate, type of site, geometric size of site, and type of rocks present in the sites, etc.) During the survey, we faced several challenges namely the fact that some tumuli were completely ransacked to the point that it was almost impossible to differentiate them from the background as we were in a rocky desert area. In addition, the sand depositions seemed to cover and hide some sites and visibility became extremely difficult in afternoon when the sun was strong.

2.2.2. Digital Elevation Model

- A SRTM DEM 1° arc-second image with a 30 m spatial resolution was downloaded from the United States Geological Survey (USGS) website http://earthexplorer.usgs.gov/
- An ASTER GDEM V2 images was downloaded from ASTER (Advanced Space-borne Thermal Emission and Reflection Radiometer) global DEM website http://www.jspacesystems.or.jp/ersdac/GDEM/E/4.html
- A DEM of the study area with 10 m cell size was generated through the digitization of contours and elevation points of a 1:50,000 scale topographic map.

2.2.3. Landsat ETM+ Multi-Spectral Image

A Landsat 7 ETM+ multi-spectral image, with 30 m spatial resolution of the study area, was obtained freely from the National Aeronautics and Space Administration (NASA) database http://earthexplorer.usgs.gov/, with many processing parameters including the supervised classification and band ratio technique.

2.3. Methodology

In this study, the methodology, which integrates GIS and multi-criteria evaluation, was used and applied to the Awsard area (Figure 3). The data analyses and processing in this study were based on both archaeological and geo-environmental information collected on 233 sites successively identified during first field survey campaign carried out on the sampling zone. At the end of the model set up, we realized a second field survey to collect data used to validate the models.

2.3.1. DEM Quality Assessment

However, before proceeding with the geo-environmental parameters analysis of Awsard, the reliability and accuracy of the DEM created by digitalization of contour lines from the topographic map was evaluated and compared to the Shuttle Radar Topography Mission (SRTM) DEM. The 30 m resolution DEMs were provided by ASTER. We used the equation of Schumann [45] to check the validity of different DEMs by comparing a number of reference points (datum points) (Table 1). The DEMs were validated by reference elevation data distributed across the study area. The height precision or quality of each DEM is expressed by Equation (1).

$$\text{RMSE}_{\text{DEM}} = \sqrt{\frac{\sum_i^n (E_R - E_{\text{DEM}})^2}{n}} \tag{1}$$

where RMSE is Root mean-square error, E_R is the elevation of reference, E_{DEM} is the elevation as provided by the different DEM data and n is the total number of points used as reference.

Table 1. DEM quality assessment.

E_R (m)	$E_{\text{DEM TOPO}}$ (m)	$E_{\text{DEM ASTER}}$ (m)	$E_{\text{DEM SRTM}}$ (m)	$(E_R - E_{\text{DEM TOPO}})^2$ (m)	$(E_R - E_{\text{DEMA STER}})^2$ (m)	$(E_R - E_{\text{DEM SRTM}})^2$ (m)
338	329	333	330	81	25	64
346	343	368	321	9	484	625
286	280	275	277	36	121	81
283	268	273	253	225	100	900
277	279	284	276	4	49	1
276	272	261	271	16	225	25
289	291	274	277	4	225	144
302	318	290	293	256	144	81
334	356	337	334	484	9	0
283	274	272	272	81	121	121
			$\frac{\sum_i^n (E_R - E_{\text{DEM}})^2}{n}$	119.6	150.3	204.2
			$\sqrt{\frac{\sum_i^n (E_R - E_{\text{DEM}})^2}{n}}$	10.936	12.260	14.290

The RMSE for the DEM created through the digitized contour lines of the topographic maps was the lowest at 10.936 m, followed by the RMSE of the ASTER GDEM (global digital elevation model) with 12.26 m and the RMSE of SRTM DEM was the highest with 14.29 m. Moreover, from the results above, the DEM, which was produced by digitization of the topographic maps, became the basic layer of the terrain for further processing.

2.3.2. Geo-Archaeological Variables

As noted by Clement et al. [46], "The premise behind modelling of archaeological sites is that historic and prehistoric peoples were closely tied to their natural and cultural environment, and that these environments were a significant determinant in their choice of site location." Predictive modelling examines soils (geology), distance to water sources, aspect, altitude and slope as potential environmental parameters or variables. GIS tools could be necessary for mapping and additional analysis such as buffer zoning, neighbouring computation and overlay analysis were used to construct a predictive model. The specific predictive model was built based on the use of a multi-parametric spatial analysis method of geographic elements, statistical and archaeological information in order to construct partial maps of archaeological interest [45,47,48]. The environmental factors, namely elevation, aspect, slope, distance from water sources, geology or rock types, were noted to have an influence on the choice of habitation in the Neolithic period in Awsard. These factors were statistically examined and assigned weight constraints according to expert knowledge. The cultural variables such as distance to the main settlement and/or ancient road proximity were used based on their availability for better accuracy of the model. However, as highlighted by Verhagen et al. [33] "cultural variables are rarely included in the models, thus resulting in predictions based on physical properties of the current landscape". This may be explained by the fact that social and cognitive factors seem to be difficult to model as they are regarded as being too abstract and intangible for use in a predictive mode. Nevertheless, there is a lack of cultural data in the study area and archaeological researchers in the Saharan areas are only limited to rock arts and tumuli recording, notwithstanding settlements [49].

At the first stage of this analysis, we evaluated the functional relations between the locations of the archaeological sites and the following environmental parameters: elevation, slope, aspect, distance from water streams, and geology. Since the model variables change from one location to another, there are no fixed ranges and corresponding coefficients. Only statistical analysis of data collected in sampling zone allows coefficient assignment. Therefore, inside every variable, based on statistical analysis and number of expected and observed sites in an area, many classes were created with a corresponding coefficient consideration. The coefficient assignment is ruled by the following formula in Equation (2):

$$Coef = 5 \times \left(1 + \frac{Obs - Exp}{Obs + Exp}\right) \tag{2}$$

where Coef is the coefficient, Obs are the observed sites and Exp are the expected sites.

The theory behind this formula is to compare homogenous distribution of sites (expected according to percentage of the areas) against the observed distribution of sites. If observed sites are more than expected sites, the coefficient will be higher than 5, otherwise the coefficient will be less than 5.

Slope

Slope is a direct function of the topography of a region [50]. Land slope is basic parameter for the construction of a predictive model. The basic argument here is that it is difficult to maintain a funeral mound on ground that is too steep since the steepness will increase the erodibility of the mound. Tumuli erection becomes impossible as a slope reaches certain values [7,51]. A layer consisting of slope (degree) data was generated by using the elevation grid as an input layer. The appropriate slope for tumuli is about 3–10%, the areas of this slope range were assigned a coefficient of 8. Therefore, steep areas with a slope of >20% were assigned a coefficient of 1, and gentle slope areas (0–3%) with a coefficient of 4 (Table 2). Slope seems to be a variable with little real predictive power, while almost 98% of all sites within the training area are located on terrain where the slope is between 0 and 10%, over 99% of the training area showed a slope of between 0% and 10% (Figure 3a,b). The vast majority of sites (47.64%) were found to be laid on plain areas (0–3% slope) and 50% of sites on 3–10% of terrain, thus it seems like Neolithic people occupied areas frequently located on landforms displaying low and moderate slopes.

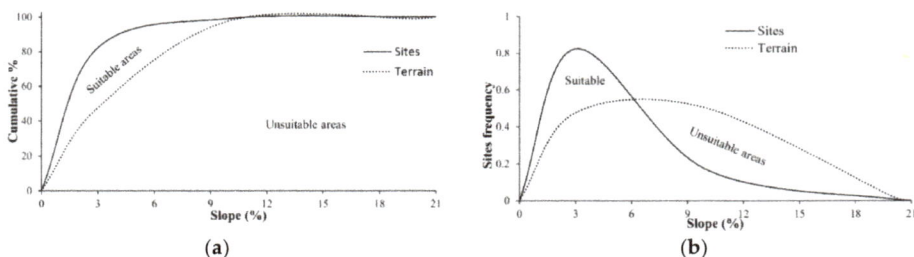

Figure 3. Sites repartition in relation to slope: (**a**) cumulative repartition; (**b**) site frequency.

Geology and Lithology

Geological formations and sediments bearing raw materials were identified using existing geological maps (Figure 1) improved by thematic maps generated from Landsat ETM+ images. Landsat 7 ETM+ band 1, band 3 and band 5 were used to produce a colour composite by applying the image-processing procedure known as "bands ratio". Crósta and Moore [52] showed that this band combination is the most effective to discriminate rock types in semi-arid regions. The combination of spectral and texture characteristics mixed with supervised classification using previous field investigations (training sites) were used to identify and map rock types in the Awsard area. Five geological units including quartzite, schist, granites, sandstones, and sandy plain were identified. The accuracy assessment of the classified image was done through a matrix confusion approach [53,54]. Thereby overall accuracy found was 89%.

During the field survey, it turned out that the spatial distribution of archaeological sites is linked to the geomorphological setting and petrographic nature of the rocks. Thus, several types of geomorphological substrates of tumuli were distinguished. The most important are:

- the alluvial cones
- the pediments;
- the plains and river terraces;
- the cuesta cuff trays;
- the benches of the slopes of the valleys and below;
- the dykes through the vast flat surface;
- the precambrian plan crystalline basement.

Petrographic types of mound materials vary from one place to another and heavily depend on the lithology of the geological substrates and their neighbourhoods. When substrate belongs to the Precambrian, the most used stones are magmatic rocks (granitites, dolerites, gabbros and syenites) and other associated metamorphic rocks (serito-schist, light-grey quartzite, gneiss, migmatites, etc.); by contrast, when it consists of Paleozoic cover, the most used stones are quartzite and sandstones which are cut into slabs or plates that were used as steles in many tumuli. The availability of basic raw materials for use in erection of tumuli could be important [26]. The consolidated, hard and resistant materials are required for the construction of tumuli [55], which is why the areas with hard and resistant geological units including quartzite, schist and granite receive high coefficients, respectively 7, 6 and 5, while areas with soft and less resistant geological units receive low coefficient (1 for sandy plain and 4 for sandstones (Table 2)).

Aspect

There is a little pattern in the sites' data to suggest that western aspect areas were less preferred than other aspects or exposures. A coefficient of 4 was affected to western exposure areas while northern aspect exposure areas received a coefficient of 6, and eastern and southern aspect exposure areas obtained a coefficient of 5 (Table 2).

Topography (Altitude)

A greater concentration of archaeological sites has been recorded at lower altitudes in the study area (Figure 4). The spatial distribution of known sites in the model development zone known as the sampling zone permitted the range of 250–300 m to be defined as the most suitable altitude for site occurrence. Consequently, a coefficient of 6 was assigned to areas between 250 and 300 m (Table 2).

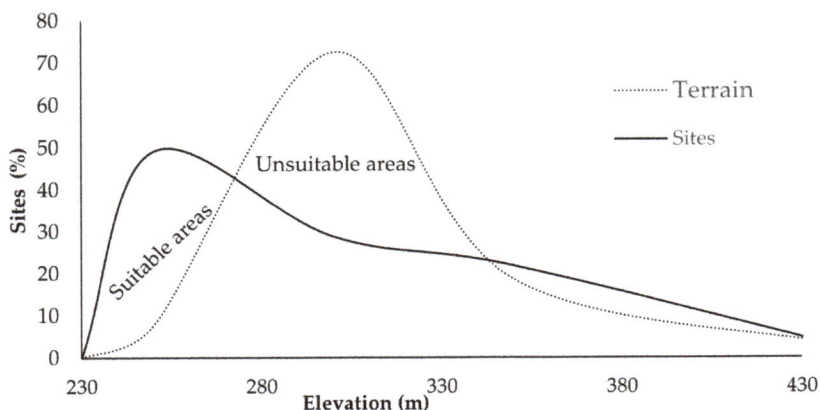

Figure 4. Sites' location in relation with elevation.

Distance to Water Source

Fresh water has always been a main factor for human occupation of any territory through history. The prehistoric population was set to occupy the areas within moderate distance close from the closest water sources [56]. In the study area and particularly in training site, 75% of sites are located in the areas less than 500 m from streams or lakes, and 98% of sites are located within 1 km distance close to water points. A coefficient of 7 has been given to areas within 100 and 250 m of a watershed, and a low coefficient of 1 to areas beyond 1 km from a watershed (Table 2).

2.3.3. Dependency between Variables and Sites

The geo-archaeological variables at the location of known sites are tested against the background environment as a whole to identify whether there are statistically significant differences between the distributions of sites against a particular geo-environmental variable [57]. Layers of geo-archaeological data are stored in 10 × 10 m raster grids in the GIS. Each grid cell receives a value for each geo-archaeological variable. The values for each of the geo-archaeological variables are also recorded for known archaeological site locations.

The values of these variables at known sites locations and their values across the entire study area are compared through the use of the Kolmogorov–Smirnov goodness-of-fit test [57]. This test examines the goodness-of-fit of a single sample to a referent population, and is appropriate for continuously distributed data, such as geo-environmental data [57]. The test is performed by plotting the cumulative distributions of the sample and population, respectively, and obtaining the difference (D) between these two curves. For a level of significance alpha = 0.05, for example, and in a two-tailed test, the value of d can be calculated from Equation (3):

$$d = \frac{1.36}{\sqrt{n}} \tag{3}$$

Table 2. Kolmogorov–Smirnov goodness-of-fit test for some variables.

	Class	Expectations (Sampling Zone)				Observations (Sampling Zone)			Coefficient	D
		Area km^2	Area %	Expected	Cumulative Frequency	Sites	% Sites	Cumulative Frequency		
Slope in %	0–3	17.40	82.29	192	0.823	111	47.64	0.476	4	0.347
	3–10	3.57	16.91	39	0.992	119	51.07	0.987	8	0.005
	10–20	0.17	0.80	2	1	3	1.29	1.000	6	0.000
	>20	0.00	0.00	0	1	0	0.00	1	1	0.000
	$n =$	21.14	100	233		233	100		$D_{max} =$	**0.347**
Geological unit	Quartzite	1.42	10.23	24	0.102	48	20.60	0.206	7	0.104
	Schist	2.19	15.83	37	0.261	54	23.18	0.438	6	0.177
	Granite	2.84	20.49	48	0.466	57	24.46	0.682	5	0.217
	Sandstone	6.68	48.27	112	0.948	74	31.76	1.000	4	0.052
	Sandy Plain	0.72	5.18	12	1	0	0.00	1	1	0.000
	$n =$	13.84	100	233		233	100		$D_{max} =$	**0.217**
Dist. water (m)	<100	4.32	26.50	62	0.265	40	17.17	0.172	4	0.093
	100–250	3.62	22.20	52	0.487	66	28.33	0.455	6	0.032
	250–500	4.30	26.38	61	0.751	68	29.18	0.747	5	0.004
	500–1000	3.53	21.68	51	0.968	54	23.18	0.979	5	0.011
	>1000	0.53	3.24	7.56	1	5	2.15	1	4	0.000
	$n =$	16.29	100	233		233	100		$D_{max} =$	**0.093**
Elevation (m)	<250	1.62	7.66	17.85	0.08	116	49.79	0.498	9	0.421
	250–300	15.30	72.37	168.62	0.80	66	28.33	0.781	3	0.019
	300–350	3.90	18.45	42.99	0.98	51	21.89	1.000	5	0.015
	>350	0.32	1.52	3.54	1.00	0	0.00	1.0	0	0.000
	$n =$	21.14	100	233		233	100		$D_{max} =$	**0.421**
Aspect	North	4.09	19.35	45.08	0.19	55	23.61	0.24	5	0.043
	East	4.53	21.42	49.91	0.41	61	26.18	0.50	5	0.090
	South	4.91	23.24	54.15	0.64	53	22.75	0.73	5	0.085
	West	7.61	35.99	83.86	1.00	64	27.47	1.00	4	0.000
	$n =$	21.14	100	233		233	100		$D_{max} =$	**0.090**

In this research, 233 archaeological sites have been recorded in the study area, so $n = 233 > 50$; the formula above could be applied for determination of d which is equal to 0.089.

Here two hypotheses have been proposed:

- Null hypothesis (H_0): that the sites are distributed irrespective of a given geo-environmental variable.
- Alternative hypothesis (H_1): that the sites are distributed respective of a given geo-environmental variable.

If D_{max} exceeds a critical value d (0.089), then H_0 may be rejected and the sample distribution can be said to be different from the population distribution [34]. The D_{max} for every geo-archaeological variable exceeds the critical value of 0.089, thus hypothesis H_0 is rejected and consequently the sites are located with respect to slope, aspect, elevation, distance to watershed and geological units (Table 2).

The statistical analysis has been used in the coefficient assignment of each class of every model variable. As mentioned above, a zone of 21 km^2 has been selected as the sampling zone in the predictive model's development. All sites within this zone were recorded during the field survey. This zone contains 233 sites, and their spatial distribution with respect to environmental variables was statistically analysed and it has been proven that their distribution is governed by particular rules. Therefore, they are not randomly distributed, after all no population can colonize a given territory in a hazardous way. There are always determining factors. The coefficient given to each class indicates the level of potential to carry the sites of that given class. The higher value means many sites have been found within it.

2.3.4. Calculation of Class Coefficient and Weighting Variables

The method used to generate the archaeological predictive model (APM) is a multi-criteria evaluation (MCE), also known as multi-criteria decision analysis (MCDA). This method is based on various criteria combinations. Those criteria mainly correspond to the environmental variables that govern the suitability and unsuitability of a given area to host archaeological sites. The known sites' location is not required as an input in this methodology but it has been used to strengthen the expert opinion of the Institut National de Science de l'Archéologie assigning weights to each variable before running the final model. MCE can be Boolean or categorical, weighted or unweighted. ArcGIS 10.2 (Esri), the program used to generate the APM, allows the choice of one of two methods: Boolean aggregation or weighted linear combination.

The Boolean aggregation method involves a binary division of each variable layer into suitable and unsuitable areas. These variables are then combined either inclusively (logical AND) or exclusively (logical OR). In the inclusive result, only the areas that are suitable in every layer become suitable in the final image. This is a risk-adverse strategy that ensures all qualities are high. In the exclusive result, an area that is suitable in only one layer will be suitable in the final result. This is a risk-taking strategy that ensures that only one quality is high. The weighted linear combination method is performed in two main steps. Firstly, each variable is classified on a continuous or categorical scale. A weight is then assigned to each individual variable with respect to how important it is compared to each of the other variables. Personal knowledge or expert opinion combined with statistical analysis of known sites' locations are used to assign the weights. The final aggregation of data allows a trade-off between qualities, if necessary, so the final suitable area is not necessarily highly ranked with respect to every variable [58].

2.3.5. Principal Component Analysis

Principal components analysis (PCA) has been used in predictive models to eliminate multi-collinearity among variables. Multi-collinearity arises where there are very strong correlations among independent variables and consequently the independent variables have no significant impact on the dependent variable [59]. According to the Esri ArcGIS Resources Center website [60], the purpose of this spatial statistical technique (PCA) tool found in the Spatial Analyst toolbox of ArcGIS 10.2 is to eliminate multi-collinearity; for example, since elevation, slope, and aspect are derived from a DEM, from which most of the variance can be explained. The tool then creates a multiband raster that contains the same amount of bands as the specified number of components. The first principal component raster band contains the greatest amount of variance, the second contains the second greatest variance, and so on [60]. The first principal component rasters containing 95% of the variance were utilized in building the model.

2.3.6. Analytic Hierarchy Process

The AHP technique has used MCE approach in predictive modeling to locate archaeological sites. The AHP combines the criteria weights and the class scores, consequently determining a global score for each class, and a consequent ranking. The global score for a given class is a weighted sum of the scores it obtained with respect to all the criteria. Table 3 shows the method to assign weight to each parameter and classes in that parameter according to their importance. Value 9 shows high importance while value 1/9 shows low importance.

The AHP technique is based on different levels of analysis: level 1 is the goal of the analysis, level 2 is multi-criteria that consist of several factors influencing archaeological sites location in our case. Several other levels of sub criteria and sub-sub criteria can be added but in this current study only one level of sub-criteria has been used. The last level is the alternative choices (Figure 5). The lines between levels indicates the relationship between factors, choices and goal. In level 2, only one comparison matrix corresponds to pair-wise comparisons between 5 factors with respect to the goal (archaeological predictive model). Thus, the comparison matrix of level 2 has the size of 5 by 5, because each choice is connected to each factor.

Table 3. Pair-wise comparison matrix formulation.

1/9	1/7	1/5	1/3	1	3	5	7	9
Extremely Less important	Very Strongly	Strongly	Moderately	Equally	Moderately	Strongly	Very Strongly	Extremely Very important

Source: Saaty [61]. Note: 1/8, 1/6, 1/4, 1/2, 2, 4, 6 and 8 can be used if more classes exist.

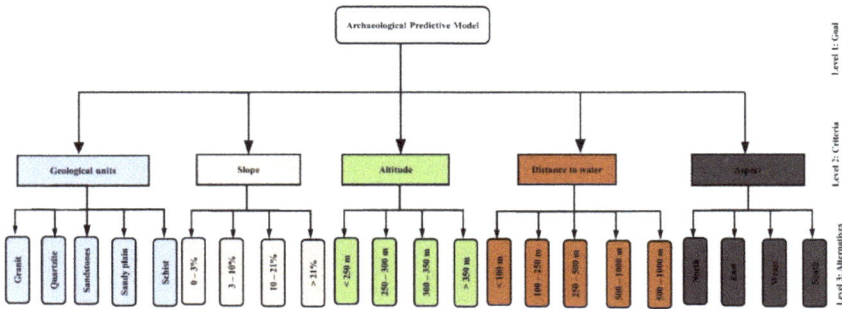

Figure 5. Different level of analytic hierarchy process (AHP).

A ratio matrix has been formed by the pairwise matrix comparison method. Criteria and sub-criteria in each level are in pairs with their importance to an element (criteria and sub-criteria) in the next higher level. Elements go to a downward level from a higher level [62]. The pairwise matrix $A = [\alpha_{ij}]_{n \times n}$ can be expressed mathematically as follows:

$$A = \begin{bmatrix} a_{11} & a_{12} & a_{13} & a_{1n} \\ a_{21} & a_{22} & a_{23} & a_{2n} \\ a_{31} & a_{32} & a_{33} & a_{3n} \\ a_{n1} & a_{n1} & a_{n1} & a_{nn} \end{bmatrix} \tag{4}$$

The matrix has a reciprocal property which is expressed mathematically as:

$$\alpha_{ji} = \frac{1}{\alpha_{ij}} \tag{5}$$

Table 4. Pair-wise comparison matrix.

	Geology	Slope	Elevation	Distance Water	Aspect
Geology	1.000	1.500	2.000	6.000	6.000
Slope	0.667	1.000	1.330	4.000	4.000
Elevation	0.500	0.752	1.000	3.000	3.000
Distance water	0.167	0.250	0.333	1.000	1.000
Aspect	0.167	0.250	0.333	1.000	1.000

After forming pairwise matrices as detailed in Table 4, weight vectors $w = [w_1, w_2, w_3, \ldots, w_n]$ were calculated according to Satty's eigenvector method. Weighting process consist of following two stages. First, the pairwise comparison matrix $A = [\alpha_{ij}]_{n \times n}$ is normalized by Equation (6) and then the weights are computed by Equation (6) resulting in values in the table below (Table 5).

$$\alpha_{ij} = \frac{\alpha_{ij}}{\sum_{i=1}^{n} \alpha_{ij}}, \tag{6}$$

For all $j = 1, 2, \ldots, n$

$$w_i = \frac{\sum_{i=1}^{n} \alpha_{ij}}{n} \tag{7}$$

For all $i = 1, 2, \ldots, n$.

Table 5. Normalized pair-wise comparison matrix.

	Geology	Slope	Elevation	Distance Water	Aspect	Priority
Geology	0.400	0.400	0.400	0.400	0.400	0.400
Slope	0.267	0.267	0.266	0.267	0.267	0.267
Elevation	0.200	0.200	0.200	0.200	0.200	0.200
Distance water	0.067	0.067	0.067	0.067	0.067	0.067
Aspect	0.067	0.067	0.067	0.067	0.067	0.067

The consistence index is determined using the the formula bellow:

$$CI = \frac{\lambda_{max} - n}{n - 1} \tag{8}$$

whereas λ_{max} is the highest or principal eigenvalue of the matrix.

In this study, λ_{max} is equivalent to 5.324 according to computations, and n is the order of the matrix which is equal to 5, consequently CI calculated using Equation (8) is 0.081. The consistency ratio (CR) is an indicator of the degree of consistence in the pair-wise comparisons matrix. It is used as an indicator of establishment [63]. It can be calculated using the the formula of Equation (9):

$$CR = \frac{CI}{RI} \tag{9}$$

where RI is the random consistency index which is equal to 1.12 as $n = 5$ according to Saaty [61].

A value of 0.072 has been obtained as the consistency ratio. Since this value of 0.072 for the proportion of inconsistency CR is less than 0.10, the judgments matrix is reasonably consistent and, therefore, can be used in the process of decision-making using AHP [64]. Otherwise, if $CR \geq 0.10$ it means that pairwise consistency is inadequate, thus the original weights in the pairwise comparison matrix must be redone.

After obtaining the weight of every criterion and its respective classes, the predictive model was produced in GIS using spatial overlay tools.

3. Results and Discussion

This predictive model for archaeological sites had interesting results. It showed that the areas with high potential of hosting archaeological sites are located in the southern zones of the study area. Other areas with a high probability of having archaeological sites are located near small isolated mountains that occupy the central part of the study area from north-east to south-west (Figure 6). The model provides a perspective on past occupation in the south of Morocco. Any populations always occupied the zones with favorable environments. The proximity to water sources played a major role in the determination of land suitability for Neolithic settlement and then in tumuli distrubution in the study area.

A sum of 365 out of 815 archaeological sites (45%) lay within 500 m distance of water sources, and 82% of sites are 1 km distance closer to paleorivers, while 284 sites (35%) are located around the paleolake in the south-western part of the study area, life always depends on water availability and accessibly since water reflects the image of society [65]. However, it must be mentioned that few sites are very close to water streams with only 14% of sites (115 tumuli) within a 100 m buffer zone of water bodies (Figure 6d). This can be explained by the fact that populations did not inhabit so close to rivers just to avoid flooding, because in the Neolithic period the Sahara was very humid with many river systems, lakes and marshes before it began to dry out during the long period of increasing aridity around 4000 BCE [66].

Lower slope terrains (Figure 6b) have as high a potential as archaeological sites as lower elevations (Figure 6c). This is normal since most people would not live on steepy and mountainious areas (Figure 7b). In general, flat areas tend to have the maximum of sites. Sites' distribution also takes in consideration geological components (Figure 7a), after all certain materials are essential for tumuli construction. It turned out that 90% of sites (731 out of 815) were built on concrete rocks (granite, quartzite, schist and sandstones) against 10% of tumuli built on soft rocks especially in the sandy plain, but there are also some tumuli made of concrete raw materials from the neighbourhood (Figure 6a).

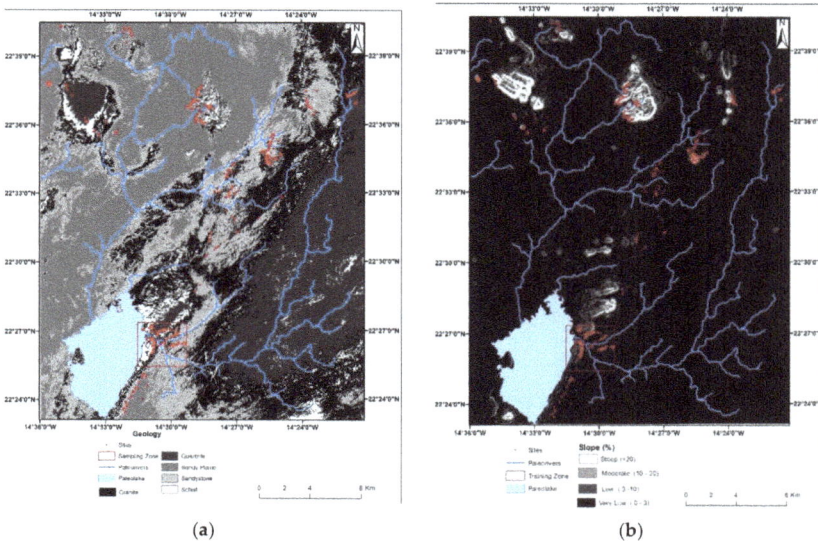

(a)   (b)

Figure 6. *Cont.*

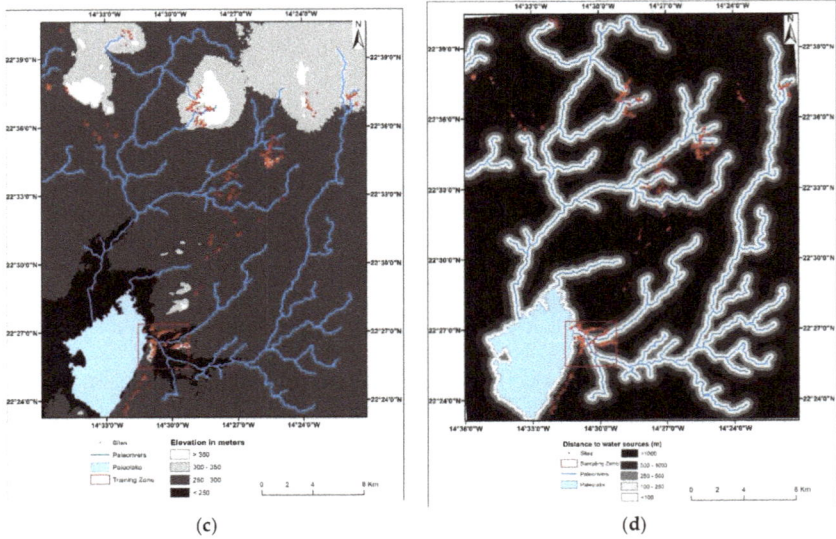

(c) (d)

Figure 6. Partial maps of different geo-environmental variables: (**a**) geology; (**b**) slope; (**c**) elevation; (**d**) distance to water sources.

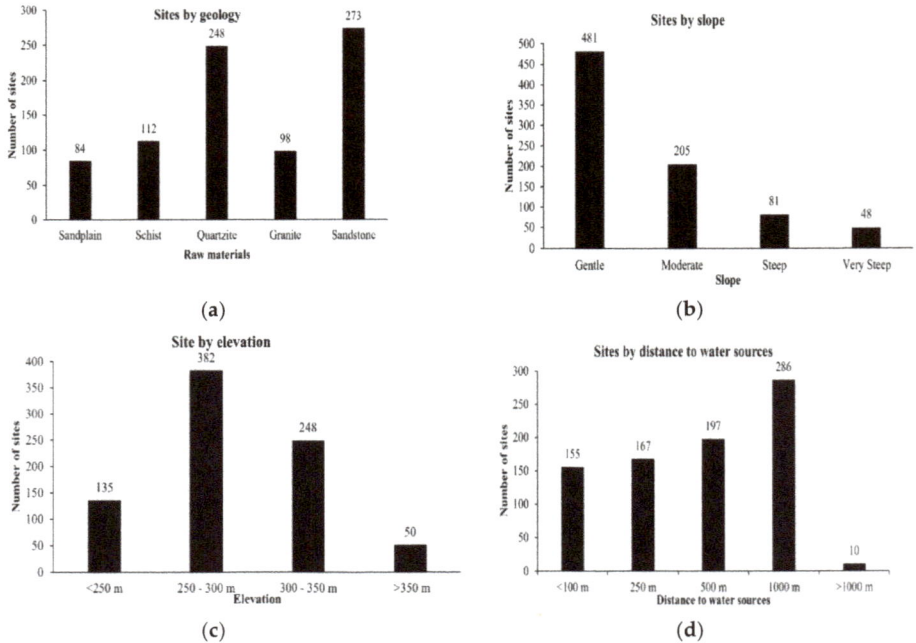

(a) (b)

(c) (d)

Figure 7. Sites' distribution in study area. (**a**) In respect to geological unit; (**b**) In respect to slope; (**c**) In respect to elevation; (**d**) In respect to distance to water sources.

The predictive accuracy of any model must be verified to validate its efficacy. The formulation of the predictive model does not guarantee the accuracy of its predictions; our model was developed based on statistics of 233 known sites in the sampling zone (data collected during first field survey). After model realization, a second field campaign was carried out in order to test and validate the model efficiency, and for this focus more than 582 supplement sites were recorded; this means that a total of 815 sites were used in the development and validation of this predictive model.

Prediction accuracy assessment of this model expressed by gain estimation method according to Kvamme [67], is given by the Equation (10).

$$G = 1 - \left(\frac{\% \, PS}{\% \, OS} \right) \qquad (10)$$

where PS is the total area where sites are predicted and OS are the observed sites within the area.

It has also been found that 56.87% of all sites are located in only 4.04% of the total area, and therefore this zone is qualified as a high potential zone for archaeological sites (Table 6). A gain of 92.87% was confirmed for this model, which is sufficient since archaeological prediction gains range from 1 (high predictive model) to 0 (no predictive model).

Table 6. Predictive gain computation.

Prediction	Area (%)	Sites (Number)	Sites (%)
Low	40.50	56	9.62
Medium	55.46	195	33.50
High	4.04	331	56.87

4. Conclusions

This study used a global methodology to create a predictive model that indicates areas with archaeological potentiality. The methodology involves archaeological research and geospatial techniques combined with multi-criteria decision making using AHP. The derivered predictive model (Figure 8) seems to be the best alternative and a useful tool for achaeological investigations for local archaeologists who usually have to monitorilarge areas with limited resources. The GIS has a huge range of applications for archaeology from recording a simple archaeological site, to managing thousands of sites and creating predictive models, etc. In this study the predictive model was used to predict archaeological sites' locations, on the basis of observed patterns, on assumptions about human behavior. It was constructed through the application of GIS tool, remote-sensing data and archaeological field research data. The methodology has been applied successfully in the Moroccan Sahara to identify the locations of funeral monuments, where it provides accurate predictions.

The evaluation of the model demonstrated the validity of the proposed methodology. The prediction results, in the form of a prediction map with a gain of 92.8%, is sufficient to consider the model as reliable. When used as a contribution to archaeological field research, the aim is to identify small areas with high probability of hosting archaeological sites, and therefore further field investigations are required to confirm site presence. This study showed that a well-built predictive model can provide reliable predictions of where archaeological sites should and should not be located in a given landscape.

The use of GIS and remote sensing could considerably change the way archaeological survey and heritage management are done in Morocco. This paper showed how effective predictive modelling could enrich archaeological knowledge about ancient cultures. This paper highlighted the way geospatial technology applied in archaeological research could be an efficient solution to the limited funds involved in archaeology by minimizing and optimizing the area of interest for archaeological expeditions, since specific areas with a high probability of undiscovered archaeological remains are already known through the model.

This predictive model is recommended for Moroccan archaeologists to search other archaeological sites, and will make their work easier. Moreover, this tool can be used to promote geo-tourism by guiding tourists to areas with archaeological sites.

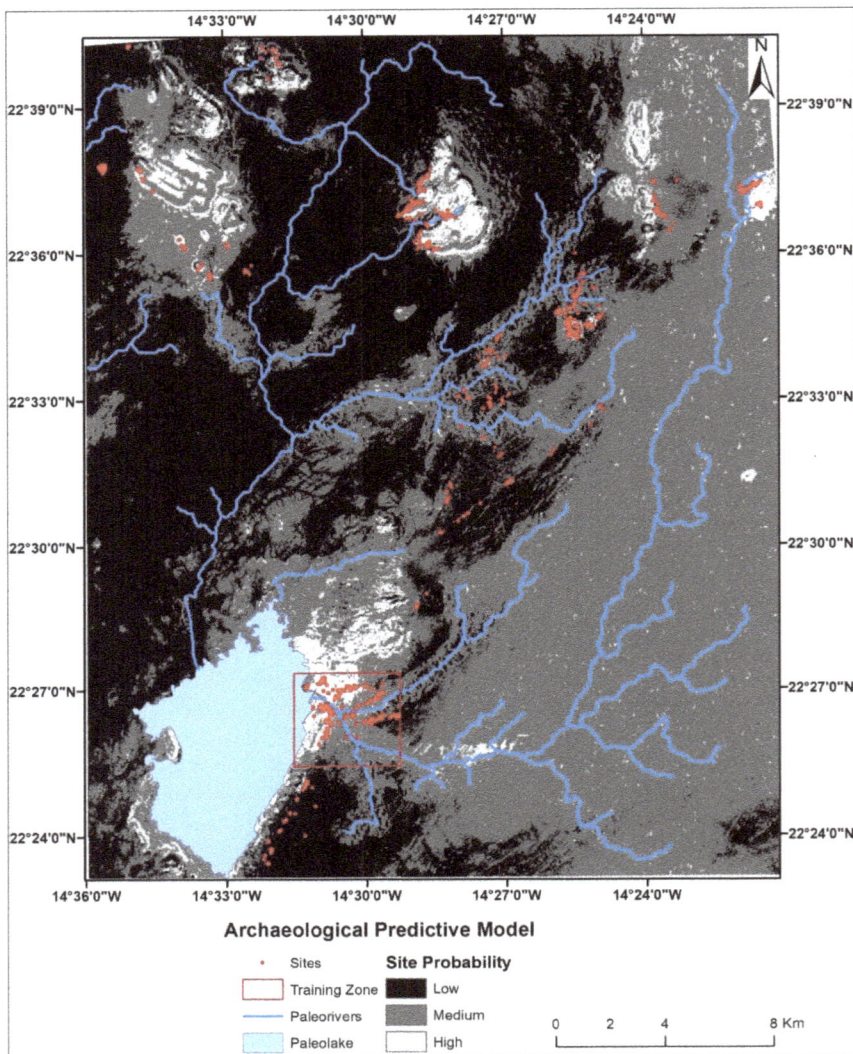

Figure 8. Archaeological predictive map of sites in Awsard.

Author Contributions: Different authors has contributed to this article and their respective contribution is detailed as follow: Conceptualization, A.F.N. and H.R.; Methodology, A.F.N. and H.R.; Software, A.F.N. and H.R.; Validation, A.F.N. and H.R.; Formal Analysis, A.F.N. and H.R.; Investigation, A.F.N. and A.O.; Writing-Original Draft Preparation, A.F.N. and H.R.; Writing-Review & Editing, A.F.N., H.R. and K.M.; Visualization, A.F.N.; Supervision, H.R. and A.O.; Project Administration, A.O.; Funding Acquisition, A.O.

Funding: This research was partially funded by Awsard Province Administrative Council provided funding for this research, as part of its support of Ange Felix Nsanziyera PhD programme.

Acknowledgments: This study is a part of multidiscipline project named "Awsard Natural and Cultural Heritage Atlas" that aims to make a full inventory of all natural and archaeological heritage all-over the Saharan provinces of Morocco. The project was directed by the Moroccan Association of Rock Art (AMAR) and Association Nature-Initiative (ANI) in Collaboration with Directorate of Cultural Heritage of Morocco. Authors wish to thank all teams including diverse scientists, Association Nature Initiative of Dakhla (ANI), Awsard Province Administrative Council and other person who participated in the realization of this study.

Conflicts of Interest: The authors declare no conflict of interest. The founding sponsors had no role in the design of the study; in the collection, analyses, or interpretation of data; in the writing of the manuscript, and in the decision to publish the results.

References

1. Souville, G. Les principaux types des tumuli marocains. *Bulletin de la Société Préhistorique de France* **1959**, *56*, 394–402. [CrossRef]

2. Nami, M. L'épopée de la pierre à travers les temps préhistoriques, La pierre et son usage à travers les âges: Jardins des Hespérides. *Bulletin Semestriel de La Société Marocaine d'Archéologie et du Patrimoine* **2008**, *4*, 18–24.

3. Nami, M.; Moser, J.; Mikdad, A.; Eiwanger, J. Quelques aspects de l'Iberomaurisien du Rif Oriental (Maroc). *Bulletin d'Archéologie Marocaine* **2012**, *4*, 16–33.

4. Bokbot, Y. Neolothique et protohistoire du Maroc. In Proceedings of the Actes du IIIème Colloque International sur l'Histoire et l'Archéologie de l'Afrique du Nord, Tabarka, Tunisie, 8–13 May 2000; Edition de l'Institut National du Patrimoine: Tunis, Tunisia, 2000; pp. 35–45.

5. Haidar-Boustani, M.; Iban, J.J.; Al-Maqdissi, M.; Armendari, A.; Gonzalez Urquijo, J.; Teira, L. Prospections Archaeologiques a l'ouest de la ville de homs: Rapport preliminaire campagne 2004. *Annales D'Histoire et D'Archaeologie* **2004**, *16–17*, 9–38.

6. Rowlands, A.; Sarris, A. Detection of exposed and subsurface archaeological remains using multi-sensor remote sensing. *J. Archaeol. Sci.* **2007**, *34*, 795–803. [CrossRef]

7. Siart, C.; Eitel, B.; Panagiotopoulos, D. Investigation of past archaeological landscapes using remote sensing. *J. Archaeol. Sci.* **2008**, *35*, 2918–2926. [CrossRef]

8. Kamermans, H. Smashing the crystal ball. A critical evaluation of the Dutch national archaeological predictive model (IKAW). *Int. J. Hum. Arts Comput.* **2007**, *1*, 71–84. [CrossRef]

9. McCauley, J.F.; Schaber, G.G.; Breed, C.S.; Grolier, M.J.; Haynes, C.V.; Issawi, B.; Elachi, C.; Blom, R. Subsurface valleys and geoarchaeology of the eastern Sahara revealed by shuttle radar. *Science* **1982**, *281*, 1004–1020. [CrossRef] [PubMed]

10. Bewley, R.; Donoghue, D.; Gaffney, V.; Van Leusen, M.; Wise, A. *Archiving Aerial Photography and Remote Sensing Data: A Guide to Good Practice*; Archaeology Data Service; Oxbow Books: Oxford, UK, 1999.

11. Wynn, J.C. Applications of high-resolution geophysical methods to archaeology. In *Archaeological Geology of North America Centennial Special Volume*; Lasca, N.P., Donahue, J., Eds.; Geological Society of America: Boulder, CO, USA, 1990; pp. 603–617.

12. Theocaris, P.S.; Liritzis, I.; Lagios, E.; Sampson, A. Geophysical prospection, archaeological excavation, and dating in two Hellenic pyramids. *Surv. Geophys.* **1996**, *17*, 593–618. [CrossRef]

13. Gaffney, C.; Gaffney, V. Non-invasive investigations at Wroxeter at the end of the twentieth century. *J. Archaeol. Prospect.* **2000**, *7*, 65–67. [CrossRef]

14. Jones, R.E.; Sarris, A.J. Geophysical and related techniques applied to archaeological survey in the Mediterranean: A review. *J. Mediterr. Archaeol.* **2000**, *13*, 3–75.

15. Neubauer, W. Images of the invisible-prospection methods for the documentation of threatened archaeological sites. *Naturwissenschaften* **2001**, *88*, 13–24. [CrossRef] [PubMed]

16. Kvamme, K.L. Geophysical surveys as landscape archaeology. *Am. Antiq.* **2003**, *68*, 435–457. [CrossRef]

17. Montufo, A.M. The use of satellite imagery and digital image processing in landscape archaeology: A case study from the island of Mallorca, Spain. *Geoarchaeology* **1997**, *2*, 71–85. [CrossRef]

18. Powlesland, D.; Lyall, J.; Donoghue, D. Enhancing the record through remote sensing: The application and integration of multi-sensor, non-invasive remote sensing techniques for the enhancement of the Sites and Monuments Record. Heslerton Parish Project, N. Yorkshire, England. *Internet Archaeol.* **1997**, *2*. [CrossRef]

19. Fowler, M.J.F. Satellite remote sensing and archaeology: A comparative study of satellite imagery of the environs of Figsbury Ring, Wiltshire. *J. Archaeol. Prospect.* **2002**, *9*, 55–69. [CrossRef]

20. Banes, I. Aerial remote-sensing techniques used in the management of archaeological monuments on the British Army's Salisbury Plain Training Area Wiltshire, UK. *J. Archaeol. Prospect.* **2003**, *10*, 83–90. [CrossRef]

21. Devereux, B.J.; Amable, G.S.; Crow, P.; Cliff, A.D. The potential of airborne lidar for detection of archaeological features under woodland canopies. *Antiquity* **2005**, *79*, 648–660. [CrossRef]

22. Challis, K. Airborne laser altimetry in alluviated landscapes. *J. Archaeol. Prospect.* **2006**, *13*, 103–127. [CrossRef]

23. Lasaponara, R.; Masini, N. *Satellite Remote Sensing: A New Tool for Archaeology*; Springer Science & Business Media: Cham, The Netherlands, 2012.

24. Chen, R.; Lasaponar, R.; Masini, N. An overview of satellite synthetic aperture radar remote sensing in archaeology: From site detection to monitoring. *J. Cult. Herit.* **2015**, *23*, 5–11. [CrossRef]

25. Tapete, D.; Cigna, F. Trends and perspectives of space-borne SAR remote sensing for archaeological landscape and cultural heritage applications. *J. Archaeol. Sci. Rep.* **2017**, *14*, 716–726. [CrossRef]

26. Van der Meer, F.D.; De Jong, S.M. Introduction. In *Imaging Spectrometry: Basic Principles and Prospective Applications*; Van der Meer, F.D., de Jong, S.M., Eds.; Kluwer Academic Publishers: Dordrecht, The Netherlands, 2001; pp. XXI–XXIII.

27. Kumar, L.; Schmidt, K.; Dury, S.; Skidmore, A. Imaging spectrometry and vegetation science. In *Imaging Spectrometry: Basic Principles and Prospective Applications*; Van der Meer, F.D., de Jong, S.M., Eds.; Kluwer Academic Publishers: Dordrecht, The Netherlands, 2001; pp. 111–155.

28. Lucas, R.M.; Rowlands, A.P.; Niemann, O.; Merton, R. Hyperspectral sensors: Past, present and future. In *Advanced Image Processing Techniques for Remotely Sensed Hyperspectral Data*; Varshney, P.K., Arora, M.K., Eds.; Springer: Berlin/Heidelberg, Germany, 2004; pp. 11–40.

29. Huggett, J. Looking at intra-site GIS. In *CAA96 Computer Applications and Quantitative Methods in Archaeology*; Lockyear, K., Sly, T.J.T., Mihailescu-Birliba, V., Eds.; BAR International Series 845; Archaeopress: Oxford, UK, 2000; pp. 117–122.

30. Vullo, N.; Fontana, F.; Guerreschi, A. The application of GIS to intra-site spatial analysis: Preliminary results from Alpe Veglia (VB) and Mondeval de Sora (BL), two Mesolithic sites in the Italian Alps. In *New Techniques for Old Times: CAA98*; Barcelo, J.A., Briz, I., Vila, A., Eds.; British Archaeological Reports International Series 757; Archaeopress: Oxford, UK, 1999; pp. 111–116.

31. Robinson, J.; Zubrow, E. Between space space: Interpolation in Archaeology. In *Geographical Information Systems and Landscaoe Archaeology*; Gillings, M., Mattingly, D., Dalem, J.V., Eds.; Oxbow Books: Oxford, UK, 1999; pp. 65–84.

32. Wheatley, D.; Gillings, M. Spatial technology and archaeology. In *The Archaeological Applications of GIS 2002*; Taylor and Francis: London, UK, 2002. [CrossRef]

33. Verhagen, P. *Case studies in Archaeological Predictive Modelling*; Archaeological Studies Leiden University; Leiden University Press: Leiden, The Netherlands, 2007.

34. Wheatley, D. Cumulative viewshed analysis: A GIS-based method for investigating intervisibility, and its archaeological applications. In *Archaeology and Geographical Information Systems: A European Perspective*; Lock, G., Stančič, Z., Eds.; Taylor & Francis: London, UK, 1995; pp. 171–185.

35. Wheatley, D. The use of GIS to understand regional variation in Neolithic Wessex. In *New Methods, Old Problems: Geographic Information Systems in Modern Archaeological Research*; Maschner, H.D.G., Ed.; Occasional Paper No. 23; Center for Archaeological Investigations, Southern Illinois University: Carbondale, IL, USA, 1996; pp. 75–103.

36. Wheatley, D. Making space for an archaeology of place. *Internet Archaeol.* **2004**, *15*. [CrossRef]

37. Roeloffs, A.; Wiatr, T.; Reicherter, K.; Museum, S.N. Prospection of Karstic Caves Using Gis and Remote Sensing. In Proceedings of the ISPRS WG VII/5 Workshop, Cologne, Germany, 18–19 November 2011; pp. 107–115.

38. Belmonte, J.; Esteban, C.; Cuesta, L.; Perera Betancort, M.; Gonzarez, J. Pre-Islamic burial moniments in Northern and Saharn Morocco. *Archaeoastronomy* **1999**, *30*, 21. [CrossRef]

39. Linstädter, J.; Blatt, M. Sea, slopes and shelters: Archaeological Surveys along the Mediterranean Coast, west of the Melilla Peninsula (Morocco). In *Pleistocene Foragers: Their Culture and Environment*; Pastoors, A., Aufferman, B., Eds.; Wissenschaftliche Schriften des Nneanderthal Museums: Mettmann, Germany, 2013; pp. 27–32.

40. Di Lernia, S. Places, monuments, and landscape: Evidence from the Holocene central Sahara, Azania. *Archaeol. Res. Afr.* **2013**, *48*, 173–192. [CrossRef]

41. Galan, E.; Torres, J.; Senoran, J.M. Archaeological Interventions. In Search of Traces of the Human Presence in the Valley. *Complutum* **2014**, *25*, 45–76.
42. Pappu, S.; Akhilesh, K.; Ravindranath, S.; Raj, U. Applications of satellite remote sensing for research and heritage management in Indian prehistory. *J. Archaeol. Sci.* **2010**, *37*, 2316–2331. [CrossRef]
43. Brooks, N.; Clarke, J.; Garfi, S.; Pirie, A. The archaeology of Western Sahara: Results of environmental and archaeological reconnaissance. *Antiquity* **2009**, *83*, 918–934. [CrossRef]
44. Kohler, T.A. Predictive locational modelling: History and current practice. In *Quantifying the Present and Predicting the Past: Theory, Method and Application of Archaeological Predictive Modeling*; Judge, W.L., Sebastian, L., Eds.; US Bureau of Land Management: Denver, CO, USA, 1988; pp. 19–59.
45. Alexakis, D.; Sarris, A.; Astaras, T.; Albanakis, K. Integrated GIS, remote sensing and geomorphologic approaches for the reconstruction of the landscape habitation of Thessaly during the neolithic period. *J. Archaeol. Sci.* **2011**, *38*, 89–100. [CrossRef]
46. Clement, O.C.; Sahadeb, D.; Kloot, R. Using GIS to Predict Likely Archaeological Sites. Available online: https://digitalcommons.du.edu/cgi/viewcontent.cgi?article=1056&context=geog_ms_capstone (accessed on 22 June 2017).
47. Makepeace, G.A. *The Prehistoric Archaeology of Settlement in South-East Wales and the Borders*; Archaeopress: Oxford, UK, 2006; pp. 25–26.
48. Van Leusen, M.; Deeben, J.; Hallewa, D.; Zoetbrood, P.; Kamerman, H.; Verhagen, P. *Predictive Modelling for Archaeological Heritage Management: A Research Agenda*; van Leusen, P.M., Kamermans, H., Eds.; Rijksdienst voor het Oudheidkundig Bodemonderzoek: Amersfoort, The Netherlands, 2004.
49. Di Lernia, S. The archaeology of rock art in Northern Africa. In *The Oxford Handbook of the Archaeology and Anthropology of Rock Art*; David, B., McNiven, I., Eds.; Oxford University Press: Oxford, UK, 2017. [CrossRef]
50. Canning, S. "BELIEF" in the past: Dempster-Shafer theory, GIS and archaeological predictive modelling. *Aust. Archaeol.* **2005**, *60*, 6–15. [CrossRef]
51. Mink, P.B.; Ripy, J.; Bailey, K. Predictive archaeological modeling using gis-based fuzzy set estimation: A case study in Woodford County, Kentucky. In Proceedings of the 88th Annual Meeting on Transportation Research Board, Washington DC, USA, 11–15 January 2009.
52. Crósta, A.P.; Moore, J.M. Geological mapping using Landsat Thematic Mapper imagery in Almeria Province, south-east Spain. *Int. J. Remote Sens.* **1989**, *10*, 505–514. [CrossRef]
53. Congalton, R.G. A review of assessing the accuracy of classifications of remotely sensed data. *Remote Sens. Environ.* **1991**, *37*, 35–46. [CrossRef]
54. Foody, G.M. Status of land cover classification accuracy assessment. *Remote Sens. Environ.* **2002**, *80*, 185–201. [CrossRef]
55. Diggs, D.M.; Brunswig, R.H. Weights of Evidence Modeling in Archaeology in Rocky Mountain National Park. In Proceedings of the ESRI International User Conference, San Diego, CA, USA, 12–16 July 2010; Available online: http://www.unco.edu/geography/faculty/diggs/docs/diggs_brunswig_esri_2009b_ppoint2003.ppt (accessed on 21 July 2014).
56. Yang, L.; Pei, A.; Guo, N.; Liang, B. Spatial Modality of Prehistoric Settlement Sites in Luoyang Area. *Sci. Geogr. Sin.* **2012**, *32*, 993–999.
57. Kvamme, K.L. One-Sample Tests in Regional Archaeological Analysis: New Possibilities through Computer Technology. *Am. Ant.* **1990**, *55*, 367–381. [CrossRef]
58. Eastman, J.R. *IDRISI Kilimanjaro Guide to GIS and Image Processing*; Clark Univesity: Worcester, MA, USA, 2003.
59. Fattah, M.A. Multicollinearity. Central Michigan University, 2010. Available online: http://www.chsbs.cmich.edu/fattah/courses/empirical/multicollinearity.html (accessed on 15 May 2014).
60. Esri. How Principal Component Analysis Works. Retrieved from ArcGIS Resource Center. Available online: http://help.arcgis.com/en/arcgisdesktop/10.0/help/index.html#/How_Principal_Components_works/009z000000qm000000/ (accessed on 7 May 2014).
61. Saaty, T.L. *The Analytical Hierarchy Process*; McGraw Hill: New York, NY, USA, 1980.
62. Malczewski, J. *GIS and Multicriteria Decision Analysis*; John Wiley & Sons, Inc.: Toronto, ON, Canada, 1999.
63. Chandio, I.A. GIS-basedland suitability analysis of sustainable hillside development. In Proceedings of the Fourth International Symposium on Infrastructure Engineering in Developing Countries, Karachi, Pakistan, 11–13 December 2013. [CrossRef]

64. Park, S.; Jeon, S.; Kim, S.; Choi, C. Prediction and Comparaison of Urban Growth by Land Suitability Index Mapping Using GIS and RS in South Korea. *Landsc. Urban Plan.* **2011**, *99*, 104–114. [CrossRef]

65. Mithen, S. The domestication of water: Water management in the ancient world and its prehistoricorigins in the jordan valley. *Philos. Trans. R. Soc. Lond. A Math. Phys. Eng. Sci.* **2010**, *368*, 5249–5274. [CrossRef] [PubMed]

66. Joanne, C.; Brooks, N.; Banning, E.B.; Matthewsd, M.; Campbell, S.; Clare, L.; Cremaschi, M.; di Lernia, S.; Drake, N.; Gallinaro, M.; et al. Climatic changes and social transformations in the Near East and North Africa during the 'long' 4th millennium BC: A comparative study of environmental and archaeological evidence. *Quat. Sci. Rev.* **2016**, *136*, 96–121. [CrossRef]

67. Kvamme, K.L. Development and testing of Quantitative models. In *Quantifying the Present and Predicting the Past: Theory, Method and Application of Archaeological Predictive Modelling*; Judge, W., Sebastian, L., Eds.; Bureau of Land Management: Denver, CO, USA, 1988; pp. 325–428.

geosciences

MDPI

Article

The Application of Freely-Available Satellite Imagery for Informing and Complementing Archaeological Fieldwork in the "Black Desert" of North-Eastern Jordan

Stefan L. Smith [1,*] and Marie-Laure Chambrade [2]

[1] Department of Archaeology, University of Gent, Campus Ufo, Sint-Pietersnieuwstraat 35, 9000 Gent, Belgium

[2] University of Lyon, CNRS, Archéorient UMR 5133, Maison de l'Orient et de la Méditerranée, 5/7 rue Raulin, CEDEX 07, 69365 Lyon, France; marie.chambrade@gmail.com

* Correspondence: stefan.smith@dunelm.org.uk; Tel.: +32-9-33-10162

Received: 30 November 2018; Accepted: 12 December 2018; Published: 15 December 2018

Abstract: Recent developments in the availability of very high-resolution satellite imagery through platforms like GoogleEarth (Google, Santa Clara County, CA, USA) and Bing Maps (Microsoft, Redmond, WA, USA) have greatly opened up the possibilities of their use by researchers. This paper focusses on the exclusive use of free remote sensing data by the Western Harra Survey (WHS), an archaeological project investigating the arid "Black Desert" of north-eastern Jordan, a largely impenetrable landscape densely strewn with basalt blocks. The systematic analysis of such data by conducting a holistic satellite survey prior to the commencement of fieldwork allowed for the precise planning of ground surveys, with advanced knowledge of which sites were vehicle-accessible and how to efficiently visit a stratified sample of different site types. By subsequently correlating the obtained ground data with this analysis, it was possible to create a typological seriation of the site forms known as "wheels", determine that at least two-thirds of sites are within 500 m of valleys or mudflats (highlighting these features' roles as access routes and resource clusters) and identify numerous anthropogenic paths cleared through the basalt for site access and long-distance travel. These results offer new insights into this underrepresented region and allow for supra-regional comparisons with better investigated areas by a method that is rapid and cost-effective.

Keywords: remote sensing; free satellite imagery; GoogleEarth; Bing Maps; archaeological fieldwork; arid environments; basalt desert; landscape accessibility; Harra; Jordan

---

1. Introduction

The "Black Desert" of north-eastern Jordan, known locally as the Harra, has been subject to varied but intermittent research over the last century. Though the first use of aerial photography for archaeological research commenced in the 1920s, identifying a dense distribution of prehistoric stone structures, ground investigations did not occur until the 1970s, mainly due to the extreme inaccessibility of the terrain. Meanwhile, large-scale analyses of satellite imagery to enable a holistic coverage of the region have only been possible for the last few years, and with the recent release of very high-resolution imagery through platforms like GoogleEarth and Bing Maps, this has become freely available to all projects. This has not only greatly aided the archaeology of this region, as indeed similar developments in the last decade have all over the world, but has actually enabled the commencement of holistic studies of the type that already exist in more accessible parts of Mesopotamia and the Levant, such as the Jazira in northern Syria and the Shamiya in central Syria. While the capacity of free satellite imagery to enable the recognition of sites in this region by remote sensing has been investigated by other authors, e.g., Reference [1], it is by the integration with ground survey data that its full potential is unlocked. This paper examines this process in the context of its benefit to the Western Harra Survey (WHS), an archaeological, geographical and geological investigation of a section of this region (see Figures 1 and 2).

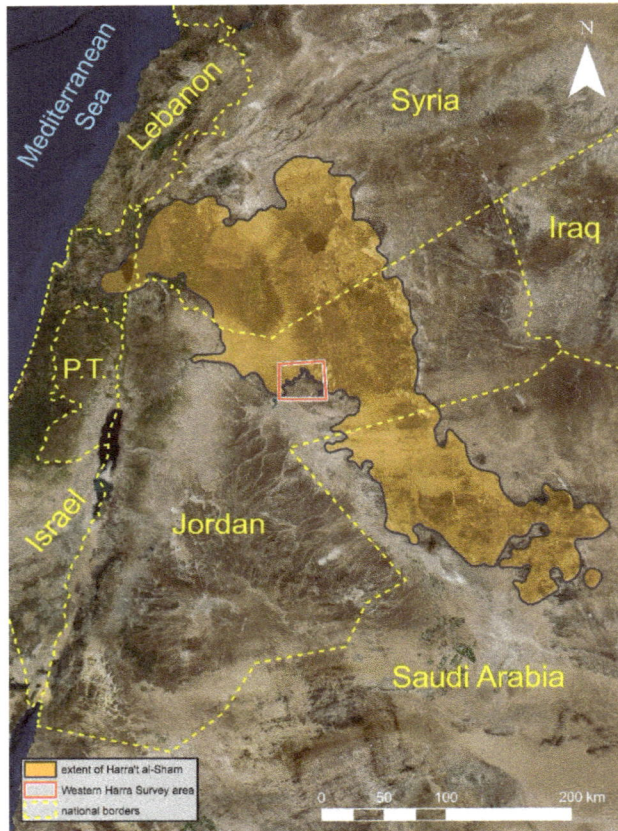

Figure 1. Map of the location of the Western Harra Survey (WHS) and the extent of the Harra't al-Sham, based on Reference [2] (32°0' N, 37°6' E). Sources: Esri, DigitalGlobe, GeoEye, i-cubed, USDA FSA, USGS, AEX, Getmapping, Aerogrid, IGN, IGP, swisstopo and the GIS User Community.

The Harra is part of the Harra't al-Sham, the largest (50,000 km^2) of a succession of basaltic plateaus from Syria in the north to Yemen in the south (Figure 1). The Jordanian Harra covers 11,400 km^2. It features volcanoes with lava flows dating from the Oligocene to the Quaternary (most recently ca. 400 ka) and is covered with a silty, carbonate-and-quartz-containing loess, on top of which a paving of basalt blocks partly protects the sediment from erosion and deflation [1]. The process of "stone heaving", which brings these blocks to the surface, is still poorly understood [2]. The basalt blocks are of variable dimensions but have always made traversing this region extremely difficult, often being impossible except along seasonal rainfall valleys or mudflats (Figure 3). Since the lowest stone course of every prehistoric structure visited in the WHS corresponds to the present-day ground level, it can be stated with confidence that at no time was the loess cover significantly higher so as to ease human travel.

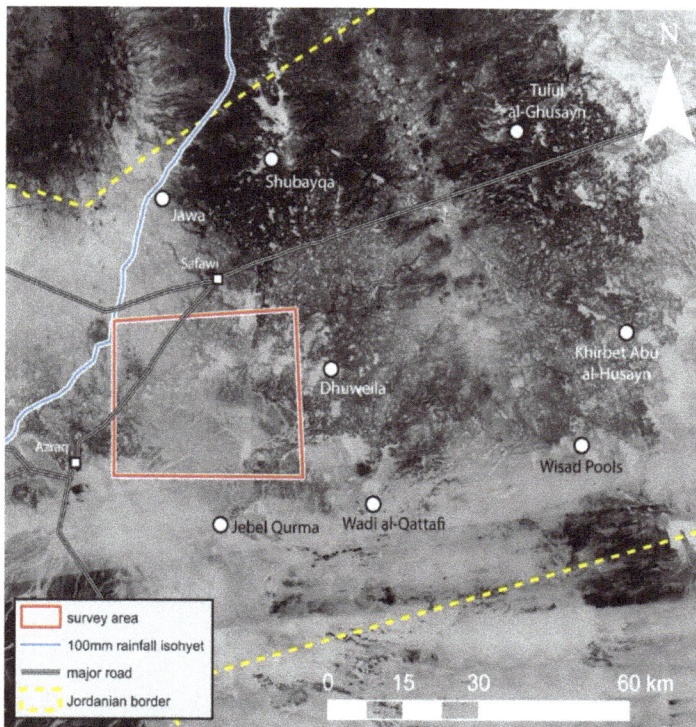

Figure 2. Map of the WHS in the context of other past and present archaeological fieldwork projects in the Harra, with the modern towns of Azraq and Safawi indicated (32°0′ N, 37°24′ E). CORONA satellite data available from the US Geological Survey.

Figure 3. A view of the typical landscape of dense basalt blocks that makes up the majority of the Harra.

In the area under study, the altitude of the plateau varies between ca. 500 and 800 m a.s.l. The hydrographic system is endorheic, and the relatively dense hydrographic networks are composed of valleys with temporary or sporadic flow during rainy seasons. Pluvial and thermal regimes have been Mediterranean since the beginning of the Holocene [3], with hot, dry summers and cold, wet winters. The current mean annual precipitation in north-eastern Jordan is between less than 100 mm and 300 mm [4], resulting in arid bioclimatic conditions. Pans or mudflats, locally called qe'an (singular: qa'a), of different sizes and types collect water flows and sediments. Together with the seasonal rainfall valleys (called awdiya; singular: wadi), qa'a edges seem to be the main areas suitable for the development of substantial vegetation. During the rainy season, qe'an can fill with water, thus forming an ephemeral lake ecosystem (Figure 4).

Figure 4. Photograph taken after seasonal rainfall in late October 2015, showing how qe'an fills with water to become temporary lakes.

The existence of countless structures built from the local basalt stone has been known to the Bedu nomads of the region for centuries, who termed them the "Works of the Old Men" [5,6]. Their academic study properly commenced in the 1970s with excavations at the site of Jawa and surveys in its vicinity by Svend Helms [7], followed in the succeeding decade by Alison Betts' "Black Desert Survey" east of the WHS area and excavations at the site of Dhuweila [8]. Since then, a handful of surveys and limited excavations have slowly brought to light, necessarily in a keyhole fashion, evidence for widespread prehistoric occupation of the area (see Figure 2). While there is as yet no consensus on whether the majority of structures represent settlements of permanent occupation, e.g., Reference [9], or seasonal camps, e.g., Reference [10], evidence based on lithic material and, more recently, Optically-Stimulated Luminescence (OSL) [11] and Accelerator Mass Spectrometry (AMS) [12,13] dating indicates a long-term occupation chronology lasting from at least the Epipalaeolithic (Late Natufian) period onwards (ca. 12,650 cal BC at Shubayqa [13]; ca. 9000 BC at Dhuweila [14]). Human presence in the Harra is subsequently attested to for all periods up to and including the Early Bronze Age (early 3rd millennium BC), but is particularly well-represented between the 7th and 4th millennium BC [9,10,15,16]. This data does not necessarily indicate permanent occupation, but it speaks against any prolonged period of abandonment.

The WHS [17,18] is a multidisciplinary project co-directed by the authors in collaboration with their respective institutions, the Institut Français du Proche-Orient (IFPO) and the Department of Antiquities of the Hashemite Kingdom of Jordan. Its primary goals are to explore and study the edges and interior of the Jordanian Harra, with a focus on the geography, geology and prehistoric human occupation (emphasising the 7th to early 3rd millennium BC), using a 36 by 32 km survey area located in the western portion of the basalt region as a proxy [19–21]. Commencing in 2015, it has thus far comprised two fieldwork seasons and one artefact study season. The survey area was selected for both its location between previously and currently investigated areas (e.g., Jebel Qurma, Dhuweila, Jawa (see Figure 2)) as well as its representation of all types of landscapes typical of the region (Figure 5).

Figure 5. Photographs illustrating the four typical landscapes of the Harra, all of which are represented in the WHS area: (**A**) Undulating steppe carpeted by a dense layer of basalt blocks, extremely difficult to traverse; (**B**) large traversable wadi systems surrounded by pockets of dense to medium-dense basalt outcrops; (**C**) large qe'an, often adjoining each other over many tens of kilometres, providing access into areas otherwise similar to A; (**D**) hilly areas crossed by small awdiya that are not easy to travel along, with medium-dense basalt coverage right up to the wadi edges.

Furthermore, this area is at least partially accessible by a combination of asphalt roads, the so-called "TAP line", a road constructed in the 1940s to follow the course of the Trans-Arabian oil pipeline (see Figure 6), and bulldozed tracks constructed for quarry vehicles and oil prospection routes from the 1980s. Though none but the first of these access routes are easily traversed, compared to the complete vehicle inaccessibility of much of the Harra, they provide vital links to its largely unexplored interior. This region therefore represents a good compromise between a desire to penetrate deep into the "Black Desert" and the practicalities of access and time constraints.

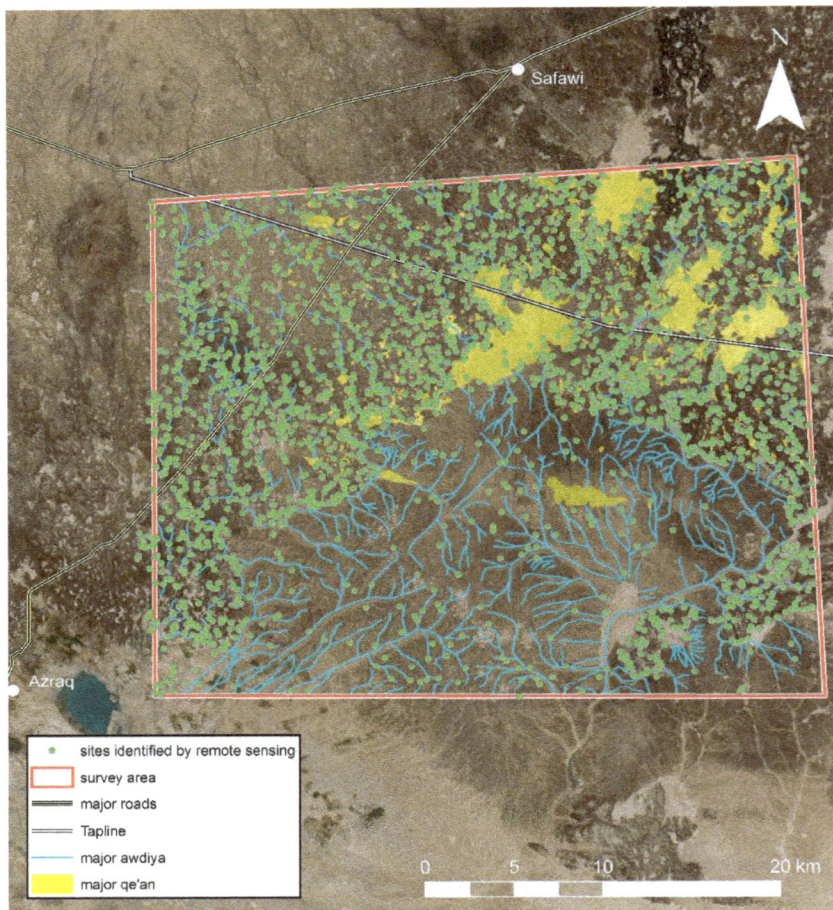

Figure 6. Map showing all sites within the WHS identified by the satellite imagery survey, with main roads and natural features highlighted (32°0′ N, 37°7′ E). Sources: Esri, DigitalGlobe, GeoEye, i-cubed, USDA FSA, USGS, AEX, Getmapping, Aerogrid, IGN, IGP, swisstopo and the GIS User Community.

2. Materials and Methods

Aerial flights across the "Black Desert" commenced with the establishment of the Airmail route from Cairo to Baghdad in the early 1920s, and it was not long before pilots noted the structures visible in this landscape and began taking photographs of them [5,22]. These were, however, focused on individual features and thus are very localised; meanwhile, subsequent more widespread aerial photography in Jordan during and after the Second World War was limited to the western part of the country [23]. The Harra was first holistically covered by US military CORONA satellite imagery (now declassified) during the 1960s and 1970s. However, it was with the commencement of flights for explicit archaeological purposes by the Aerial Photographic Archive for Archaeology in the Middle East (APAAME) [24] in 1997 that the region began to be systematically photographed from the air at a resolution that could be used for accurate interpretations of its archaeology. Their database, consisting of tens of thousands of images, allowed projects working in the Harra and elsewhere to gain an unprecedented view of the regional contexts of their keyhole ground investigations, paving the way for the use of high-resolution satellite imagery from the 2010s onwards.

When the WHS project began in early 2015, free very high-resolution remote sensing data was only available for individual pockets of this region, with the GoogleEarth platform releasing small sections of GeoEye imagery for the Harra in 2013, up to a resolution of ca. 0.5 m/pixel [1]. Therefore, a preliminary remote sensing analysis was first carried out using CORONA imagery. These have been used to great effect in parts of Syria and northern Iraq to identify sites, especially as they were mostly taken before the commencement of modern urban expansion and widespread mechanised agricultural practices [25,26]. Due to the relatively small nature of sites in the Harra, however, with most features comprising individual structures no more than 70 m in diameter (and often much smaller), this proved to be an insufficiently precise dataset, with many structures appearing unclearly or not at all. This however changed when full very high-resolution GoogleEarth coverage was released in 2017, with further GeoEye (Herndon, VA, USA) followed by CNES/Airbus (Paris, France) satellite imagery becoming available, as well as DigitalGlobe (Westminster, CO, USA) products appearing in Bing Maps, fundamentally altering the possibilities for remote sensing analyses and by extension, ground investigations in this region.

With all but the smallest structures (see Section 3.1.2) clearly visible and definable on this imagery, it could be used to conduct a remote sensing survey. The area in question was thus systematically analysed, latitudinal line by latitudinal line, at an "eye altitude" of 1 km in the GoogleEarth platform, and all likely features of archaeological interest were recorded in a GIS database. This process took 4 months, gradually compiling a KML-file that was then converted into an ArcGIS (ESRI, Redlands, CA, USA) shapefile before being integrated into a geodatabase for further analysis within the ArcMap program. This process allowed for the identification of nearly 3000 individual sites or features of all morphologies, assigned numbers in the order in which they were recorded; this then became their site number in the field also. Each feature was also given a preliminary site type definition (see Section 3.1.2) and a clarity rating of "definite" (almost certainly a site), "probable" (potential site with a greater than 50% chance of actually being one) and "tentative" (potential site with a lesser than 50% chance of actually being one). This survey of the satellite imagery allowed for both direct analyses of the archaeological landscape to be conducted and for the subsequent fieldwork to be efficiently planned. Methodologies similar to this have been used by numerous contemporary archaeological projects, as recently illustrated by Ansart et al. [27]. To locate and map features in the field, the database was loaded onto both a tablet computer and a handheld GPS device. The former was used to locate sites in the field using the tablet's built-in GPS function, with the points shapefile overlaid on pre-downloaded offline imagery from both GoogleEarth and Bing Maps. The latter was used to record the form of specific structures and parts of structures with more precise locational positioning.

3. Results

3.1. Individual Archaeological Sites

3.1.1. Site Locations and their Relationships to Natural Features

The satellite imagery survey allowed for numerous results to be obtained directly from the remote sensing analysis, independent of fieldwork data. Most simply, the relative locations of each archaeological feature allowed for their distributions, densities, areas of concentration and spatial relationships to natural features to be determined (Figure 6). In total, 2770 individual features were identified within the WHS area. This wealth of data already led to some interesting results that impacted upon later interpretations.

Most visible archaeological features appear to be located by areas of relatively dense basalt. This is an expected correlation given the fact that it is anthropogenic basalt structures that are the easiest to identify in this region on satellite imagery, and that for any structures not located within or near basalt raw material, a less prominently visible material is likely to have been used. However, a further noticeable pattern is that there is a concentration of features along the edges of basalt fields and qe'an or awdiya. A number of factors could be involved here, including areas of easiest accessibility in the Harra, access to collection of seasonal rainfall and access to the best areas for vegetation for herd grazing. Furthermore, the practicalities of erecting buildings are easier in these border regions, where basalt stones are available for construction, but areas do not have to be first cleared of basalt to create habitable space for humans and livestock.

However, the satellite survey also revealed a large number of features located deep within dense basalt fields. These would have been extremely difficult to access but for human modification of the landscape (see Section 3.2.2), and at some distance away from necessary natural resources. The effort entailed in such endeavours, which does not appear strictly necessary from a geographical viewpoint, indicates that significant advantages must have existed for settlements thus located. While a selection of all site types were visited in the field, this study focussed on "wheels", and their identified subdivisions and associations, which forms the basis of the following discussions.

3.1.2. "Wheels" and "Encircled Enclosure Clusters"

It has long been recognised by explorers and researchers in the Harra that its prehistoric structures can be divided into several distinct types based on their morphologies as viewed from above [5,6,28]. Using the higher-resolution free satellite imagery now available, structural differences can be defined with greater clarity, allowing for improved large-scale mapping of distinct forms that can be investigated on the ground (Figure 7). For previously known and identified site types such as enclosures, "wheels", "pendants", "kites" and meandering walls, this simply meant that they could be identified with a greater degree of certainty than previously possible. However, for certain features, the higher quality satellite imagery actually allowed for the discerning of discrete site types that have previously been grouped together. Most notably, it became clear that the sites known as "wheels" or "jellyfish" [6,28] require a typological seriation, something that has been recognised before by Rollefson et al. [16]. In the WHS area, two distinct forms could be morphologically defined by remote sensing data, which during the course of the later fieldwork were found to have impacts on their frequency and material remains.

The first form is, true to its name, indeed "wheel-like" in shape (Figure 8a). Its main features comprise a roughly circular or elliptical outline, inside which enclosures are divided by mostly straight walls, arranged like the spokes of a wheel. Though these "spokes" sometimes come to a central point, they often converge around one to three sub-circular central enclosures, from which the other arc-shaped enclosures emanate. Occasionally, such sites are encircled by a series of very small enclosures, no more than 2 m across. In the survey area, 70 of these true "wheels" were identified with a certainty of "definite" or "probable", 77% of which are located on the outskirts of the basalt terrain; within 1 km of the edge of the Harra, a qa'a, or a major wadi (Figure 9).

The second form, which the authors have termed "encircled enclosure clusters" are each comprised of a randomly clustered set of at least four sub-circular or sub-elliptical enclosures (Figure 8b). This creates an irregular external outline, sometimes with one or two additional protruding enclosures. Few, if any, of the internal walls are straight, and there is no clear central enclosure. As their name suggests, they are always encircled by a series of very small enclosures, which vary in clarity on remote sensing and on the ground. Over three times as many "encircled enclosure clusters" as "wheels" were identified in the survey area; a total of 226 sites ("definite" or "probable"). Almost the same proportion of these sites (80%) is located on the edge of the basalt desert (Figure 9).

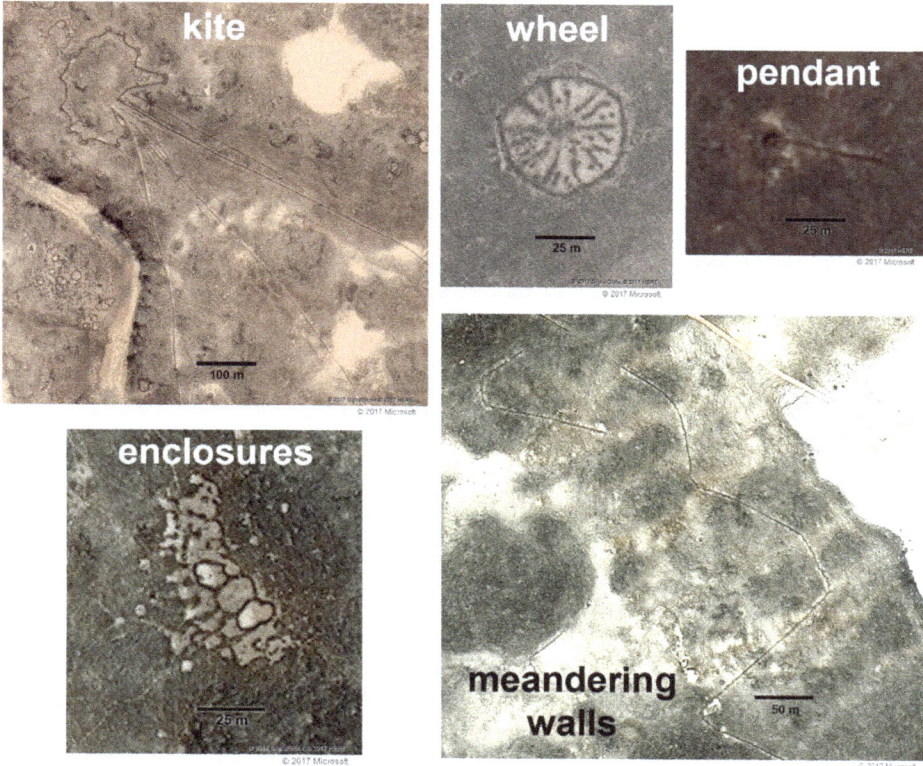

Figure 7. Representative satellite images of the main previously-identified prehistoric site forms found in the Harra. Microsoft product screen shots reprinted with permission from Microsoft Corporation.

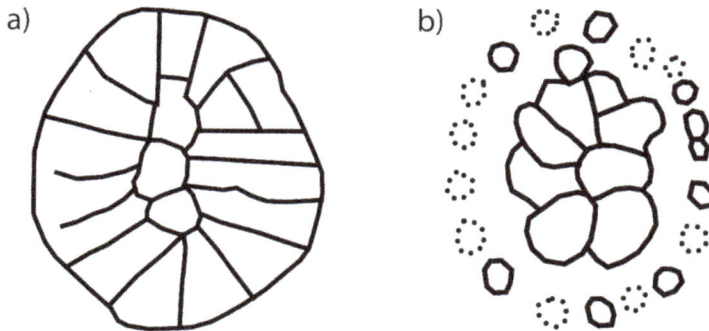

Figure 8. Representative line drawings highlighting the differences between (**a**) "wheels" and (**b**) "encircled enclosure clusters".

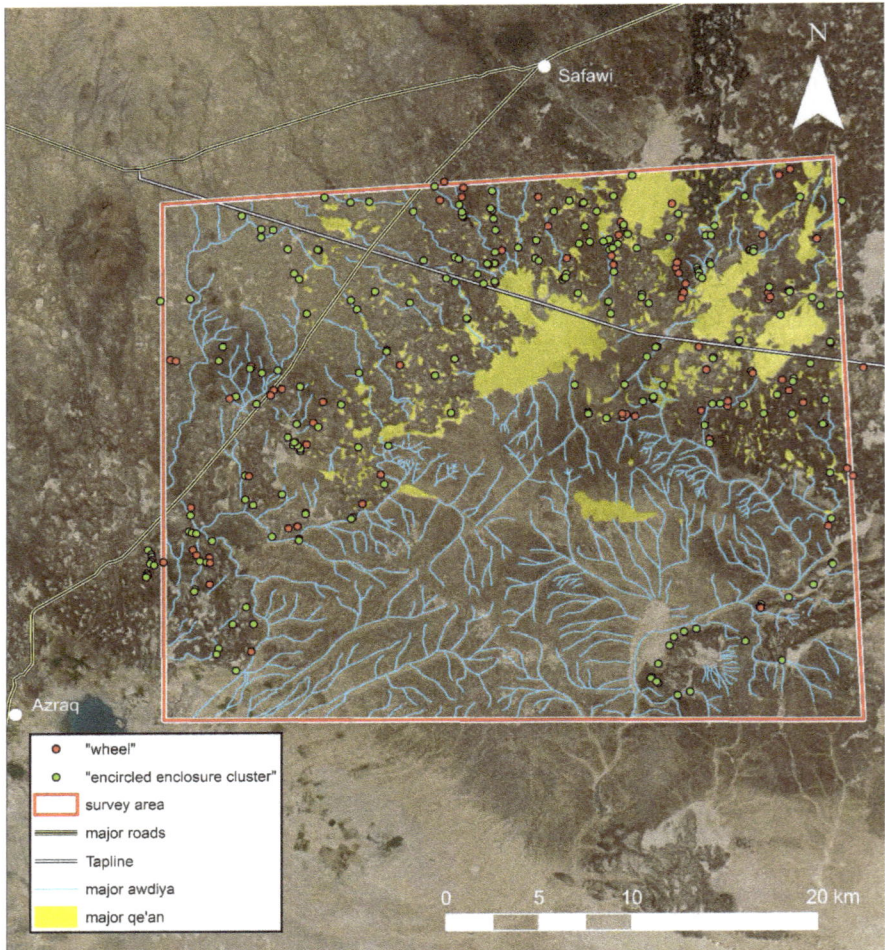

Figure 9. Map showing the distribution of "wheels" and "encircled enclosure clusters" in the WHS, with natural features highlighted (32°0′ N, 37°7′ E). Sources: Esri, DigitalGlobe, GeoEye, i-cubed, USDA FSA, USGS, AEX, Getmapping, Aerogrid, IGN, IGP, swisstopo and the GIS User Community.

This definition of the distinctions between "wheels" and "encircled enclosure clusters" also highlights an example of the bidirectional relationship between remote sensing surveys and ground-based fieldwork. The very small encircling enclosures that are occasionally present at "wheels" and always present at "encircled enclosure clusters" initially appeared as black "dots" on satellite imagery, indicating possible basalt cairns. However, in some regions with particularly good structure preservation and high contrast between the basalt and the underlying silt they appeared more as enclosures, with a small amount of light-coloured loess in their centres (Figure 10). Investigating these features on the ground revealed that these are in fact square structures with right-angled corners, and thus are unique compared to other enclosures in the region (Figure 11). This fieldwork-derived data can now be fed back into the satellite imagery survey, helping us to interpret what we see in remote sensing more accurately for future investigations.

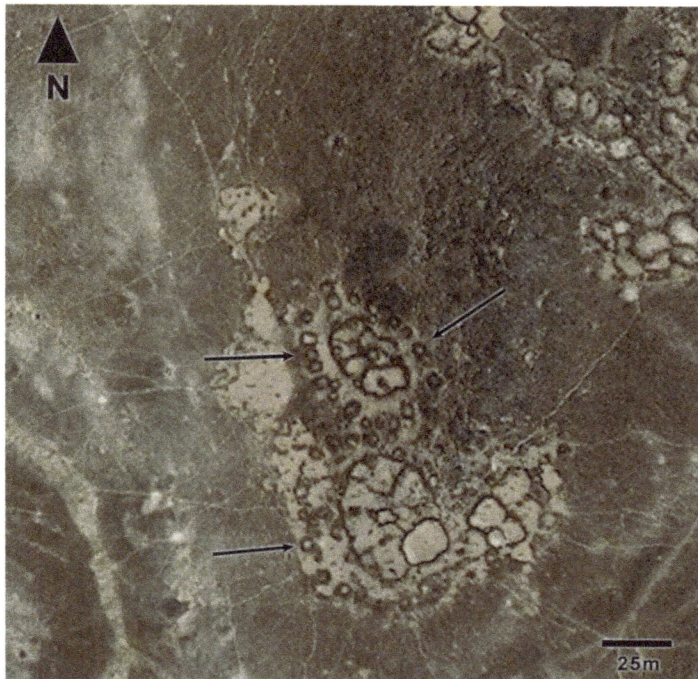

Figure 10. Satellite image showing a particularly distinct view of two "encircled enclosure clusters", with clear examples of the associated small square structures indicated by arrows. Microsoft product screen shot(s) reprinted with permission from Microsoft Corporation.

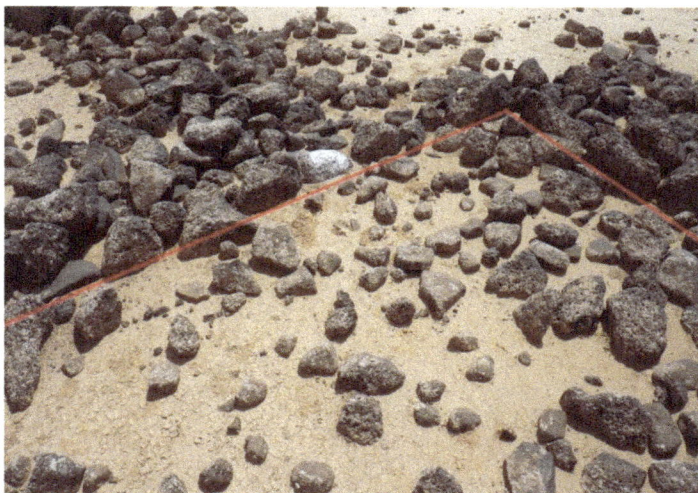

Figure 11. Ground view of one of the small square structures that surround "encircled enclosure clusters", with an overlaid red line drawn to indicate its inner edge.

3.2. Regional Human Circulation

3.2.1. Access via Natural Features

Travelling in this region has always been difficult, whether on foot or by wheeled vehicle, due to the dense cover of basalt boulders. Yet the resilience of subsistence in this region depends on access within it, even in the present day. Two solutions exist to allow for easy and speedy travel, and thus for communication, trade and the exploitation of resources. Awdiya, as natural corridors, and qe'an, as wide open spaces, offer natural routes since they can be followed or crossed easily. As detailed above regarding "wheels" and "encircled enclosure clusters", the majority of all sites are in fact in close proximity to or within these natural access features. However, even though only ca. 20% of all sites are located in the deep basalt (over 1 km from a qa'a, wadi, or the edge of the Harra), this still accounts for 559 individual features. Moreover, even just 1 km is a long distance to traverse on a regular basis across the Harra. Decreasing the distance from the nearest natural access point to a more manageable 500 m also significantly increases the number of basalt-located sites to 33%, or 926 individual features.

3.2.2. Anthropogenic Pre-Modern Paths

For these sites, the creation of access paths is practically a requirement. Such routes, found in association with prehistoric sites and clearly deliberately arranged by the clearing of a path through the basalt boulders, were first identified on the ground; this was then confirmed by their visibility on satellite imagery. Their existence and association with prehistoric time periods have also been noted by several other researchers in the region [10,29], however, they have not yet been the focus of systematic analyses across a large area. Furthermore, there are several other types of access routes that are visible in remote sensing throughout the Harra, complicating identification. Apart from the asphalt road, the "TAP line" and the tracks opened by bulldozers mentioned earlier, many narrow "sheep tracks" crisscross the region, which are created by stones shifting naturally by the continued kicking of sheep and goat feet as they take the same routes on a regular basis in single file.

It thus becomes necessary to define these different routes based on their appearances in satellite imagery. While the asphalt roads and the cobbled "TAP line" are self-evident, all other routes are simply variations of paths cleared by moving basalt rocks. However, several clear distinctions between these route types can be made from their forms, arrangements and widths (Figure 12). Modern tracks are very straight and wide, while sheep tracks form a random network of narrow paths. Pre-modern routes, on the other hand, are unconnected to vehicle tracks but are clearly linked with ancient sites, and frequently have ancient remains located along the way. Distinctions between recent and ancient routes are also quantifiable from their widths, which, despite some overlap, vary greatly from sheep tracks to the bulldozed roads (Figure 13).

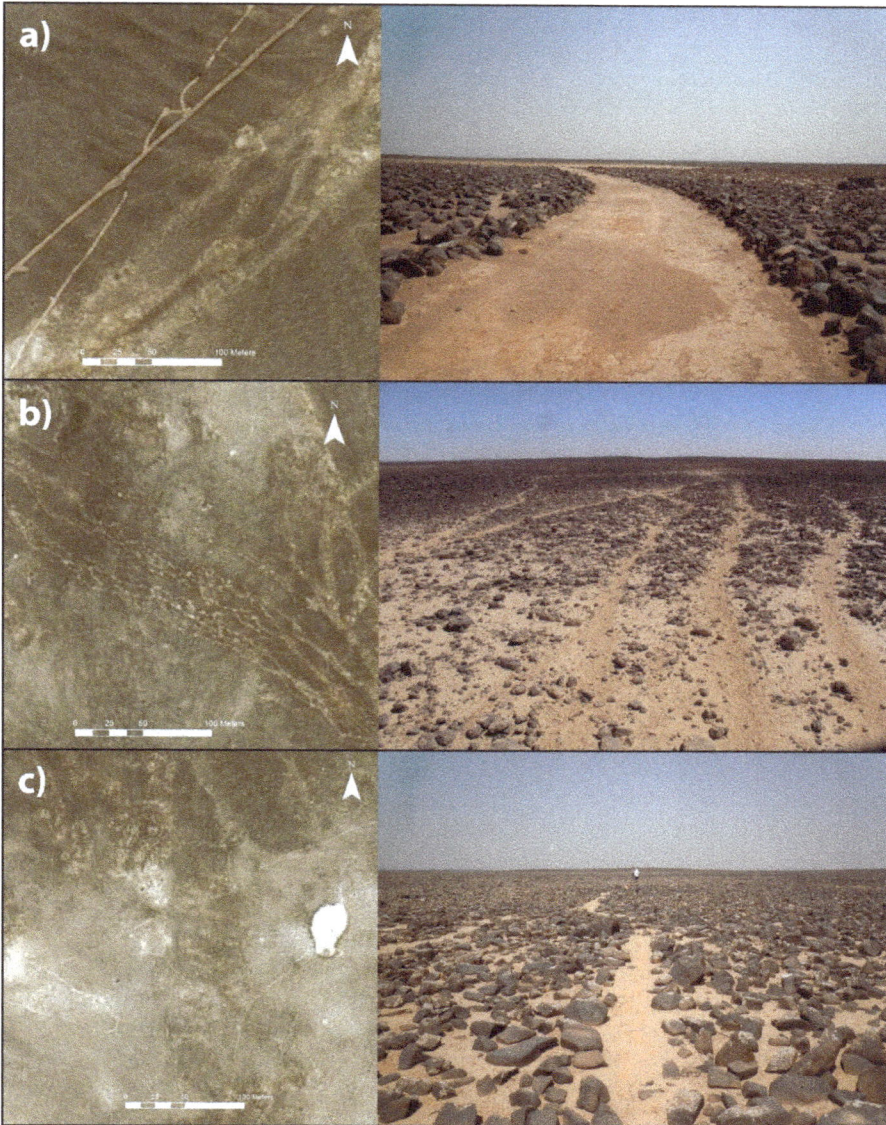

Figure 12. Complementary satellite and ground views of representative examples of (**a**) modern vehicle tracks, (**b**) "sheep tracks" and (**c**) probable pre-modern routes. Sources: Esri, DigitalGlobe, GeoEye, i-cubed, USDA FSA, USGS, AEX, Getmapping, Aerogrid, IGN, IGP, swisstopo and the GIS User Community.

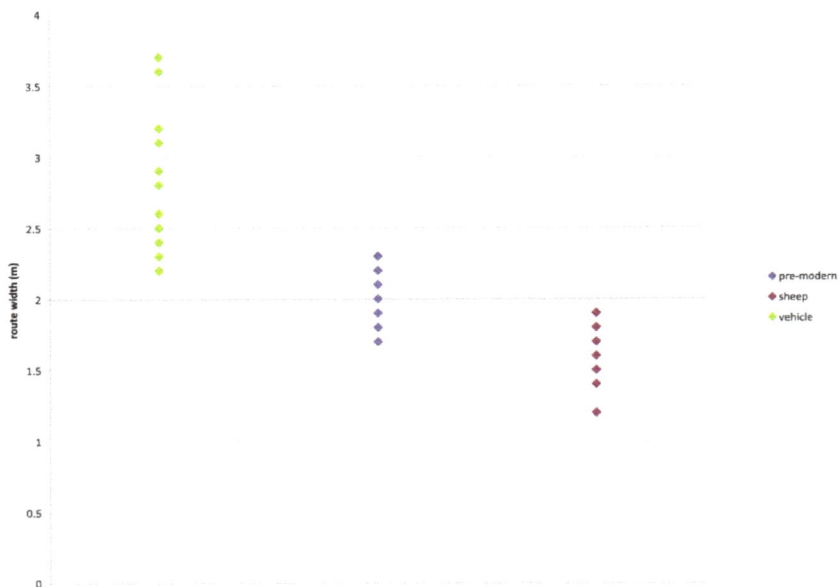

Figure 13. Graph showing the widths of paths in the analyzed area corresponding to the three types of routes present in the region.

Incorporating ground observations with remote sensing investigations to test the validity of the latter allowed for confident mapping of ancient routes from satellite images alone. For this, we chose a selected region in the northeast of the survey area to conduct a preliminary mapping analysis (Figure 14). This found that 38% of these routes identified connect a qa'a or wadi to other such natural features, while a similar number (36%) connect such features to a site. By contrast, only 16% of routes connect a site to one or more other sites. From this, it could be concluded that pre-modern routes were arranged for two main purposes: (1) Connecting multiple natural access features to provide possibilities of long-distance movement by using these as "stepping stones" across the basalt desert; and (2) connecting sites to their natural approaches for immediate access and to their local resources such as ephemeral water in awdiya and qe'an. This explains the noticeable hierarchy of route networks, ranging from paths several kilometres long to ones leading merely a few tens of meters from a site to a wadi. The most notable examples within the latter variety are path arrangements on slopes with anthropogenically constructed horizontal stone breaks to create stair treads. This leads to the artificial creation of suitable areas for natural vegetation to grow, leading in turn to their easy identification on the ground (Figure 15). By correlating such ground observations with remote sensing, we can now confidently identify these very localised and short-length features on the imagery also.

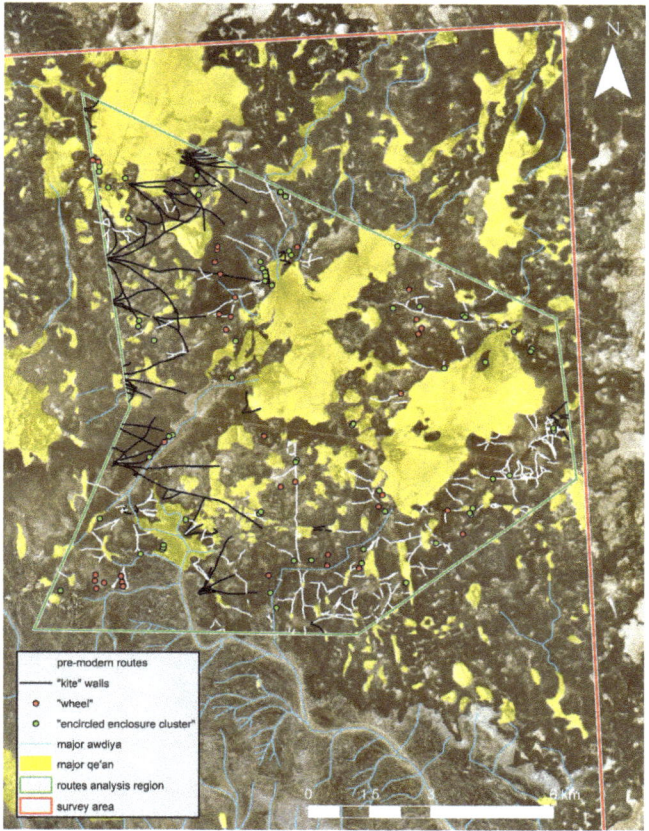

Figure 14. Map showing all probable pre-modern routes identified in the analysis area, with "wheels" and "encircled enclosure clusters" marked for locational comparison (32°0′ N, 37°7′ E). Also highlighted are natural features and "kite" walls, which can additionally be used for human circulation (the latter due to the ubiquitous linear clearings of removed construction stones adjacent to the walls). Sources: Esri, DigitalGlobe, GeoEye, i-cubed, USDA FSA, USGS, AEX, Getmapping, Aerogrid, IGN, IGP, swisstopo and the GIS User Community.

Figure 15. Complementary satellite and ground views of representative examples of pre-modern paths with "steps" built into them, indicated by arrows in the left-hand image. Sources: Esri, DigitalGlobe, GeoEye, i-cubed, USDA FSA, USGS, AEX, Getmapping, Aerogrid, IGN, IGP, swisstopo and the GIS User Community.

4. Discussion

4.1. Impacts on Ground Fieldwork Practicalities

The results obtained by the satellite imagery survey, combined with ground-based research, had numerous impacts upon the fieldwork itself. From a standpoint of pure practicality, the ability to pinpoint specific sites for ground investigation that have a good chance of being particularly beneficial to analysis is extremely useful. Our surface survey could thus target representative selections of sites of, for example, specific morphological types, various states of preservation, isolated or grouped locations and varied geographic/geological associations (e.g., close/far from a wadi or qa'a). It was therefore possible to investigate a stratified sample of archaeological remains from the outset of the first fieldwork season, something that would have been otherwise impossible.

A second practical use of the satellite imagery was the ability to determine the accessibility of features we wished to investigate. By mapping visibly cleared routes wide enough to allow the passage of vehicles (see above), a network of "roads" could be superimposed on the desired sites to be visited, and clusters of archaeological features within a reasonable walking distance (maximum ca. 1 km) of these were earmarked for targeting (Figure 16). The formerly unfeasible investigation of sites deep within the basalt was therefore in specific instances made possible.

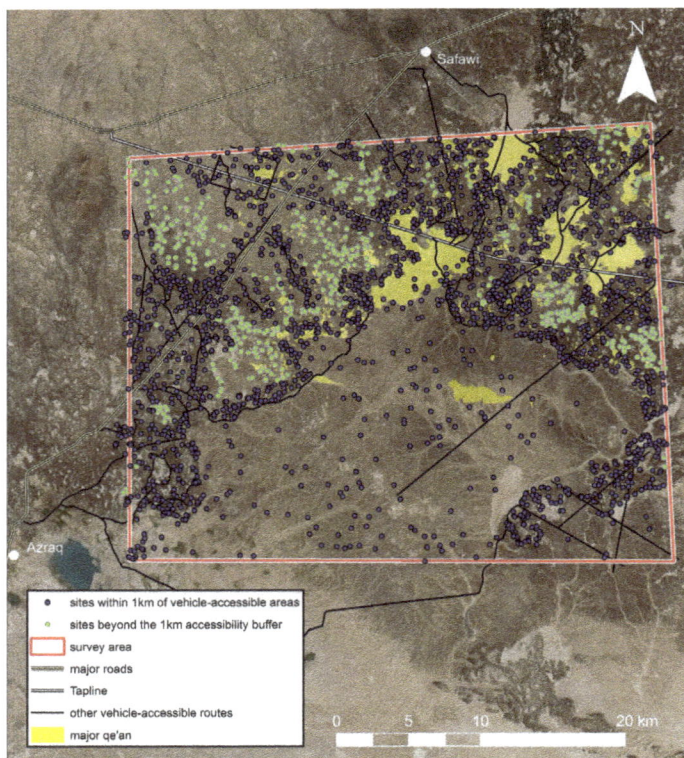

Figure 16. Map showing all routes traversable by vehicle in the WHS area (including vehicle-accessible qe'an), with sites within 1 km of such routes highlighted (32°0' N, 37°7' E). Sources: Esri, DigitalGlobe, GeoEye, i-cubed, USDA FSA, USGS, AEX, Getmapping, Aerogrid, IGN, IGP, swisstopo and the GIS User Community.

Both of these advantages, as well as other practicalities in the field, were greatly enhanced by the use of offline imagery on tablet computers. Doing so enabled direct comparisons between the remote sensing data and ground observations to be made, allowing for some immediate confirmations or denials of hypothesised features or their properties based on the satellite imagery survey. Crucially, this was achieved without the need to wait for an opportunity to access the Internet, which might not occur until the end of the season. Furthermore, any beneficial modifications to the fieldwork plan based on in situ findings, such as visiting extra locations and sites, could be made ad-hoc.

Overall, these practical advantages of freely available very high-resolution satellite imagery analyses significantly increase the efficiency of the ground fieldwork, enabling precise data collection from the outset with a much smaller initial outlay, both in terms of time and expense, than would be necessary otherwise. Given the well-known time and budget constraints of many projects within the discipline of archaeology, this provides a particularly meaningful advantage.

4.2. Impacts on Regional Analyses

The advantages of the satellite imagery upon regional analyses are even further reaching. Firstly, in an area such as the Harra, ground investigations are inevitably of a keyhole nature, with only a small percentage of the landscape being feasibly accessible. By incorporating these with the analysis of satellite imagery, however, they can be turned into holistic datasets, enabling regionwide studies that in turn can be correlated with other, more accessible areas with more intensive surface surveys such as the Greater Western Jazira, e.g., Reference [30], and Shamiya, e.g., Reference [31], regions of north-eastern and central Syria, respectively. This can be achieved by the analysis and extrapolation of data pertaining to both individual sites and inter-site features such as routes.

The ability to identify and quantify site types, and be able to precisely determine these with great accuracy is key to ascertaining their spread across a wide landscape. By correlating the visual appearance of these sites on satellite imagery with their known forms from ground investigations, they can be rapidly mapped across a large area. By relying on the surface studies to inform what can be seen on the imagery, even features which cannot be clearly seen on the latter can still be integrated into the holistic mapping once a comparative recognition link has been established. For example, once we have determined that features which look like cairns around "encircled enclosure clusters" are in fact square structures at all examples visited on the ground, it can be said with a high degree of certainty that they will also be square structures at other sites not visited in the field.

Furthermore, with the inclusion of data from fieldwork such as statistically-significant correlations between certain site types and certain human activities or occupation periods, even more detailed interpretations can be made about the landscape as a whole. In the WHS, the former has been investigated by analysing the ratio of different lithic tool varieties present at different site types. Certain tools can indicate certain specific uses for different spaces. For example, lithic scrapers are associated with the processing of animal hides, e.g., Reference [32], while a preponderance of debitage (offcut flakes or chips) might indicate a tool production site. One tentative correlation that was already identified from the first two fieldwork seasons is that "wheel" sites contain on average more scrapers and significantly more lithic cores than "encircled enclosure clusters", while flakes were found to be proportionally more numerous at the latter site type. If a simultaneous use of both site types is assumed, this could indicate complementary site functions of settlement, processing or storage of tools and/or livestock. If this hypothesis were upheld by further investigation, it would allow the mapping of aspects of socio-economic organisation across the entire landscape by satellite survey.

However, it is also quite possible that these different site types are the result of chronological variations, which ties into another important correlation that the WHS is attempting to determine: That between site form and occupation dates. In the 2017 fieldwork season, the collection of soil samples for dating by OSL, using a method detailed by Athanassas et al. [11], was commenced. Though the results of this analysis are not yet available, and more sites need to be sampled to provide a comprehensive dataset, it is hoped that in time we will be able to rapidly map the spread of human occupation across the Harra over the *longue durée*. This methodology has previously been used to great effect by the Fragile Crescent Project of Durham University, UK, in the holistic mapping of sites across Northern Mesopotamia [33,34].

Difficulties of circulation in the Harra due to the basaltic pavement, which preserve parts of this region from long-term (and especially modern) human impact, allow in turn for the recognition of ancient routes. This is a good opportunity for studying past circulation from archaeological remains, especially from pre- and protohistoric periods; a relatively rare phenomenon, though other such examples do exist in the Middle East, see e.g., References [35,36]. While pathways can be identified and studied on the ground, the contrast between linear cleared features and the surrounding basalt boulders is particularly recognisable in very high-resolution satellite images. Beyond analysing the interpreted local and daily exploitation of resources, this allows for a more comprehensive regional analysis through systematic mapping, enabling for example the posing of questions surrounding the territorial and economic impact of these routes. Circulation in the region was undoubtedly a critical issue in the past, especially during the peak of occupation between the 7th and the 4th millennium BC. It is now possible to investigate and even quantify issues of spatiality in regional and interregional trade and socio-economic networks that have been suggested for prehistoric north-eastern Jordan [37].

Together, these methods allow us to interpret the distribution of human practices and of human settlement at various time periods across a vast region much quicker than would be possible from decades of fieldwork, with the inclusion of areas that are practically inaccessible on the ground. For this process to be accurate, the quality of remote sensing data is obviously a deciding factor. It is therefore significant that, in the Harra at least, the resolution and clarity of freely-available satellite imagery is now at a level where it can be used for this methodology. Thus, while such analyses are in no way as reliable as ground truth data, they can significantly develop at least a broad understanding of the archaeology of a region by a method that is rapid, cost-effective and encompasses a large scale. Although some factors of its success pertain specifically to the Harra (such as the particularly high contrast and sharp definitions of black basalt structures atop light brown loess), the methodology of the WHS is a comprehensive example of what can be achieved through the use of freely available satellite imagery in general, and illustrates that, as long as its known limitations are kept in mind, this democratisation of remote sensing data yields significant rewards for all academic research.

Author Contributions: Conceptualization, S.L.S. and M.-L.C.; Methodology, S.L.S.; Formal Analysis, S.L.S.; Investigation, S.L.S. and M.-L.C.; Resources, S.L.S. and M.-L.C.; Writing—Original Draft Preparation, S.L.S. and M.-L.C.; Writing—Review & Editing, S.L.S. and M.-L.C.; Visualization, S.L.S. and M.-L.C.; Project Administration, S.L.S. and M.-L.C.; Funding Acquisition, S.L.S. and M.-L.C.

Funding: The Western Harra Survey fieldwork has been made possible by funding from the Council for British Research in the Levant (CBRL) (2015 season), the National Geographic Society (2015 season), the Curtiss T. Brennan & Mary G. Brennan Foundation (2017 season) and Archéorient—UMR 5133 CNRS (2017 & 2018 seasons). The 2017 and 2018 seasons as well as the project in general has received funding from the Research Foundation—Flanders (FWO) and the European Union's Horizon 2020 research and innovation programme under the Marie Skłodowska-Curie grant agreement No. 665501.

Acknowledgments: The Western Harra Survey is a co-directed project by both authors. Thank you to all additional team members: Anne Binder, Lanah Haddad, S. Nina Mann (2015 season); David Burke (2017 season); Imad Alhussain (2017 & 2018 seasons). Thanks also go to the Department of Antiquities of the Hashemite Kingdom of Jordan, in particular HE Monther Jamhawi, Aktham Oweidi, Ahmad Lash and Aisar Radaydeh. We are also grateful to IFPO and CBRL for facilitating our fieldwork and study season in numerous ways.

Conflicts of Interest: The authors declare no conflict of interest. The funding sponsors had no role in the design, execution, interpretation, or writing of the study.

References

1. Kempe, S.; Al-Malabeh, A. Hunting kites ('desert kites') and associated structures along the eastern rim of the Jordanian Harrat: A geo-archaeological Google Earth images survey. *Z. Orient-Archäol.* **2010**, *3*, 46–86.

2. Kempe, S.; Al-Malabeh, A. Desert kites in Jordan and Saudi Arabia: Structure, statistics and function, a Google Earth study. *Quat. Int.* **2013**, *297*, 126–146. [CrossRef]

3. Issar, A.S.; Zohar, M. *Climate Change—Environment and History of the Near East*; Springer: Berlin, Germany, 2007.

4. Traboulsi, M. Les précipitations au Proche-Orient: Variabilité Spatio-Temporelle et Relations avec la dynamique de l'atmosphère (1960–61/1989–90). Ph.D. Thesis, University of Bourgogne, Dijon, France, 2004.

5. Maitland, P. The 'Works of the old Men' in Arabia. *Antiquity* **1927**, *1*, 197–203. [CrossRef]

6. Kennedy, D. The "Works of the Old Men" in Arabia: Remote sensing in interior Arabia. *J. Archaeol. Sci.* **2011**, *38*, 3185–3203. [CrossRef]

7. Betts, A.V.G. (Ed.) *Excavations at Jawa 1972–1986*; Edinburgh University Press: Edinburgh, UK, 1991.

8. Betts, A.V.G. (Ed.) *The Harra and the Hamad: Excavations and Survey in Eastern Jordan*; Sheffield Archaeological Monographs 9; Sheffield Academic Press: Sheffield, UK, 1998.

9. Müller-Neuhof, B. A 'marginal' region with many options: The diversity of Chalcolithic/Early Bronze Age socio-economic activities in the hinterland of Jawa. *Levant* **2014**, *46*, 230–248. [CrossRef]

10. Akkermans, P.M.M.G.; Huigens, H.O.; Brüning, M.L. A landscape of preservation: Late prehistoric settlement and sequence in the Jebel Qurma region, north-eastern Jordan. *Levant* **2014**, *46*, 186–205. [CrossRef]

11. Athanassas, C.D.; Rollefson, G.O.; Kadereit, A.; Kennedy, D.; Theodorakopoulou, K.; Rowan, Y.M.; Wasse, A. Optically stimulated luminescence (OSL) dating and spatial analysis of geometric lines in the Northern Arabian Desert. *J. Archaeol. Sci.* **2015**, *64*, 1–11. [CrossRef]

12. Müller-Neuhof, B.; Abu-Azizeh, W. Milestones for a tentative chronological framework for the late prehistoric colonization of the basalt desert (north-eastern Jordan). *Levant* **2016**, *48*, 220–235. [CrossRef]

13. Richter, T.; Arranz-Otaegui, A.; Yeomans, L.; Boaretto, E. High Resolution AMS Dates from Shubayqa 1, northeast Jordan Reveal Complex Origins of Late Epipalaeolithic Natufian in the Levant. *Sci. Rep.* **2017**, *7*. [CrossRef] [PubMed]

14. Betts, A.V.G. The Epipalaeolithic Periods. In *The Harra and the Hamad: Excavations and Survey in Eastern Jordan*; Sheffield Archaeological Monographs 9; Betts, A.V.G., Ed.; Sheffield Academic Press: Sheffield, UK, 1998; pp. 11–35.

15. Rollefson, G.; Rowan, Y.; Wasse, A. The Late Neolithic colonization of the Eastern Badia of Jordan. *Levant* **2014**, *46*, 285–301. [CrossRef]

16. Rollefson, G.O.; Athanassas, C.D.; Rowan, Y.M.; Wasse, A. First chronometric results for 'works of the old men': Late prehistoric 'wheels' near Wisad Pools, Black Desert, Jordan. *Antiquity* **2016**, *90*, 939–952. [CrossRef]

17. Western Harra Survey—Home. Available online: http://bit.ly/WHS_Facebook (accessed on 7 December 2018).

18. Western Harra Survey—YouTube. Available online: http://bit.ly/WHS_YouTube (accessed on 7 December 2018).

19. Smith, S.L. The Late Chalcolithic and Early Bronze Age of the Badia and Beyond: Implications of the Results of the First Season of the Western Harra Survey. In *Landscapes of Survival, Proceedings of the International Conference on the Archaeology and Epigraphy of Jordan's North-Eastern Desert, Leiden University, Leiden, The Netherlands, 17–18 March 2017*; Akkermans, P.M.M.G., Ed.; Sidestone Press: Leiden, Netherlands, in press.

20. Smith, S.L. Surveying the Black Desert: Investigating Prehistoric Human Occupation in North-Eastern Jordan. ArchéOrient—Le Blog 2017. Available online: http://archeorient.hypotheses.org/7882 (accessed on 27 November 2018).

21. Chambrade, M.-L.; Smith, S.L. Western Harra Survey Project. *Archaeol. Jordan Newsl.* **2018**, 7–8. Available online: http://www.acorjordan.org/archaeology-jordan-aij (accessed on 28 November 2018).

22. Kennedy, D.L. Pioneers above Jordan: Revealing a prehistoric landscape. *Antiquity* **2012**, *86*, 474–491. [CrossRef]

23. Kennedy, D.; Bewley, R. Aerial archaeology in Jordan. *Antiquity* **2009**, *83*, 69–81. [CrossRef]

24. APAAME—Finding the Past Frame by Frame. Available online: http://www.apaame.org (accessed on 27 November 2018).

25. Philip, G.; Donoghue, D.; Beck, A.; Galiatsatos, N. CORONA satellite photography: An archaeological application from the Middle East. *Antiquity* **2002**, *76*, 109–118. [CrossRef]

26. Ur, J.A. CORONA Satellite Photography and Ancient Road Networks: A Northern Mesopotamian Case Study. *Antiquity* **2003**, *77*, 102–115. [CrossRef]

27. Ansart, A.; Braemer, F.; Davtian, G. Preparing an archaeological field survey: Remote sensing interpretation for herding structures in the Southern Levant. *J. Field Archaeol.* **2016**, *41*, 699–712. [CrossRef]

28. Betts, A.V.G. "Jellyfish": Prehistoric desert rockshelters. *ADAJ* **1982**, *26*, 183–188.

29. Huigens, H.O. The identification of pathways on *harra* surfaces in north-eastern Jordan and their relation to ancient human mobility. *J. Arid Environ.* **2018**, *155*, 73–78. [CrossRef]

30. Smith, S.L. The View from the Steppe. Using Remote Sensing to Investigate the Landscape of "Kranzhügel" in Its Regional Context. In *Studies in Honor of Tony J. Wilkinson. New Agendas in Remote Sensing and Landscape Archaeology in the Near East*; Oriental Institute Publications; Lawrence, D., Altaweel, M., Philip, G., Eds.; University of Chicago Press: Chicago, IL, USA, in press.

31. Geyer, B.; Chambrade, M.-L.; Coqueugniot, E. Water quest and subsistence strategies during the Neolithic in the arid margins of Northern Syria. In *Settlement Dynamics and Human-Landscape Interaction in the Dry Steppes of Syria*; Studia Chaburensia 4; Morandi Bonacossi, D., Ed.; Harrassowitz Verlag: Wiesbaden, Germany, 2014; pp. 1–31.

32. Yerkes, R.; Barkai, R.; Gopher, A.; Zutovski, K. The use of fan scrapers: Microwear evidence from Late Pottery Neolithic and Early Bronze Age, Ein Zippori, Israel. *J. Lithic Stud.* **2016**, *3*, 185–205. [CrossRef]

33. Galiatsatos, N.; Wilkinson, T.J.; Donoghue, D.; Philip, G. The Fragile Crescent Project (FCP): Analysis of Settlement Landscapes using Satellite Imagery. In Proceedings of the Computer Applications to Archaeology 2009 Conference, Williamsburg, VA, USA, 22–26 March 2009; Available online: http://dro.dur.ac.uk/6909 (accessed on 27 November 2018).

34. Lawrence, D.; Bradbury, J.; Dunford, R. Chronology, Uncertainty and GIS: A Methodology for Characterising and Understanding Landscapes of the Ancient near East. In Proceedings of the 2nd International Landscape Archaeology Conference, Berlin, Germany, 6–8 June 2012; Bebermeier, W., Hebenstreit, R., Kaiser, E., Krause, J., Eds.; Excellence Cluster Topoi: Berlin, Germany, 2012; Volume 3, pp. 353–359.

35. Wilkinson, T.J.; French, C.; Ur, J.A.; Semple, M. The Geoarchaeology of Route Systems in Northern Syria. *Geoarchaology* **2010**, *25*, 745–771. [CrossRef]

36. Zboray, A. Prehistoric trails in the environs of Karkur Talh, Jebel Uweinat. In *Desert Road Archaeology in Ancient Egypt and Beyond*; Africa Praehistorica, 27; Förster, F., Reimer, H., Eds.; Heinrich-Barth-Institut: Köln, Germany, 2013; pp. 381–389.

37. Müller-Neuhof, B. Chalcolithic/Early Bronze Age Flint Mines in the Northern Badia. *Syria* **2013**, *90*, 177–188. [CrossRef]

geosciences

MDPI

Article

Mapping Archaeology While Mapping an Empire: Using Historical Maps to Reconstruct Ancient Settlement Landscapes in Modern India and Pakistan

Cameron A. Petrie [1,2,*], Hector A. Orengo [3], Adam S. Green [2], Joanna R. Walker [1], Arnau Garcia [2], Francesc Conesa [2], J. Robert Knox [4] and Ravindra N. Singh [5]

[1] Department of Archaeology, University of Cambridge, Cambridge CB2 3DZ, UK; jrw95@cam.ac.uk
[2] McDonald Institute for Archaeological Research, University of Cambridge, CB2 3ER, UK;
     ag952@cam.ac.uk (A.S.G.); ag2023@cam.ac.uk (A.G.); fcic2@cam.ac.uk (F.C.)
[3] The Catalan Institute of Classical Archaeology, 43003 Tarragona, Spain; hao23@cam.ac.uk
[4] Formerly of the Department of Asia, British Museum, London WB1B 3DC, UK; jrknox57@googlemail.com
[5] Department of AIHC and Archaeology, Banaras Hindu University, Varanasi 221005, India;
     drravindransingh@gmail.com
*   Correspondence: cap59@cam.ac.uk; Tel.: +44-(0)1223-338-582

Received: 10 November 2018; Accepted: 19 December 2018; Published: 25 December 2018

Abstract: A range of data sources are now used to support the process of archaeological prospection, including remote sensed imagery, spy satellite photographs and aerial photographs. This paper advocates the value and importance of a hitherto under-utilised historical mapping resource—the Survey of India 1″ to 1-mile map series, which was based on surveys started in the mid–late nineteenth century, and published progressively from the early twentieth century AD. These maps present a systematic documentation of the topography of the British dominions in the South Asian Subcontinent. Incidentally, they also documented the locations, the height and area of thousands of elevated mounds that were visible in the landscape at the time that the surveys were carried out, but have typically since been either damaged or destroyed by the expansion of irrigation agriculture and urbanism. Subsequent reanalysis has revealed that many of these mounds were actually the remains of ancient settlements. The digitisation and analysis of these historic maps thus creates a unique opportunity for gaining insight into the landscape archaeology of South Asia. This paper reviews the context within which these historical maps were created, presents a method for georeferencing them, and reviews the symbology that was used to represent elevated mound features that have the potential to be archaeological sites. This paper should be read in conjunction with the paper by Arnau Garcia et al. in the same issue of *Geosciences*, which implements a research programme combining historical maps and a range of remote sensing approaches to reconstruct historical landscape dynamics in the Indus River Basin.

Keywords: archaeological landscapes; settlements; historical maps; Survey of India; Archaeological Survey of India; heritage; colonial studies

---

1. Introduction

Landscape archaeology is fundamentally concerned with recording of the distribution of ancient settlement and the environments within which they were situated, and makes use of a range of different sources of data to aid the multi-faceted and multi-phased process of landscape prospection [1]. Remote assessment is important for directing effective survey on the ground, but it also provides insight into the distribution of archaeological sites in areas and/or regions that are otherwise inaccessible (e.g., areas of Afghanistan [2–5]; Syria [6]; Thessaly [7]). Satellite imagery and aerial photography now play an indispensable role in the landscape archaeology of many regions, but their usefulness is curtailed by their relatively restricted time depth. In contrast, for some parts of the world there are historical map series that document the landscape long before major intensive agricultural practices and development programmes removed heretofore preserved vestiges of past landscapes. These historic maps provide an important additional source of data for landscape archaeology that is complementary to remote sensing imagery and aerial photography, and has a similar but relatively under-utilised potential to generate major new insights into past landscapes. This paper argues that the digitisation and analysis of historic maps has unique potential for the landscape archaeology of South Asia, where there are a range of historical map resources and modern development has profoundly obscured complex archaeological landscapes.

The Survey of India 1″ to 1-mile (1:63,360 scale) map series originated in a very specific imperial context [8], and were based on surveys carried out from the mid–late nineteenth century onwards. These surveys were started in the wake of the conquest of Punjab and the then North-West Frontier Province and the Indian rebellion of 1857, which is also known as the First War of Independence. The resulting maps were published from the early twentieth century onward as part of series that were issued progressively and updated incrementally, and they were primarily designed to present information that was relevant to the military, such as the location of villages, roads, irrigation canals and the nature of land-use [9]. The surveys and the maps they produced also incidentally documented the locations and, to some extent, the height and area, of thousands of elevated mounds that were visible in the landscape at the time that the surveys were carried out. It appears that in most instances the surveyors did not recognise these mounds as anything unusual. A significant proportion of these mounds were actually the remains of ancient settlements, some of which built up during the processes of the formation and abandonment of ancient settlements millennia ago.

While archaeologists have long been aware of the potential of the Survey of India 1″ to 1-mile maps, and made use of them as early as the late 1940s/early 1950s [10], their use has been limited in terms of the areal extent. J.R. Knox carried out a systematic assessment of a significant number of map sheets as part of graduate research in the 1970s, but these maps have not been used to guide more recent archaeological surveys or as a data source in combination with remote sensing data. As such, the significance of these maps as a data source for large-scale systematic mapping of archaeological sites has not been explored. This paper considers the historical context within which the Survey of India 1″ to 1-mile map series was created, its incidental documentation of archaeological sites, and the degree to which surveyors knew what they were recording. To highlight the utility of this rich data source, this paper also outlines a systematic method for (a) georeferencing these maps and (b) identifies symbols that represent features and/or archaeological site locations.

2. Reconstructing Archaeological Landscapes Remotely

The declassification of spy satellite photographs (e.g., CORONA, GAMBIT, HEXAGON keyhole series) and the availability of Open Access remote sensing data (e.g., Landsat, Aster, Copernicus; [1,11,12]) has meant that a wide range of imaging system data sources are now used (both extensively and intensively) for prospection. Furthermore, a substantial (and growing) body of literature now highlights the importance of historic satellite and remote sensing imagery for documenting ancient landforms and settlements, particularly those at risk of being modified, obscured and/or obliterated in the process of modern urban and rural development [1,3–5,13–37].

Such high-resolution imagery has had a profound impact upon archaeological knowledge, but it does have limitations. Declassified spy satellite imagery only post-dates the 1950s and though it saw markedly improved resolution in the 1970s [1], it has issues with distortion [38]. Furthermore, not all parts of the world have coverage, let alone high-quality coverage, as much spy satellite imagery acquisition focused on geopolitical "hot spots". While China, the former USSR, and parts of the Middle East are covered extensively in US imagery, and Soviet imagery covered parts of Europe, imagery for other areas is either poor or non-existent. Coverage for much readily available remote sensing imagery is global, but is often limited in terms of its resolution and chronological range, which means that many features are not visible because the imagery was acquired after instances of modern disturbance [1,15].

The use of aerial photography for archaeology has a long history, beginning with the use of balloons in the nineteenth century and being advanced through the methods of reconnaissance and documentation using aeroplanes developed during the First World War [1,39,40]. Aerial archaeology in many parts of the world is now well developed (e.g., UK, USA, France, Italy, Jordan), and important historic imagery is available via substantial archives (e.g., National Collection of Aerial Photography, which contains images from the UK and much of the world). Historic aerial photos provide an invaluable record of many landscapes before they were disturbed by modern activities, and our understanding of the archaeology of some regions has been completely transformed through their use (e.g., Jordan; [1,7,40–44]). In relative terms, however, aerial photo coverage is again limited and/or unsystematic. Vertical and oblique images also have specific advantages and disadvantages related to distortion and visibility of features, with the latter being affected the timing of the photographs (time of day and time of year etc. [45–48]).

Maps have a longer history, and thus considerable potential for documenting lost or disappearing landscapes. While early maps contain limited topographic information and lack accuracy, the information documented by mapping projects in various nation states and former imperial dominions from the eighteenth and nineteenth century AD onwards was of a different order of accuracy due to advances in methods and the scale of the endeavour. Starting from the Cassini maps of France and the maps created of the UK prior to the establishment of the Ordnance Survey, surveyors set out to document landscapes and topography systematically. As these maps were created at particular points in time, it is also possible to monitor change in landscape over time by comparing different maps, particularly where individual maps are updated and reissued [49]. The maps produced by these systematic surveying projects have considerable potential for documenting archaeological sites, and it is notable that in the UK, there was clear interaction between the Ordnance Survey and the various Royal Commissions for Historic Monuments, such that easily visible archaeological sites are clearly marked on the maps and their legends [49,50].

There is growing interest in the history and execution of the mapping projects carried out in various imperial dominions, particularly those of the UK [8,51–54], but also the former USSR. The inclusion of archaeological sites on Soviet era maps of parts of the former USSR and Afghanistan produced by the Soviet Military Topographic Service has been noted [5,28,55], but these maps have been used in relatively limited areas and/or on a relatively small scale. It is notable that the types of sites that are easy to distinguish on these maps are clearly ancient sites with distinctive morphologies.

3. Mapping Territory and Surveying Archaeology: The Survey of India, the Archaeological Survey of India, and the 1″ to 1-Mile Map Series

3.1. Systematic Mapping in Nation States and Imperial Dominions: Trigonometry, Topography and Archaeology

The mapping of nation states and imperial dominions is a deeply political act, typically intertwined with military objectives. For instance, the roots of the UK's Ordnance Survey lie in the commencement of a systematic attempt to map the highlands of Scotland in 1747 after the Jacobite rising [49,56]. The establishment of the Survey of India soon followed in 1767 [51,52,57–59]. These august institutions are probably best seen as siblings—developed in parallel by government agencies that were interacting, and with objectives and innovations being transferred quickly between the two.

Accurate mapping soon became a key objective, and the Principal Triangulation of Britain, carried out between 1791 and 1853, was the first high-precision trigonometric survey of the whole of Great Britain and Ireland [56,59]. Shortly thereafter, the Great Trigonometrical Survey (GTS) set out to complete the more formidable task of measuring the entire South Asian Subcontinent between 1802 and 1871 [51,52]. This work of the GTS was inherently scientific and profoundly significant, but in many ways, it was controversial, particularly its cost in life and materials [51]. It nonetheless provided the framework for the more detailed mapping of topography that started 1831 [8,51], and was accompanied by the Revenue Survey, and occurred concurrently with the compilation of the *Gazetteers of British India: District Series* (1833 onwards [60]) and *The Imperial Gazetteer of India* [61].

The history of the Survey of India is a rich vein for scholarly enquiry, particularly as it was extensively documented [57,58,62–64]. Beyond consideration of its contribution to geodesy and cartography, it is ripe for the deconstruction of imperial ideology, the political impact of map making, and the conception of borders and frontiers [8,51–53]. Most scholarly attention has been directed to the period before 1843, which Edney [51], also [8] has referred to as a phase of "cartographic anarchy", when colonial surveyors struggled with the practical complications and epistemological challenges of surveying such extensive and (geographically and culturally) complex territory. In contrast, the period from 1843 to 1904 is regarded as one of consolidation and comprehensive survey [52,57,58], which was marked by progressive expansion of the territory controlled by the colonial authorities and the implementation of measures to systematically document the subcontinent district by district (see above).

The Survey of India 1" to 1-mile maps (Figures 1–3) were intended to document the landscape of British dominions in the Subcontinent systematically, and there are also 1" to 2-mile and 1" to 4-mile series. Complaints from the military authorities in 1905 resulted in the establishment of a committee to formulate a policy on map production to meet military and civil needs, and this Committee of 1905–06 devised a specific layout for these maps using contours and colours that continued in use up until the Second World War when the civil survey and publication programme was curtailed [57,58]. Comprehensive historical information about the progress of the survey and the publication of the 1" to 1-mile and associated series is restricted and not yet in the public domain. These maps were, however, issued on the basis of a grid of quadrangles that spanned the entire South Asian subcontinent and the surrounding regions (Figure 2).

The production of these maps in the early twentieth century occurred on the back of the topographic surveys that had started over 50 years before, and whose execution was documented in two editions of *The Manual of Surveying in India* [9,65]. The importance of these maps for both administrative and military purposes meant that specific guidance was given for the types of features that were essential to document, particularly rivers, streams, canals, populated areas, roads and areas suitable for military camps or positions ([9], pp. 441–448). As Thuillier and Smyth ([9], p. 281) noted "The conspicuous objects and all geographical items thus laid down, but all the different descriptions of land separated, viz. cultivation, waste, fallow, sites of villages and land fit for cultivation, the area of which being required for settlement purposes, is found by triangulation on the map". The geographical and topographic features were supposed to be documented using the same system and/or a consistent set of principles, and reference was made to a number of core European military surveying manuals ([9], p. 285).

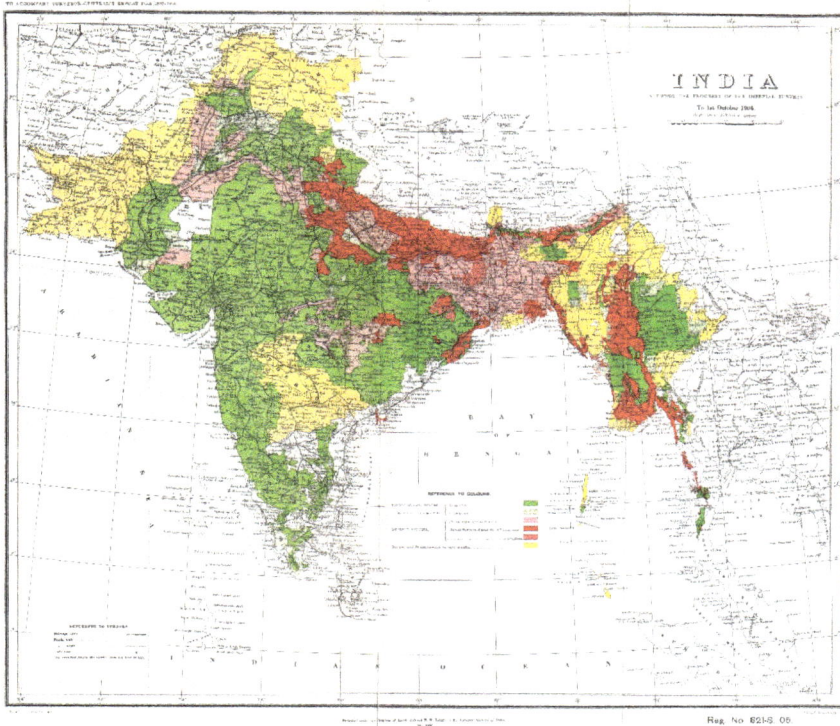

(a)

REFERENCE TO COLOURS.

(b)

Figure 1. (**a**) Map showing progress of imperial surveys up to 1st October 1904 (Image: Pahar.in; Public Domain [66]). (**b**) Close-up of the legend from the same map (Image: Pahar.in; Public Domain [66]).

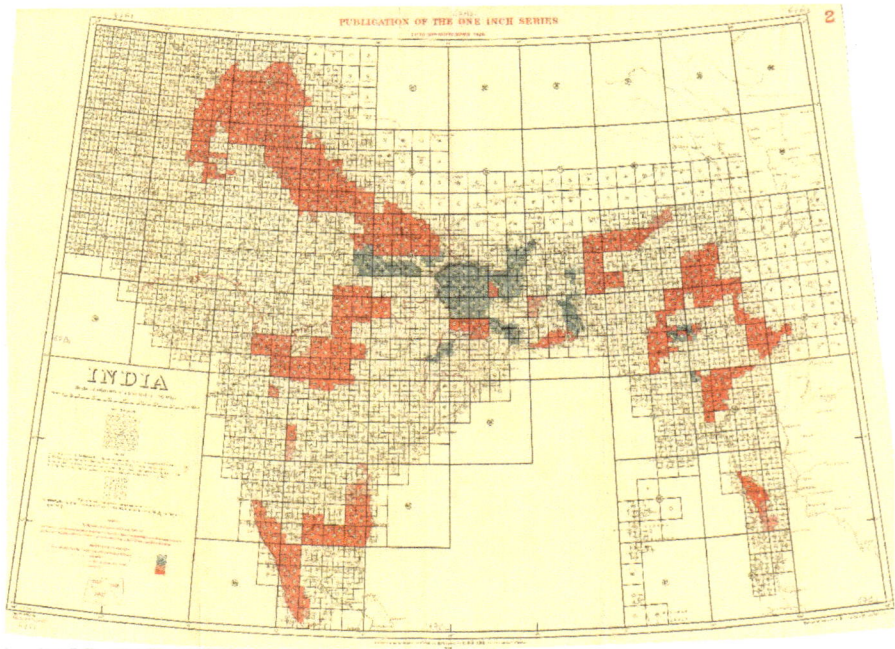

(a)

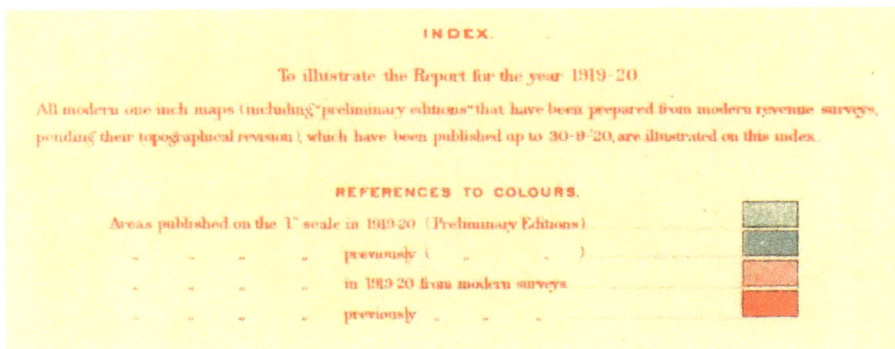

(b)

Figure 2. (**a**) Map showing structure of 1″ to 1-mile map series and its publication as at 1920 (Image: Pahar.in; Public Domain [66]). (**b**) Close-up of the legend from the same map (Image: Pahar.in; Public Domain [66]).

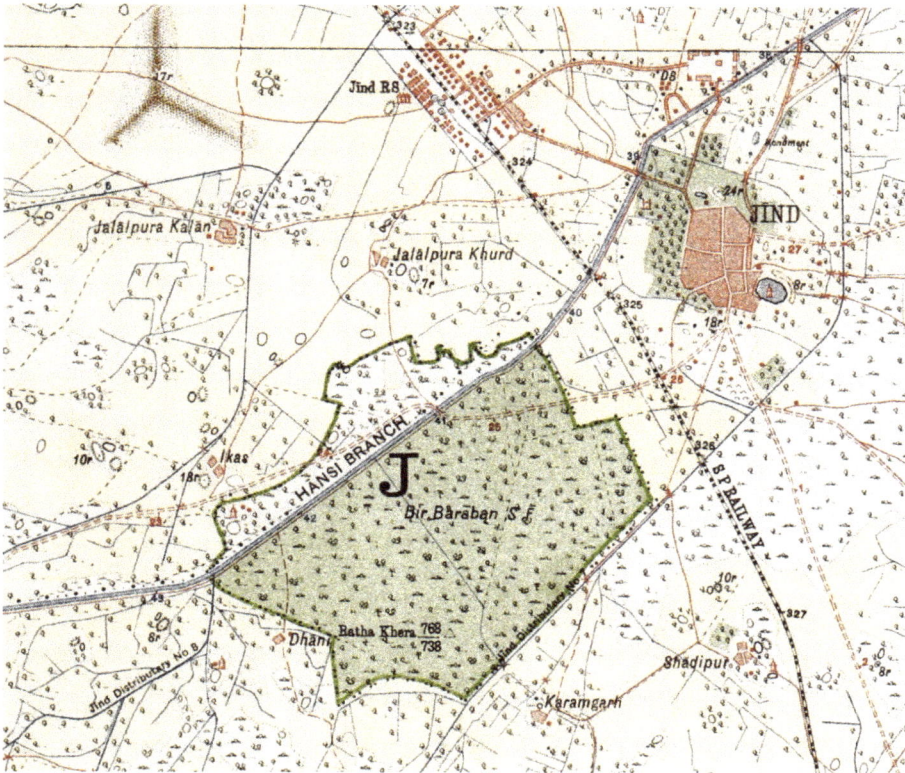

Figure 3. Detail from 1″ to 1-mile map sheet 53/C7/1915, showing the types and range of information being documented (e.g., built-up areas, railways, roads, ponds, irrigation canals, cultivated areas, forest, temples) (Image courtesy of Cambridge University Library).

Specific choices were made about line style and thickness to ensure that photography could be used for reproducing the maps, and certain types of land were not mapped, including plots of barren waste and jungle of less than 40 acres in area ([9], p. 312). For reasons that will become clear below, it is notable that specific guidelines were in place for documenting hills and elevated areas, and reference was made to 'vertical' and 'horizontal' styles:

"In the first method the shade is formed by strokes of the pencil or pen radiating from or converging into any curved part of a hill, according as it projects or re-enters:—they are supposed to describe the same course as water would do, if it would trickle in streams down the slopes, and are darker or lighter according to the steepness of the slope. The other method has the shade formed by lines parallel, or nearly so, to the horizon. It is much more easy to apply, and more natural than the former, and has some claim to particular notice from its easy application in sketching, and the facility with which it may be demonstrated and acquired. The horizontal manner marks the contours of hills by waving lines, each line continuing on the same level, while following every undulation of the ground. In practice, either or both of the styles may be used at the pleasure of the draftsman, or as may be best suited to the nature of the ground he wishes to portray" ([9], pp. 297–300)

Surveyors were thus granted a degree of agency in how they chose to depict hills or elevated ground, and they presumably adapted their symbology to suit the needs of depicting elevations on the relatively flat plains that predominate in many areas. Accuracy and precision were nonetheless critical, and "hilly features should invariably be put in, *by the Surveyor* himself, after a careful study of the ground, and without this personal examination in the field, it must be vain to attempt to give even an approximation to truth. It is evident that a map, to be anything, ought to be precise; it is otherwise worse than useless" ([9], p. 300).

During the period in which the systematic topographic mapping of the subcontinent was underway, Colonel Alexander Cunningham was appointed as the Archaeological Surveyor of the Government of India in November 1861. Cunningham had made the case that the government was honour bound to understand and preserve the ancient monuments of the British dominions in India. He argued that they must be "preserved by the accurate drawings and faithful descriptions of the archaeologist", and that it "would redound equally to the honour of the British Government to institute a careful and systematic investigation of all existing monuments of ancient India" ([67], pp. iii, iv). It is notable that the Archaeological Survey of India (ASI) was established over twenty years before the UK's Ancient Monuments Protection Act 1882, and 47 years before the various Royal Commissions on the Historic Monuments of Scotland, Wales and England (1908).

A major limitation of the early work of the ASI under Cunningham was that he and his subordinates largely documented settlements that were known from historical documents. Cunningham ([67], p. iv; [68], pp. xxi–xxxiv) started by following in the footsteps of the Chinese pilgrim Xuanzang, who travelled throughout South Asia in the seventh century AD. Cunningham also took account of other 'foreign' accounts of South Asia made by other Chinese pilgrims (e.g., Faxian, Song Yun), and the works of the Alexander historians (Arrian, Quintus Curtius) and their followers (Strabo, Pliny the elder, Ptolemy etc.). Furthermore, he also made use of the range of indigenous sources in various forms, including Vedic texts and the epic Ramayana and Mahabharata, as well sutras and astronomical works ([68], pp. xxxiv–lii). These approaches were paralleled in Arabia and the Levant, where the documentation of archaeological sites mentioned in the Bible, Classical, and Mesopotamian sources was a clear objective of the surveying that was carried out under the influence of the Survey of India ([54], pp. 42, 64, 113).

3.2. Parallel Developments in the Ordnance Survey and the Survey of India

Given that the ASI set out to complete a "systematic investigation of all existing monuments of ancient India", it is instructive to consider the interconnection between the work of the ASI and that of the Survey of India. It is also important to consider contemporary developments in the mapping of archaeological sites that were taking place elsewhere.

The early twentieth century Ordnance Survey maps of the UK typically documented visible archaeological features such as earthworks and tumuli [50,69,70]. There were, however, several occasions (e.g., Haverfield's 1906 lecture to the Royal Geographical Society on Hadrian's Wall) when it was made clear that there was a disconnect between the work of the Ordnance Survey and their depiction of archaeological sites on maps, and the professional knowledge of the archaeology that was being depicted ([50], pp. 22–23; [71]). Major change in the mapping of ancient sites in the UK only occurred when OGS Crawford (who was, interestingly, born in India) was appointed the first Archaeology Officer of the Ordnance Survey in 1920 ([50], p. 23). It is also notable that during this period, an Empire Conference of Survey Officers was inaugurated to bring together surveyors from across the British Empire [71,72], and it was just before the second event that Crawford [73] published a paper on mapping 'primitive English landmarks'. Survey of India representatives were present at the first Empire Conference [74], and it is almost certain that Crawford's paper would have been read by representatives of the Survey of India. Although knowledge of archaeology in the UK increased exponentially during the twentieth century, it is notable that the majority of what is now depicted on maps are features visible above the ground, and subtle and/or sub-surface features are not typically displayed. In the UK context, such subtle features make up a substantial proportion of the preserved archaeological landscape, which has effectively left them 'invisible' on maps, even though they are well understood through the extensive use of aerial archaeology and geophysical prospecting.

While a substantial proportion of the archaeology in the UK is not visible on the surface, this is not the case for much of the Subcontinent. Much like areas in the ancient Near East, ancient settlements in India and Pakistan were typically made of mud-brick or fired brick, resulting in the creation of mounded sites. Such sites often have variations of local terms in their names that reflect that fact (e.g., khera, tibba, tibbi, tepe, tul, theh, tol, tell), and there are instances where such features are labelled on maps (e.g., Ratha Khera, which is indicated on Figure 4, and is also the location of a trigonometric point). There were almost certainly other types of archaeological sites that do not preserve as mounds, but mounds are the most visible type of archaeological remains.

3.3. Incidental Documentation of Archaeology in the Survey of India 1" to 1-Mile Maps

There has been a long tradition of archaeological survey in South Asia, which has recorded numerous site locations [75–78]. While substantial numbers of 'proto-historic' sites are now known, up until the large scale survey projects started in the 1970s, the extent of the distribution of these elevated mounds was largely unrecognised. Of the large number of settlements now known to relate to the Indus Civilisation (*c*.3000–1500 BC), for example, it is notable that only the ancient city at Harappa was known in the initial period of topographic survey in the nineteenth century AD, and that was largely because it had been described by Charles Masson and Alexander Burnes in the earlier nineteenth century ([79], p. 6). Rather than being able to conceive that this mound might be pre- or proto-historic in date, Cunningham ([68], p. 241) suggested that it may have been the location of "another city of the Malii" mentioned by Arrian. A number of other early Indus settlements were also discovered at that time (e.g., Sutkagen-dor, Dabar Kot, Periano Ghundai, Rana Ghundai ([76], pp. 53–56), but they are situated in areas that were only subjected to preliminary reconnaissance or low-resolution survey prior to the Second World War.

There was some general awareness amongst the surveyors of the Survey of India of major archaeological sites, as the Survey of India 1" to 1-mile series maps document the archaeological sites that had been documented by the Archaeological Survey of India and/or were known at the time of the production of the maps of the during the early decades of the twentieth century. There appears to have been a consistent approach to the symbology that was used to depict the mound sites, which was presumably a result of the actions of the Committee of 1905–06. There was, however, a degree of evolution in the way that legends were depicted, with specific differences evident in maps from 1915 (Figure 4a) and those from 1936 (Figure 4b), where 'Antiquities' are indicated at the top right of the latter. These same differences are evident in the legends of the Ordnance Survey maps from the same period. Such sites are indicated on the maps through the use of gothic script (e.g., Harapâ [80], pp.105–108); Sites of Ancient Cities of Harappa, 44/B14/1936; Figure 5a), which is again something that appears in Ordnance Survey maps of the same period and up to the present. Furthermore, archaeological sites are mentioned in a number of District Gazetteers, and the Director General of the ASI and other subordinates are thanked for providing relevant information [81]. Significantly, the well-known archaeological sites that were comprised of elevated mounds were not typically depicted through the use of contours, but through the use of 'form-lines', which are 'lines drawn on a map to indicate the estimated configuration or elevation between the contour lines' (Oxford English Dictionary) (Figure 5a). Furthermore, elevated mounds that were not then known as archaeological sites, including the major Indus Civilisation city of Rakhigarhi, were also documented using 'form lines' (Figure 6a). Although there has been considerable development adjacent to both sites, they are still visible in modern satellite photos (Figures 5b and 6b).

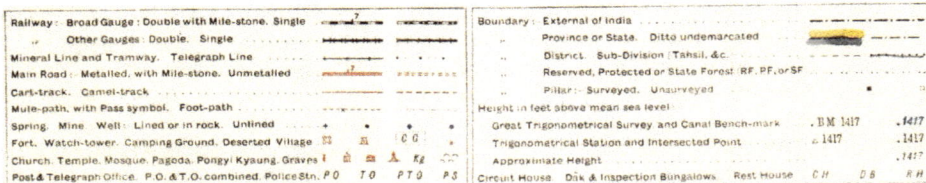

(a)

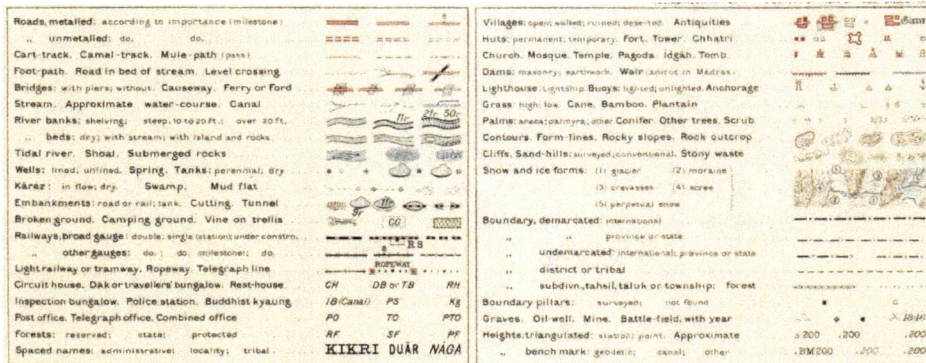

(b)

Figure 4. (**a**) Reproduction of legend from a 1915 map (53/C3/1915); (**b**) Reproduction of legend from a 1936 map (53/D9/1936) (Images courtesy of Cambridge University Library).

Examination of a selection of Survey of India 1″ to 1-mile maps from the region of northwest India has shown that a range of symbols were used to depict the elevated ground that has the potential to be archaeological mound sites. In addition to 'form-lines', these symbols include 'sand-hills' (either 'surveyed' or 'conventional'), vertical hachures, which are not shown on the standard legend, and also areas of 'graves', which are often visible on top of areas indicated with 'form-lines' (Figure 4b). These graves are typically historic and/or contemporary Muslim cemeteries and not particularly ancient. There is also a distinctive 'X' symbol used to depict 'deserted villages', and these are sometimes accompanied by a note stipulating '(old site)'—which have been observed on the 1″ to 1-mile maps sheets for Gujarat, and are also seen on the 1955 US Topo Map series, which were based on the Survey of India maps and produced in black-and-white.

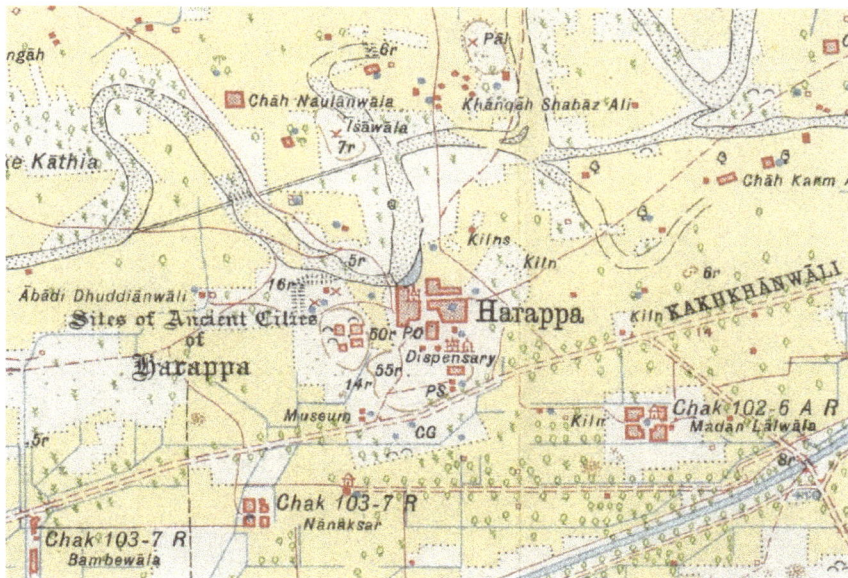

(a)

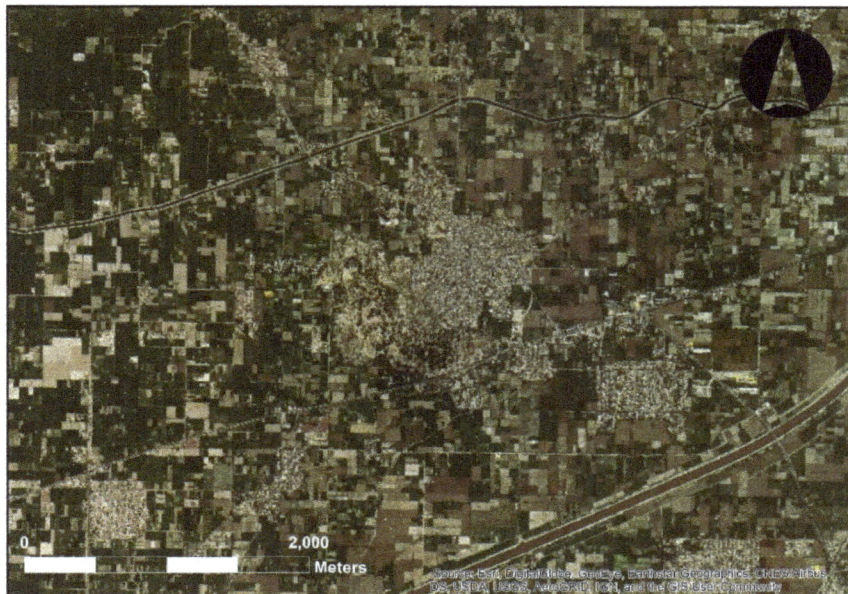

(b)

Figure 5. (a) Detail showing documentation of the "Sites of Ancient Cities of Harappa", including depiction of mounds delineated by 'form-lines', graves, and 'x' signifying deserted villages, and their relationship to the modern town (44/B14/1933) (Image courtesy of Cambridge University Library); (b) ESRI world imagery of the same region today.

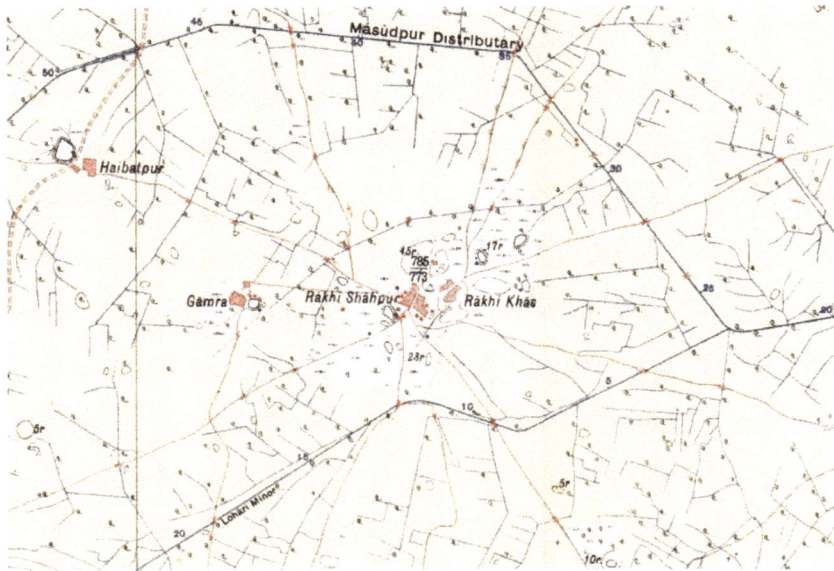

(a)

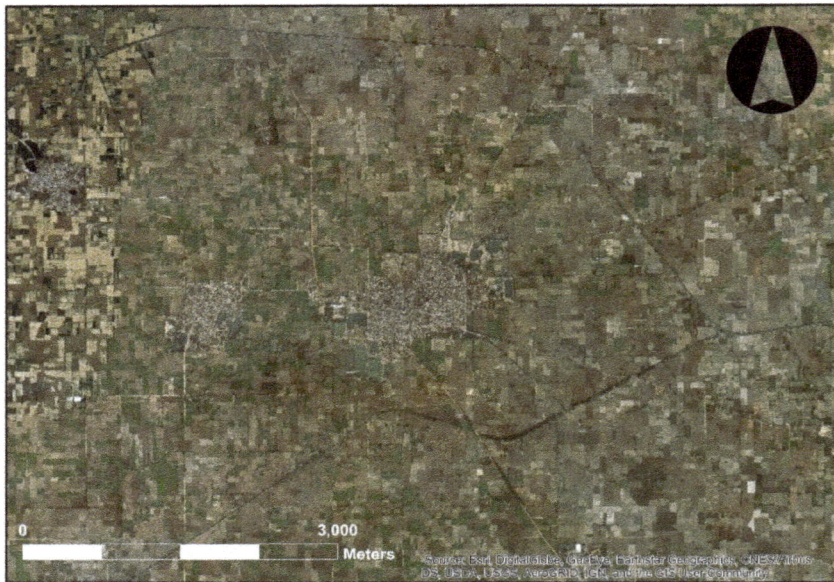

(b)

Figure 6. (a) Detail showing documentation of mounds of the Indus city of Rakhigarhi and their relationship to the modern towns of Rakhi Shapur and Rakhi Khas (53/C3/1915) (Image courtesy of Cambridge University Library); (b) ESRI world imagery of the same region today.

Although the work of the Survey of India in the early twentieth century was taking place at an enormous scale and with considerable speed, the work was systematic and the survey methods that were used resulted in the incidental location and documentation of thousands of features that may have an archaeological origin. As far as we can ascertain, it is likely that the surveying teams, and the administrators who superintended them, had limited to no awareness of the archaeological significance of these small, elevated features that were being recorded with considerable fidelity. However, it is notable that Rondelli et al. [28] have argued that Soviet surveyors knew that they were documenting archaeological sites when they were working in Central Asia. The Survey of India currently regards the sites marked with an 'X' as being "A site that was inhabited in the past, but found uninhabited during field survey" (Surveyor General of India, personal communication 2015). It is thus possible that the Survey of India teams had a vague awareness that what they were documenting was historically significant.

Irrespective of whether the surveyors knew that they were recording archaeological sites, it is clear that the Survey of India 1" to 1-mile maps provide a major data source for identifying and locating mound shaped features that might be archaeological mound sites in advance of on-the-ground survey projects. There are however, a number of challenges and limitations to using this resource, and to overcome these we have developed methods to (a) geo-reference the 1" to 1-mile maps accurately, and (b) extract location information that can be ground-truthed to assess whether or not an archaeological site is present and ascertain its date.

4. Geo-Referencing Survey of India 1" to 1-Mile Maps

In order to make use of the Survey of India 1" to 1-mile maps as a prospecting resource we have established methods for georeferenciation using both ArcMap and QGIS georeferenciation tools (a plugin using GDAL in the case of QGIS). The maps themselves, and the processes of reproduction, were comparably accurate with the Ordnance Survey maps of the same period, and likely used the Everest 1830 spheroid. They were, nonetheless, printed on paper and have also subsequently been held in library collections where they are often folded, which creates the potential for distortion, particularly in digitisation processes that involve photography or scanning, and both of these can also introduce distortion. All of these factors affect the process of georeferenciation.

Using WGS84 as the geodetic datum, the first step in the georeferenciation process consisted of the selection of two points from opposite corners at the graticule of meridians and parallels available at the frame of the map (the corner of the map frames are always coincidental with every 15 minutes on a lat/long grid, so each map covers $\frac{1}{4}$ of a degree). The coordinate values of these points transformed to decimal degrees were then inserted manually into the georeferenciation tool. These steps produced a rough initial georeferenciation that made it possible to scale, orientate and situate the map in its approximate geographical location. It is notable that the use of coordinates from the map graticule (even when employing four or six points geometrically distributed) were not accurate enough when compared with independently geo-referenced imagery, and produced large deviations across the map. After this first step, a georeferenciation method using target data from an already georeferenced source was adopted. Typically, ground control points (GCPs) were obtained in ArcGIS and QGIS using their world imagery map services. Both feature high-resolution aerial imagery that can be employed to identify features that were extant at the time that the initial surveying for the maps was carried out.

A minimum of 20 GCPs distributed evenly across each map was considered necessary for a reliable georeferenciation. GCPs were obtained from clearly delimited structures and features visible on each map sheet and modern remote sensing imagery, particularly canals, old structures and intersections in villages. Given the degree of modifications that have affected the whole of the study during the last hundred years and the scale of the maps, it was often difficult to find reliable GCPs. Many elements of the landscape have disappeared since the late nineteenth century AD, and most urban areas have changed dramatically, such that their layout as shown in the maps is no longer recognisable. The most useful elements for the retrieval of GCPs were crossings within the water channel network. Channels were carefully mapped during the nineteenth century AD, due to a combination of their lineal nature, the detailed surveying that went into their construction, and the core objectives of the original surveys, all of which mean that they are a source of reliable GCPs. Much of the large network of channels that is still being used has existed since the late nineteenth and early twentieth centuries, including many crossings and junctions. These offer well-distributed and reliable GCPs that can be correlated to the features still preserved today that are clearly visible and easy to identify in high-resolution aerial and satellite imagery. For similar reasons regional roads and railroads were preferred elements for the extraction of GCPs. For areas that lacked these features, elements such as crossings of local roads or the central points of villages were selected.

After the selection of GCPs and the creation of links between map features and their real coordinates in high-resolution satellite imagery, an evaluation process for ascertaining the reliability of the georeferenciation followed for each map was developed. The initial four points that were used to position in the map frame were eliminated as they introduced large errors, and the residual error values for each of the remaining GCPs' (the difference between the location provided as a result of the transformation and the actual coordinates of the GCP) were calculated. The root mean square error or RMSE is a measure of how well GCPs conform to the real coordinates using the root mean square sum of all GCPs' residual error values. RMSE values are thus dependent on the transformation method employed, that is, the mathematical approach adopted to deform the image to adapt it to the distribution of real coordinates. The coordinates of the selected GCPs and those of the corresponding points from the original image map file were stored in a table. This table was used to establish the links between GCPs and original points and, during the same process, to calculate residual error for each GCP and RMS value for the whole map using different transformation methods.

The transformation methods chosen for the rectification and georeferenciation process took into account the characteristics of the original 1″ to 1-mile maps. Zero and first order polynomials, Helmert transformation and other similar methods were not adequate given the internal distortions in the maps. These distortions were not just due to the quality or accuracy of the maps but, more importantly, to the types of affects that the paper have suffered over time (all maps presented folding marks) and in the digitisation process (maps digitised in 2008–2009 were photographed in keeping with University Library practices of the time, and those digitised from 2010 onwards were scanned using a drum scanner).

Thankfully, the maps maintained a high degree of integrity and their high quality and precision advised against the use of higher transformation orders that can produce important deformations in the margins and in areas where no GCPs are available. Consequently, two transformation methods were tested: a second order polynomial, which is a standard method in the georeferenciation of maps, and the *Adjust* transformation, which is an algorithm implemented in ArcMap that combines a polynomial transformation and triangulated irregular network interpolation approaches [82]. In most cases, the *Adjust* transformation was superior to the results of a second order polynomial and, therefore, this method was employed to generate rectified georeferenced maps.

As shown in Table 1, the georeferenciation process produced maps with an average RMSE of 0.000430° (equivalent to around 47 m at this latitude) using a second order polynomial with values between 0.000641° and and 0.000141°. Using the *Adjust* transformation the average RMSE was 0.000102357° (*c*.10.3 m) with values between 0 and 0.000244° (*c*.26.8 m). The rectified maps produced using the *Adjust* transformation were considered to have enough spatial accuracy to extract features that could be checked during fieldwork. With a maximum RMS value of 26.8 m (which could have been larger depending on the area explored in each individual map), these provided accurate locations for the central point of mounds. Mounds, being systematically much larger than these error values, would consistently fall within the location suggested in the rectified maps.

Table 1. Calculation of the root mean square error for a selection of georeferenced Survey of India 1″ to 1-mile maps using second order polynomials and the *Adjust* transformation. GCPs: ground control points.

Map	Number of GCPs	Second Order Polynomial	Adjust Transformation	Mound Features Identified
44J8 (1914)	26	0.000311	0.000194	✓
44J12 (1914)	22	0.000498	0.000218	✓
44J15 (1914)	29	0.000311	0.000109	✓
44J16 (1914)	28	0.000501	0.000165	✓
44K14 (1915)	26	0.000463	0.000051	✓
44N3 (1913)	22	0.000430	0.000215	✓
44N4 (1913)	22	0.000354	0.000151	✓
44N6 (1913)	18	0.000469	0.0000927	✓
44N7 (1914)	18	0.000616	0.000201	✓
44N8 (1913)	19	0.000442	0.000244	✓
44N10 (1913)	21	0.000500	0.000139	✓
44N11 (1913)	19	0.000503	0.000141	✓
44N12 (1913)	20	0.000641	0.000175	✓
44N14 (1913)	17	0.00063	0.00021	✓
44N15 (1913)	25	0.000375	0.00015	✓
44N16(1913)	20	0.000504	0.000171	✓
44O1 (1913)	27	0.000299	0.000166	✓
44O2 (1915)	39	0.000472	0.000148	✓
44O3 (1915)	34	0.000421	0.0000401	✓
44O4 (1916)	31	0.000406	0.000147	
44O5 (1915)	40	0.000636	0.000181	✓
44O6 (1915)	38	0.000447	0.000113	✓
44O7 (1915)	40	0.000354	0.000130	✓
44O8 (1916)	31	0.000514	0.000151	
44O9 (1915)	31	0.000514	0.000151	✓
44O10 (1913-14)	30	0.000361	0.0000978	✓
44O11 (1913-14)	30	0.000393	0.00016	✓
44O12 (1915)	34	0.000636	0.000219	✓
44O13 (1913-14)	34	0.000565	0.000122	✓
44O14 (1913-14)	34	0.000470	0.0000497	✓
44O15 (1913-14)	34	0.000503	0.0000944	✓
44O16 (1915)	33	0.000537	0.000147	✓
44P9 (1917)	26	0.000416	0.0000210	
44P13 (1912)	28	0.000333	0.0000886	
53B2 (1914)	24	0.000325	0.000049	✓
53B3 (1914)	22	0.000382	0.000115	✓
53B4 (1914)	19	0.000531	0.000244	✓
53B7 (1914)	34	0.000485	0.000108	
53B8 (1914)	20	0.00078	0.00022	
53B11 (1914)	32	0.000159	0.000382	
53B12 (1914)	31	0.000425	0	
53C1 (1912-13)	34	0.000532	0.000109	✓
53C2 (1912-13)	36	0.000493	0.0000907	✓
53C3 (1915)	36	0.000502	0.000201	✓
53C4 (1937)	34	0.000349	0.000119	✓
53C6 (1914-15)	33	0.000350	0.0000607	✓
53C7 (1915)	36	0.00058	0.000127	✓
53C8 (1937)	28	0.000302	0.0000676	
53C9 (1914)	26	0.000382	0.0000610	
53C10 (1914)	40	0.000420	0	
53C11 (1914)	32	0.000382	0.000159	
53C11 (1915)	31	0.000559	0.000106	
53C12 (1937)	29	0.000303	0	
53C14 (1914)	32	0.000482	0.000180	
53C15 (1915)	37	0.000500	0	
53C16 (1937)	37	0.000279	0	
53D5 (1937)	30	0.000208	0.0000327	
53D9 (1936)	35	0.000183	0.0000260	
53D13 (1936)	34	0.000141	0.0000283	
53C5 (1914)	32	0.00045	0.0000116	✓
53G2 (1915)	27	0.000355	0.000111	
53G3 (1915)	26	0.000421	0.000137	
53G4 (1937)	34	0.000403	0.0000776	
53H1 (1937)	33	0.000299	0.000202	

Each individual map sheet in the Survey of India 1″ to 1-mile series covers an area of 683.610 km², and the resolution of the images used meant that there was a pixel size of between 2.69 m × 2.68 m to 2.65 m × 2.73 m, though for some images this reached 5 m. The area covered by the 64 maps shown in Table 1 is approximately 27,000 km², and spans large parts of modern Haryana and Punjab in northwest India (Figure 7). This area is archaeological significant and was selected as it spans a variety of different climate zones and distinct ecological contexts [83].

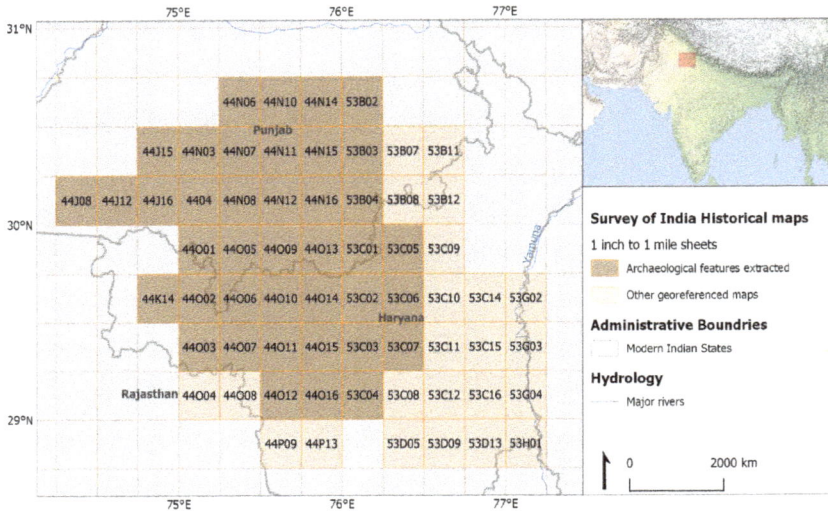

Figure 7. Geographical location of the georeferenced Survey of India 1″ to 1-mile map sheets used in this study.

5. Identifying mound features on Survey of India 1″ to 1-mile maps

With the maps georectified, it is possible to extract features of archaeological interest with a high degree of accuracy, and the manual detection and digitisation was carried out by several of the co-authors. Two GIS shapefiles were created to store the results of the digitisation, one for points and another for linear features. Both shapefiles had an associated table with columns destined to store information about the digitised features. These information categories included location coordinates, feature type, relief height and size, which were chosen to offer the maximum possible detail about the original features to assist future analysis (see below).

The point layer shapefile was dedicated to the digitisation of mound features depicted in the maps and recorded a single point per mound located in the geometrical centre of the feature. The table associated with this layer included information about the individual maps, edition, and year of publication of the map from which each point was retrieved, but also important data about each individual feature that was recorded. The type and colour of the line used to represent the mound features was noted as it provides an important indication of how the surveyor perceived the feature while in the field. As noted above, the methods used by the Survey of India meant that features and/or mounds that were similar in terms of size, shape and height might have been drawn using different methods, including 'horizontal' continuous contour lines, discontinuous 'form-lines', shaded relief, 'vertical' hatching or using a combination of these approaches (see below).

The size of the mounds was measured using the largest axis of the feature, and we opted to divide them into three categories: (1) mound features measuring up to 200 m, which were the most common and typically measure around 100 m; (2) mound features between 200 and 400 m, and (3) mound features with diameters of more than 400 m. As many of the Survey of India 1″ to 1-mile maps provide information on the relative height of the mound features in feet, it was also often possible to include a measure of height, which in combination with the measure of sizes provides a useful way to characterise the volume and character of each mound feature. Spot heights on these mounds were taken during the process of surveying, and the elevation of the mounds may have made them suitable places for circuit stations or survey points.

The second shapefile layer aimed at recording linear features. These appear to consist mostly of earthworks and relict field systems, and lines were recorded following the axis of the features. Apart from the information relative to the map from which linear features were extracted, the table associated to the linear feature shapefile layer included information on the type of line and relative height of the feature. The tables for both layers also included a field in which notes about the digitisation process could be included.

Digitisation was carried out using both ArcMap and QGIS. The manual digitisation of map features followed a systematic grid to ensure no areas were left uninspected. Observation of individual sheets and sets of sheets showed that mounds could be represented using three main approaches:

(a) 'shaded mound features', which were delineated with graded stippling and are perhaps the most common way of representing mounds, particularly those in the small size (1) category (Figure 8a). According to the Survey of India 1″ to 1-mile map legends, these features were defined as 'sand-hills', and the darker shading around the edges and/or a spot height indicates that they were formally surveyed (Figure 8a);

(b) 'form-line mound features', which were delineated using a discontinuous horizontal brown or black line (Figure 8b) that was usually employed for medium size (2) and large size (3) mounds. The use of discontinuous 'form-lines' indicates that they represent clear areas of elevation. The use of the horizontal rather than vertical lines to depict elevation is potentially related to the shallowness of the slope. It is likely that the form-lines were used as break-lines marking the transition between the flat plain and the elevated mound. The presence of a spot height presumably marks the highest point of the mound. In the process of hill-sketching, Surveyors were advised to use 25 feet contours ([9], p. 557), so an elevation of up to 7.5 m would have only warranted one contour. The choice to use form lines rather than contours to depict such mounds is interesting, as it suggests that they were recognised as not being natural hills, but there are no clear references to methods for depicting such features in either edition of *A Manual of Surveying for India* [9,65];

(c) 'hachure mound features' were delineated using vertical lines to depict elevation, which was potentially related to the steepness of the slope. Vertical hachures were sometimes used to represent small (1) and medium size (2) mounds (Figure 8c).

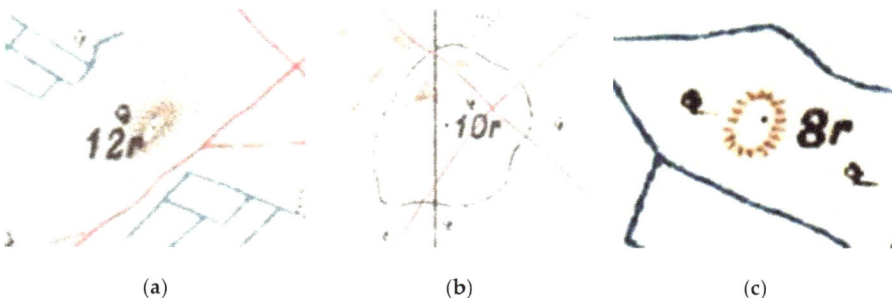

(a)   (b)   (c)

Figure 8. (**a**) Shaded; (**b**) Form-line; (**c**) Hachure mound features (Images courtesy of Cambridge University Library).

Figure 9 shows the range of mound feature types that can appear on one sheet. This includes instances where a combination of approaches was employed to represent a mound feature, and in these cases, we used a 'combined' category.

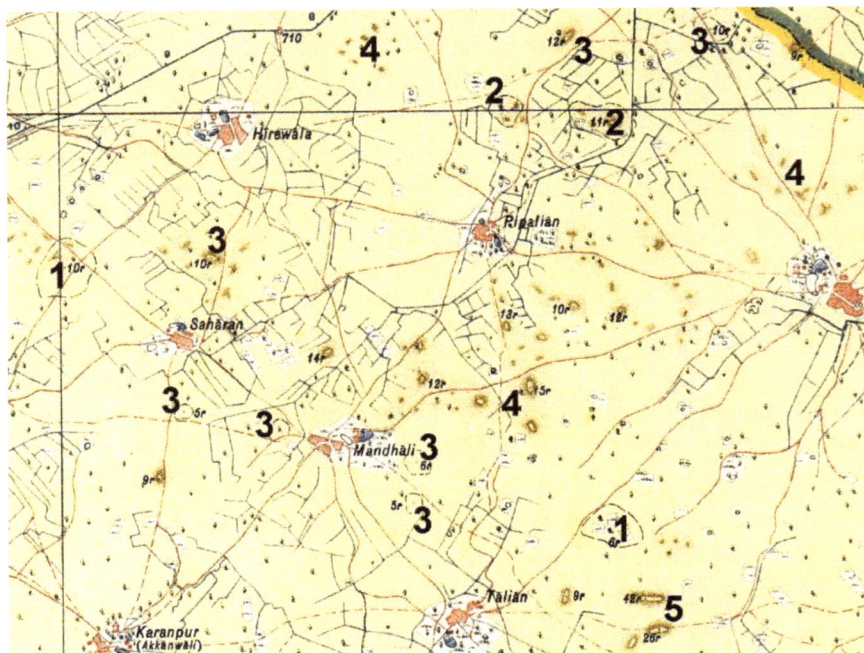

Figure 9. Survey of India 1″ to 1-mile map sheet of 44/O5/1915. Large mounds delineated by form-lines are labelled 1, and 'combined' features showing both form-lines and shaded relief are labelled 2. Medium-sized mounds delineated by either shading or form-lines are labelled 3, while clusters of small mound features delineated with shading are labelled 4. Linear mound features that may be traces of elevated roads or sand dunes are labelled 5 (Base map image courtesy of Cambridge University Library).

The use of continuous black lines to demarcate small features can be confusing, but it appears that these typically represent ponds (Figure 10a,b). When ponds contained water they were coloured blue (Figure 10a), while those that were dry at the time of survey were represented as a continuous black line without a colour fill (Figure 10b). Lines formed of points also appear, and appear to represent areas under cultivation (Figure 10c, also Figure 10a).

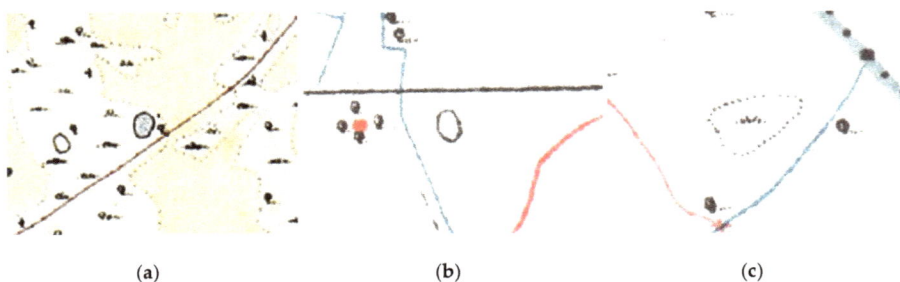

(a)   (b)   (c)

Figure 10. (a) Full pond; (b) Dry pond; (c) Dotted line feature (Images courtesy of Cambridge University Library).

Figure 11a,b show the types of relationships that existed between some linear and mound features. Mound features are visible to the top left of Figure 11a [(b) form-line] and towards the centre of Figure 11b [(c) hachure marked with a relative height of 8 feet]. It is evident that some the lineal mound features follow the orientation and direction of then-modern roads, suggesting that they potentially follow earlier routes. Figure 11b also shows how water channels go around mound features, potentially because of the topography created by the mound, which needed to be avoided for the water to flow. Figure 11c shows a combined mound feature, where a discontinuous form-line co-occurs with graves. In these instances, it is likely that the graves were relatively modern, and were excavated into an existing mound feature.

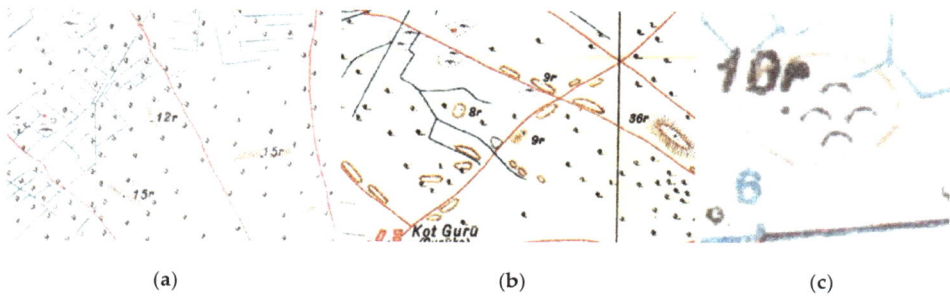

(a)   (b)   (c)

Figure 11. (**a**) Form-line and linear shaded mound features with spot heights on the same sheet; (**b**) Hachure and linear shaded mound features with spot heights and close to roads on the same sheet; (**c**) Combination of mound and graves (Images courtesy of Cambridge University Library).

Almost 9000 mound features (Table 2) have been identified on 40 of the 64 Survey of India 1″ to 1-mile map sheets listed in Table 1 (Figure 7). Within the area investigated for this study, shaded, form-line and hachure features were all common, but there is considerable variability in the number and size categories of mounds attested on individual sheets. The size 1 features in each category were the most abundant, but proportionally significant numbers of size 2 and 3 features were also located.

Table 2. The mound feature type and size categories of features recorded on the georeferenced maps.

Type	Total	Size 1	Size 2	Size 3
Shaded	3143	2792	291	57
Form-line	2056	1545	405	102
Hachure	3699	3596	90	1
Combined	32	7	17	8

The variability in mound feature occurrence indicates that it is not yet statistically robust to attempt to produce summative data on the average number of mound features and the frequency of specific types per sheet, or to consider extrapolating this across a larger area. In future, when sheets that cover the full range of climate and ecological zones have been assessed, it will be possible to carry out predictive modelling of the number of mound types and size categories that might be expected in different zones.

6. Testing the Archaeological Viability of the Identification of Mound Features

The viability of the categorisation outlined here has been tested in the field over several field seasons [84–87] (Figure 12), and the detailed results of this analysis are the subject of a separate paper [88]. To facilitate the process of ground-truthing, each feature location was included in a field survey table and assigned Historical Feature Identification Numbers (hf_id), which will be used to create a comprehensive listing of preserved and potentially lost archaeological sites. Each hf_id was accompanied by information about its feature type and size category to allow the field survey team to assess the probability that a feature identified on the historical maps remains identifiable in the contemporary landscape, and the probability that the hf_id is (or more correctly, was) an archaeological site.

A sample of mound features in the size (2) or size (3) (i.e., those greater than 200 m across on the 1″ to 1-mile maps) categories were visited and assessed to determine whether they were extant archaeological sites. This resulted in the identification of in excess of 200 archaeological sites within a delimited study area that were previously unknown [86–88]. In the early stages of this survey, it became clear that the smallest features on the historical maps were rarely preserved. As a result, in each survey unit, historical mound features in the size (1) category were visited until at least ten features tested negative. In many instances, this meant that every historical mound feature identified in a survey unit was visited, but in some, several size (1) historical mound features were not visited as their likelihood of being a preserved archaeological site was demonstrably low. It is clear that it is not feasible to simply assume that all mound features were archaeological sites, not least because there is variation in the detail and clarity of each feature, and some of these features are almost certainly geomorphological in origin. Importantly, there is significant variation in the number of mound features on each map, as some have 400+ features, while others have as few as 60. This variability by sheet no doubt reflects variation on the ground, with some areas having more features, and potentially more sites. In addition to checking mound features, it is also possible to visit locations that include topographic words within the name to ascertain whether archaeological mounds are present. It has been noted previously that a number of modern villages in this region are elevated, and are likely built on top of archaeological sites. Ground-truthing is the essential component for demonstrating the veracity of the historic map dataset. It is also important to remember that many sites have been lost as agricultural development in the region has unfolded, which is a factor that will need to be considered in future studies.

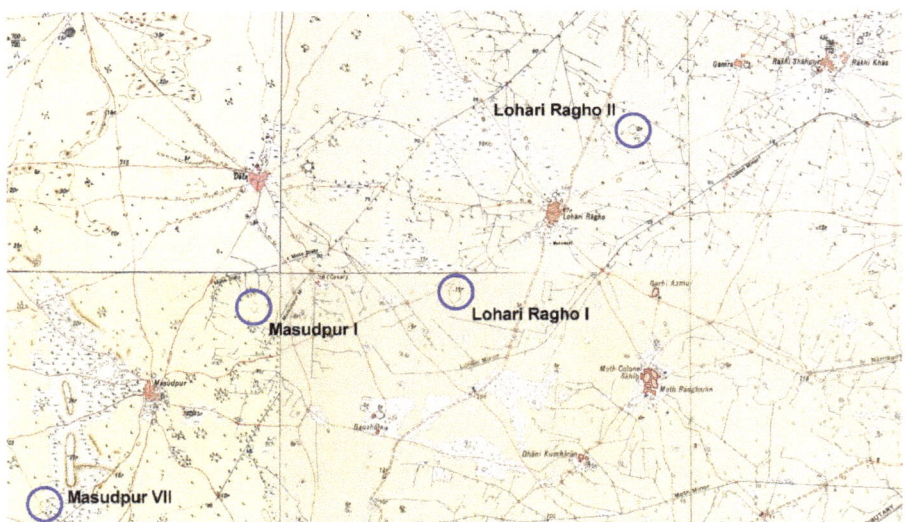

Figure 12. Four map sheets combined to show the area to the southwest of the Indus Civilisation urban site of Rakhigarhi (visible as a number of mounds under two modern villages at top right; also Figure 6), and the location of a selection of form-line mound features that correspond with surveyed archaeological sites (Base map image courtesy of Cambridge University Library).

7. Discussion and Conclusions

Imperial mapping projects in South Asia were inevitably geared towards the systematic documentation of the landscape to facilitate military domination, administrative control and economic exploitation. The Survey of India and the Archaeological Survey of India began as large-scale systematic documentation projects, and there are clear instances where information about archaeological sites that had been documented made its way onto 1" to 1-mile maps (e.g., Figure 5). There are, however, also numerous instances where otherwise undocumented archaeological sites were recorded on these maps (e.g., Figures 6 and 12), but the historical significance of these mound features does not appear to have been recognised formally at the time. It appears unlikely that the surveying teams were aware of the archaeological significance of what they were recording by and large, but the appearance of village names with some appellation of mound incorporated suggests that there was some awareness amongst local populations that these mounds were something unusual. Furthermore, many of these mounds are likely to have had significance for local populations as they were areas of sacred spaces as cemeteries, areas of economic exploitation – most particularly silt, sand and brick extraction - and their elevation means that they were not well suited for irrigation supported agriculture. This last factor also means that many mounds are currently under threat due to pressure from extensive and intensive farming and the ready availability of bulldozers.

These maps provide a new resource that can now be used to take major steps towards understanding long-term trajectories of human occupation in South Asia, and they will make it possible to develop new inclusive, comprehensive, and decolonised records of these evolving social landscapes. The systematic documentation of mound features visible on historic maps will make it possible to filter and query the data set at a later stage to select specific types of features, create thematic maps, quantify aspects of those features, and perform statistical analyses. Although substantial numbers of archaeological sites have already been documented across northwest India [75–78], attempts to conduct systematic survey using historic maps as a key data source has shown that the currently 'known' sites represent only a fraction of the actual archaeological settlements in the region [84–88]. Although a large number of the almost 9000 mound features visible in the historical maps that have been studied are likely to be natural, if even one tenth of them turn out to be archaeological sites, then the number of known sites in the region will increase dramatically. The Survey of India 1" to 1-mile maps thus have the potential to revolutionise our understanding of the archaeological landscapes of South Asia.

Beyond their use for identifying and locating hitherto unrecognised archaeological mound sites, these 1" to 1-mile maps have considerable potential for reconstructing other types of landscape features that are also often overlooked, including palaeochannel levees, relict sand-dunes, raised road-ways and relict field-systems [89]. This historical map data set can also be used to reconstruct historical landscape dynamics [90], and/or the development of land-use practices, hydrological schemes and irrigation, and urban growth from the late nineteenth and early twentieth centuries.

As with other remote sensing and aerial photography datasets, there are inevitably a range of limitations to the Survey of India 1" to 1-mile maps, not least the fact that some areas were mapped repeatedly, while others do not appear to have ever had maps produced and/or made publically available. There are, however, extant archives that contain maps of areas of archaeological interest, and further research will require the establishment of new collaborations involving scholars and governmental institutions in Pakistan, India and Afghanistan. The 1" to 2-mile and 1" to 4-mile maps typically cover the interstitial areas, and they also document important archaeological data, but these series are inevitably of lower resolution. Nonetheless, they are an additional data source that also needs to be considered.

It is important to acknowledge that the methods outlined here for making use of these Survey of India 1" to 1-mile maps are only likely to be useful for identifying mounded sites, and will not be suitable to aid detection of a wide range of other features of archaeological significance. Therefore, it is essential to integrate the use of these historical maps into comprehensive approaches that make use of the full suite of earth observation and remote sensing techniques, potentially integrating open-source multi-spectral data and the computational power of platforms like Google Earth Engine to identify hydrological and topographic features not easily visible on the surface [35,90,91]. It is also imperative that these remote prospection approaches are co-ordinated with large-scale ground-truthing surveys, that will verify which of the mound features are archaeological sites, and establish a reliable chronology for those sites and the associated landscapes. The combined analysis of these maps and the ground-truthing of the mound feature data will also make it possible to use machine learning-based approaches to carry out site detection across very large areas.

It is particularly timely to be able to add new evidence for the distribution of old settlements. The sobering truth is that in many instances, mound features in South Asia that were recorded in the early twentieth century are no longer extant, and in areas where farming is becoming increasingly mechanised and urban growth is unabated, mounds are disappearing with increasing speed. When such circumstances are in play archaeological sites and cultural heritage are clearly at-risk, and large-scale integrated mapping and survey projects cannot be commenced soon enough.

Author Contributions: Conceptualization, C.A.P., J.R.K. and R.N.S.; formal analysis, H.A.O., A.S.G., J.R.W., A.G. and F.C.; method, H.A.O.; writing—original draft, C.A.P. and H.A.O.; writing—review and editing, C.A.P., H.A.O., A.S.G., J.R.W., A.G., F.C., J.R.K. and R.N.S.

Funding: The *Land, Water and Settlement* project was primarily funded by a Standard Award from the UK India Education and Research Initiative (UKIERI) under the title "From the collapse of Harappan urbanism to the rise of the great Early Historic cities: investigating the cultural and geographical transformation of northwest India between 2000 and 300 B.C.". Smaller grants were also awarded by the British Academy's Stein Arnold Fund, the Isaac Newton Trust, the McDonald Institute for Archaeological Research and the Natural Environment Research Council (NERC). The *TwoRains* project has been primarily funded by the European Research Council under the European Union's Horizon 2020 research and innovation program (grant agreement no. 648609), but has also received support from DST/UKIERI, the British Academy and the McDonald Institute for Archaeological Research. The *WaMStrIn* and *Marginscapes* projects have been funded by the European Union's Horizon 2020 Research and Innovation programme under the Marie Skłodowska-Curie grant agreement no.s 746446 and 794711.

Acknowledgments: The research presented in this paper is the product of the long-term consideration of the importance of historic maps for understanding archaeological landscapes. Following initial steps made by J.R. Knox, the *Land, Water and Settlement* project made extensive use of these historic maps during reconnaissance and more extensive surveys, and the *TwoRains* project has developed a comprehensive method for using these maps as a large-scale data source for archaeological landscape reconstruction. The work of *TwoRains* has also been enhanced by two related projects entitled *WaMStrIn* and *Marginscapes*. The authors would like to thank the heads of the Department of AIHC and Archaeology at Banaras Hindu University and the Department of Archaeology at the University of Cambridge who have provided their support to the *Land, Water and Settlement*, *TwoRains*, *WaMStrIn* and *Marginscapes* projects. We would especially like to thank the staff of the Map Room and Imaging Services at the University Library at the University of Cambridge for providing access to and high-resolution copies of the Survey of India 1" to 1-mile maps in their collection.

Conflicts of Interest: The authors declare no conflict of interest.

References

1. Wilkinson, T.J. *Archaeological Landscapes of the Near East*; University of Arizona Press: Tucson, AZ, USA, 2003.
2. Petrie, C.A. Remote sensing in inaccessible lands: Plains and preservation along old routes between Pakistan and Afghanistan. *ArchAtlas* **2007**, *3*. Available online: http://www.archatlas.org/workshop/Petrie07.php (accessed on 8 November 2018).
3. Thomas, D.C.; Kidd, F.J. On the margins: Enduring pre-modern water management strategies in and around the Registan desert, Afghanistan. *J. Field Archaeol.* **2017**, *41*, 29–42. [CrossRef]
4. Hammer, E.; Seifried, R.; Franklin, K.; Lauricellaca, A. Remote assessments of the archaeological heritage situation in Afghanistan. *J. Cult. Herit.* **2018**, *33*, 125–144. [CrossRef]
5. Franklin, K.; Hammer, E. Untangling palimpsest landscapes in conflict zones: A "remote survey" in Spin Boldak, Southeast Afghanistan. *J. Field Archaeol.* **2018**, *43*, 58–73. [CrossRef]
6. Casana, J.; Laugier, E.J. Satellite imagery-based monitoring of archaeological site damage in the Syrian civil war. *PLoS ONE* **2017**, *12*, e0188589. [CrossRef] [PubMed]

7. Orengo, H.A.; Krahtopoulou, A.; Garcia-Molsosa, A.; Palaiochoritis, K.; Stamati, A. Photogrammetric re-discovery of the hidden long-term landscapes of western Thessaly, central Greece. *J. Archaeol. Sci.* **2015**, *64*, 100–109. [CrossRef]

8. Simpson, T. "Clean out of the map": Knowing and doubting space at India's high imperial frontiers. *Hist. Sci.* **2017**, *55*, 3–36. [CrossRef]

9. Thuillier, H.L.; Smyth, R. *A Manual for Surveying for India Detailing the Mode of Operations on the Trigonometrical, Topographical and Revenue Surveys of India*, 3rd ed.; Thacker, Spink and Co.: Calcutta, India, 1875.

10. Fairservis, W.A. *Excavations in the Quetta Valley, West Pakistan*; American Museum of Natural History: New York, NY, USA, 1956.

11. Wiseman, J.R.; El-Baz, F. (Eds.) *Remote Sensing in Archaeology*; Springer: New York, NY, USA, 2007.

12. Tapete, D. Remote Sensing and Geosciences for Archaeology. *Geosciences* **2018**, *8*, 41. [CrossRef]

13. Fowler, M.J.F.; Curtis, H. Stonehenge from 230 kilometres. *AARGnews* **1995**, *11*, 8–16.

14. Kennedy, D. Declassified satellite photographs and archaeology in the Middle East: Case studies from Turkey. *Antiquity* **1998**, *72*, 553–561. [CrossRef]

15. Kouchoukos, N. Satellite images and the representation of Near Eastern landscapes. *Near East. Archaeol.* **2001**, *64*, 80–91. [CrossRef]

16. Ogata, N. A study on cities, settlements and archaeological sites in the Silk Road region using satellite photos. *Bull. Res. Cent. Silk Roadol.* **2003**, *17*, 39–51. Available online: http://hdl.handle.net/10935/4395 (accessed on 9 November 2018).

17. Pournelle, J. Marshland of Cities: Deltaic Landscapes and the Evolution of Early Mesopotamian Civilization. Ph.D. Dissertation, University of California, San Diego, CA, USA, 2003.

18. Ur, J.A. CORONA satellite photography and ancient road networks: A northern Mesopotamian case study. *Antiquity* **2003**, *77*, 102–115. [CrossRef]

19. Fowler, M.J.F. Declassified CORONA KH-4B satellite photography of remains from Rome's desert frontier. *Int. J. Remote Sens.* **2004**, *25*, 3549–3554. [CrossRef]

20. Galiatsatos, N. Assessment of the CORONA series of satellite imagery for landscape archaeology: A case study from the Orontes valley, Syria. Unpublished. Ph.D. Dissertation, Durham University, Durham, UK, 2004.

21. Wilkinson, T.J.; Ur, J.; Casana, J. From nucleation to dispersal: Trends in settlement pattern in the Northern Fertile Crescent. In *Side-by-Side Survey: Comparative Regional Studies in the Mediterranean World*; Cherry, J., Alcock, S., Eds.; Oxbow Books: Oxford, UK, 2004; pp. 198–205, ISBN 1842170961.

22. Altaweel, M. The use of ASTER satellite imagery in archaeological contexts. *Archaeol. Prospect.* **2005**, *166*, 151–166. [CrossRef]

23. Hritz, C. The changing archaeoscape of southern Mesopotamia. In *GIS and Archaeological Site Location Modelling*; Mehrer, M., Wolcott, C., Eds.; Taylor & Francis: Boca Raton, FL, USA, 2005; pp. 413–436, ISBN 9780415315487.

24. Alizadeh, K.; Ur, J.A. Formation and destruction of pastoral and irrigation landscapes on the Mughan Steppe, north-western Iran. *Antiquity* **2007**, *81*, 148–160. [CrossRef]

25. Beck, A.; Philip, G.; Abdulkarim, M.; Donoghue, D.N. Evaluation of Corona and Ikonos high resolution satellite imagery for archaeological prospection in western Syria. *Antiquity* **2007**, *81*, 161–175. [CrossRef]

26. Casana, J.; Cothren, J. Stereo analysis, DEM extraction and orthorectification of CORONA satellite imagery: Archaeological applications from the Near East. *Antiquity* **2008**, *82*, 732–749. [CrossRef]

27. Mantellini, S.; Rondelli, B.; Stride, S. Analytical approach for representing the water landscape evolution in Samarkand Oasis (Uzbekistan). In *On the Road to Reconstructing the Past. Computer Applications and Quantitative Methods in Archaeology (CAA2008)*; Jerem, E., Redö, F., Szeverényi, V., Eds.; Archaeolingua: Budapest, Hungary, 2011; pp. 387–396, ISBN 9789639911307.

28. Rondelli, B.; Stride, S.; García-Granero, J.J. Soviet military maps and archaeological survey in the Samarkand region. *J. Cult. Herit.* **2013**, *14*, 270–276. [CrossRef]

29. Ur, J.; de Jong, L.; Giraud, J.; Osborne, J.F.; MacGinnis, J. Ancient cities and landscapes in the Kurdistan region of Iraq: The Erbil Plain Archaeological Survey 2012 season. *Iraq* **2013**, *75*, 89–118. [CrossRef]

30. Wright, R.P.; Hritz, C. Satellite remote sensing imagery: New evidence for site distributions and ecologies in the upper Indus. In *South Asian Archaeology 2007*; Frenez, D., Tosi, M., Eds.; BAR International Series 2454; Archaeopress: Oxford, UK, 2013; pp. 315–321, ISBN 9781407310626.

31. Conesa, F.C.; Madella, M.; Galiatsatos, N.; Balbo, A.L.; Rajesh, S.V.; Ajithprasad, P. CORONA photographs in monsoonal semi-arid environments: Addressing archaeological surveys and historic landscape dynamics over North Gujarat, India. *Archaeol. Prospect.* **2015**, *22*, 75–90. [CrossRef]

32. Ogata, N.; Yu, Z.; Ito, T.; Sohma, H.; Ideta, K. A study of settlement remains near the Qiemo Oasis in Northwestern China using Satellite Imagery and DEM. *Stud. Geogr. Reg. Environ. Res.* **2015**, *VIII*, 9–29. Available online: http://jairo.nii.ac.jp/0059/00005522 (accessed on 9 November 2018).

33. Hammer, E. New Research Directions in the CAMEL Lab. News and Notes, The Oriental Institute Members' Magazine. 2016, Volume 228, pp. 4–9. Available online: https://oi.uchicago.edu/sites/oi.uchicago.edu/files/uploads/shared/docs/Publications/nn228-web.pdf (accessed on 9 November 2018).

34. Hammer, E.; Lauricella, A. Historical imagery of desert kites in eastern Jordan. *Near East. Archaeol.* **2017**, *80*, 74–83. [CrossRef]

35. Orengo, H.A.; Petrie, C.A. Large-scale, multi-temporal remote sensing of palaeo-river networks: A case-study from northwest India and its implications for the Indus Civilisation. *Remote Sens.* **2017**, *9*, 735. [CrossRef]

36. Thomas, D.C. *The Ebb and Flow of the Ghūrid Empire*; Adapa Monographs; University of Sydney Press: Sydney, Australia, 2018; ISBN 9781743325414.

37. Fuentes, J.; Varga, D.; Pinto, J. The Use of High-Resolution Historical Images to Analyse the Leopard Pattern in the Arid Area of La Alta Guajira, Colombia. *Geosciences* **2018**, *8*, 366. [CrossRef]

38. Scollar, I.; Galiatsatos, N.; Mugnier, C. Mapping from CORONA geometric distortion in KH4 images. *Photogramm. Eng. Remote Sens.* **2016**, *82*, 7–13. [CrossRef]

39. Crawford, O.G.S. A century of air photography. *Antiquity* **1954**, *28*, 206–210. [CrossRef]

40. Kennedy, D.; Bewley, R. Aerial archaeology in Jordan. *Antiquity* **2009**, *83*, 69–81. [CrossRef]

41. Braasch, O. Goodbye Cold War! Goodbye Bureaucracy? Opening the skies to aerial archaeology. In *Aerial Archaeology: Developing Future Practice*; Bewley, R., Rączkowski, W., Eds.; IOS Press: Amsterdam, The Netherlands, 2002; pp. 19–22, ISBN 1586031848.

42. Bourgeois, J.; Meganck, M. (Eds.) *Aerial Photography and Archaeology 2003: A Century of Information*; Archaeological Reports Ghent University; Academia: Ghent, Belgium, 2005; ISBN 9789038207827.

43. Hanson, W.S.; Oltean, I.A. (Eds.) *Archaeology from Historical Aerial and Satellite Archives*; Springer: New York, NY, USA, 2013; ISBN 978-1-4614-4505-0.

44. Orengo, H.A.; Knappett, C. Toward a definition of Minoan agro-pastoral landscapes: Results of the survey of Palaikastro (Crete). *Am. J. Archaeol.* **2018**, *122*, 470–507. [CrossRef]

45. Grady, D. *Aerial Reconnaissance: Management of Research Projects in the Historic Environment*; MoRPHE Project Planning Notes (PPN) 5; Historic England: London, UK, 2008; Available online: https://historicengland.org.uk/images-books/publications/morphe-project-planning-note-5/heag028-morphe-ppn5/ (accessed on 9 November 2018).

46. Verhoeven, G.; Sevara, C.; Karel, W.; Ressl, C.; Doneus, M.; Briese, C. Undistorting the past: New techniques for orthorectification of archaeological aerial frame imagery. In *Good Practice in Archaeological Diagnostics: Non-Invasive Survey of Complex Archaeological Sites*; Corsi, C., Slapšak, B., Vermeulen, F., Eds.; Natural Science in Archaeology; Springer: Cham, Switzerland, 2013; pp. 31–67, ISBN 978-3-319-01784-6.

47. Doneus, M.; Geert Verhoeven, G.; Atzberger, C.; Wess, M.; Ruš, M. New ways to extract archaeological information from hyperspectral pixels. *J. Archaeol.l Sci.* **2014**, *52*, 84–96. [CrossRef]

48. McCarthy, J. Multi-image photogrammetry as a practical tool for cultural heritage survey and community engagement. *J. Archaeol. Sci.* **2014**, *43*, 175–185. [CrossRef]

49. Oliver, R. *Ordnance Survey Maps: A Concise Guide for Historians*; Charles Close Society for the Study of Ordnance Survey Maps: London, UK, 2005; ISBN 978-1-870-598-31-6.

50. Phillips, C.W. *Archaeology in the Ordnance Survey 1791–1965*; CBA: London, UK, 1980; ISBN 0900312904.

51. Edney, M.H. *Mapping an Empire: The Geographical Construction of British India, 1765–1843*; University of Chicago Press: Chicago, IL, USA, 1997; ISBN 9780226184883.

52. Barrow, I.J. *Making History, Drawing Territory: British Mapping in India, c.1756–1905*; Oxford University Press: Oxford, UK, 2003; ISBN 0195665465.

53. Chester, L.P. *Borders and Conflict in South Asia: The Radcliffe Boundary Commission and the Partition of Punjab*; Studies in Imperialism; Manchester University Press: Manchester, UK, 2009; ISBN 978-0-7190-9136-0.

54. Foliard, D. *Dislocating the Orient: British Maps and the Making of the Middle East, 1854–1921*; University of Chicago Press: Chicago, IL, USA, 2017; ISBN 9780226451336.

55. Ball, W. *Archaeological Gazetteer of Afghanistan/Catalogue des Sites Archéologiques d'Afghanistan*; Editions Recherche sur les Civilisations: Paris, France, 1982; ISBN 2-86538-040-8.
56. Hewitt, R. *Map of a Nation—A Biography of the Ordnance Survey*; Granta: London, UK, 2010; ISBN 978-1847082541.
57. Chadha, S.M. *Survey of India through the Ages*; Royal Institution of Chartered Surveyors and the Royal Geographical Society: London, UK, 1990.
58. Chadha, S.M. Survey of India through the ages. *Himal. J.* **1991**, *47*. Available online: https://www.himalayanclub.org/hj/47/2/survey-of-india-through-the-ages/ (accessed on 9 November 2018).
59. Roy, W. An account of the mode proposed to be followed in determining the relative situation of the Royal Observatories of Greenwich and Paris. *Philos. Trans. R. Soc. Lond.* **1787**, *77*, 188–226. [CrossRef]
60. Scholberg, H. *The District Gazetteers of British India: A Bibliography*; Inter Documentation: Zug, Switzerland, 1970; ISBN 9780800212650.
61. Hunter, W.W. *The Imperial Gazetteer of India*; Clarendon Press: Oxford, UK, 1881.
62. Black, C.E.D. *A Memoir on the Indian Surveys, 1875–1890*; E.A. Arnold: London, UK, 1891.
63. Phillimore, R.H. *Historical Records of the Survey of India*; Survey of India: Dehra Dun, India, 1945–1968; Volumes 1–5.
64. Kulkarni, M.N. Two centuries of surveying and mapping in India. *Surv. Rev.* **2006**, *38*, 452–458. [CrossRef]
65. Smyth, R.; Thuillier, H.L. *A Manual for Surveying for India Detailing the Mode of Operations on the Trigonometrical, Topographical and Revenue Surveys of India*, 1st ed.; Thacker and Co.: Calcutta, India, 1851.
66. Public Website. Available online: https://pahar.in/survey-of-india-report-maps/ (accessed on 8 November 2018).
67. Cunningham, A. *Four Reports Made During the Years 1862-673-64-65*; Archaeological Survey of India Reports; Archaeological Survey of India: Calcutta, India, 1871; Volume I.
68. Cunningham, A. *The Ancient Geography of India*; Chuckervertty, Chatterjee and Co.: Calcutta, India, 1871.
69. Murray, D. Ordnance survey and the depiction of antiquities on maps: Past, present and future: The current and future role of the Royal Commissions as suppliers of heritage data to the Ordnance Survey. In *Archaeology in the Age of the Internet. CAA97. Computer Applications and Quantitative Methods in Archaeology*; Dingwall, L., Exon, S., Gaffney, V., Laflin, S., van Leusen, M., Eds.; BAR International Series 750; Archaeopress: Oxford, UK, 1999; pp. 235–239, ISBN 0860549453.
70. Seymour, W.A. *A History of the Ordnance Survey*; Dawson: Folkestone, UK, 1980; ISBN 9780712909792.
71. Godwin-Austin, A. The Conference of Empire Surveyors. *Geogr. J.* **1928**, *72*, 449–453. Available online: https://www.jstor.org/stable/1783349 (accessed on 9 November 2018).
72. Evans, C.J.; Brittain, M.; Cessford, C. *Inlands and Hinterlands: The Archaeology of North West Cambridge*; CAU New Archaeologies of the Cambridge Region Series; McDonald Institute: Cambridge, UK; Volume IV, in preparation.
73. Crawford, O.G.S. Primitive English land-marks and maps. *Emp. Surv. Rev.* **1931**, *1*, 3–12.
74. Survey Review. Empire Conference of Survey Officers 1928. *Surv. Rev.* **2006**, *38*, 434–435. [CrossRef]
75. Joshi, J.P.; Bala, M.; Ram, J. The Indus Civilisation: A reconsideration on the basis of distribution maps. In *Frontiers of the Indus Civilisation: Sir Mortimer Wheeler Commemoration Volume*; Lal, B.B., Gupta, S.P., Eds.; Books & Books: Delhi, India, 1984; pp. 511–530, ISBN 9780856722318.
76. Possehl, G.L. *Indus Age: The Beginnings*; U Penn Museum: Philadelphia, PA, USA, 1999; ISBN 978-0812234176.
77. Kumar, M. Harappan settlements in the Ghaggar-Yamuna divide. In *Linguistics. Archaeology and the Human Past*; Indus Project, Research Institute for Humanity and Nature: Kyoto, Japan, 2009; Volume 7, pp. 1–75, ISBN 9784902325423.
78. Green, A.S.; Petrie, C.A. Landscapes of urbanisation and de-urbanization: Integrating site location datasets from northwest India to investigate changes in the Indus Civilization's settlement distribution. *J. Field Archaeol.* **2018**, *43*, 284–299. [CrossRef]
79. Possehl, G.L. A short history of archaeological discovery at Harappa. In *Harappa Excavations 1986–1990: A Multi-Disciplinary Approach to Third Millennium Urbanism*; Meadow, R.H., Ed.; Monographs in World Archaeology No. 3; Prehistory Press: Madison, WI, USA, 1996; pp. 5–11, ISBN 9780962911019.
80. Cunningham, A. *Report for the Year 1872–73*; Archaeological Survey of India Reports; Archaeological Survey of India: Calcutta, India, 1875; Volume V.
81. *Gazetteer of Attock District 1930*; Punjab District Gazetteers: Lahore, Pakistan, 1932.

82. ArcGIS Desktop, Fundamentals of Georeferencing A Raster Dataset. Available online: http://desktop.arcgis. com/en/arcmap/latest/manage-data/raster-and-images/fundamentals-for-georeferencing-a-raster-dataset. htm (accessed on 9 November 2018).

83. Petrie, C.A.; Singh, R.N.; Bates, J.; Dixit, Y.; French, C.A.I.; Hodell, D.; Jones, P.J.; Lancelotti, C.; Lynam, F.; Neogi, S.; et al. Adaptation to variable environments, resilience to climate change: Investigating *Land, Water and Settlement* in northwest India. *Curr. Anthropol.* **2017**, *58*, 1–30. Available online: http://www.journals. uchicago.edu/doi/full/10.1086/690112 (accessed on 9 November 2018). [CrossRef]

84. Singh, R.N.; Petrie, C.A.; Pawar, V.; Pandey, A.K.; Neogi, S.; Singh, M.; Singh, A.K.; Parikh, D.; Lancelotti, C. Changing patterns of settlement in the rise and fall of Harappan urbanism: Preliminary report on the Rakhigarhi Hinterland Survey 2009. *Man Environ.* **2010**, *35*, 37–53.

85. Singh, R.N.; Petrie, C.A.; Pawar, V.; Pandey, A.K.; Parikh, D. New insights into settlement along the Ghaggar and its hinterland: A preliminary report on the Ghaggar Hinterland Survey 2010. *Man Environ.* **2011**, *36*, 89–106.

86. Singh, R.N.; Green, A.S.; Green, L.M.; Ranjan, A.; Alam, A.; Petrie, C.A. Between the Hinterlands: Preliminary Results from the *TwoRains* Survey in Northwest India 2017. *Man Environ.* **2018**, in press.

87. Singh, R.N.; Green, A.S.; Alam, A.; Petrie, C.A. Beyond the Hinterlands: Preliminary Results from the *TwoRains* Survey in Northwest India 2018. *Man Environ.* **2019**, in press.

88. Green, A.S.; Singh, R.N.; Alam, A.; Garcia, A.; Green, L.M.; Conesa, F.; Orengo, H.A.; Ranjan, A.; Petrie, C.A. Re-discovering dynamic ancient landscapes: Archaeological survey of features from historical maps in northwest India and their implications for the large-scale distribution of settlements throughout South Asia. *Remote Sens.* in preparation.

89. Walker, J.R. Human-Environment Interactions in the Indus Civilisation: Reassessing the Role of Rivers, Rain and Climate Change in northwest India. Ph.D. Thesis, Department of Archaeology, University of Cambridge, Cambridge, UK, in preparation.

90. Garcia, A.; Orengo, H.A.; Conesa, F.C.; Green, A.S.; Petrie, C.A. Remote Sensing and historical morphodynamics of alluvial plains: The 1909 Indus flood and the city of Dera Ghazi Khan (Province of Punjab, Pakistan). *Geosciences* **2018**, *9*, in press.

91. Orengo, H.A.; Petrie, C.A. Multi-Scale Relief Model (MSRM): A new algorithm for the analysis of subtle topographic change in digital elevation models. *Earth Surf. Process. Landf.* **2018**, *43*, 1361–1369. [CrossRef] [PubMed]

geosciences

MDPI

Article

Remote Sensing and Historical Morphodynamics of Alluvial Plains. The 1909 Indus Flood and the City of Dera Gazhi Khan (Province of Punjab, Pakistan)

Arnau Garcia [1,*], Hector A. Orengo [2], Francesc C. Conesa [1], Adam S. Green [1] and Cameron A. Petrie [1,3]

1   McDonald Institute for Archaeological Research, University of Cambridge, Downing Street,
    Cambridge CB2 3ER, UK; fcic2@cam.ac.uk (F.C.C.); ag952@cam.ac.uk (A.S.G.); cap59@cam.ac.uk (C.A.P.)
2   Catalan Institute of Classical Archaeology, Plaça Rovellat, s/n 43003 Tarragona, Spain; horengo@icac.cat
3   Department of Archaeology, University of Cambridge, Downing Street, Cambridge CB2 3DZ, UK
*   Correspondence: ag2023@cam.ac.uk

Received: 15 November 2018; Accepted: 21 December 2018; Published: 29 December 2018

Abstract: This paper explores the historical inundation of the city of Dera Ghazi Kkan (Punjab, Pakistan) in 1909. The rich documentation about this episode available—including historic news reports, books and maps—is used to reconstruct the historical dynamics between an urban settlement and the river morphodynamics in the Indus alluvial plain. Map and document-based historical regressive analysis is complemented with the examination of images obtained through different Remote Sensing techniques, including the use of new algorithms specifically developed for the study of topography and seasonal water availability which make possible to assess long-term changes in the Indus River basin. This case of study provides an opportunity to examine: (1) how historical hydrological dynamics are reflected in RS produced images; (2) the implications of river morphodynamics in the interpretation of settlement patterning; and (3) the documented socio-political responses to such geomorphological change. The results of this analysis are used to consider the long-term dynamics that have influenced the archaeo/cultural landscapes of the Indus River basin. This assessment provides critical insights for: (1) understanding aspects of the formation, preservation of representation of the archaeological record; (2) identifying traces of morphodynamics and their possible impact over the cultural heritage; and (3) offering insights into the role that recent historical documents can have in the interpretation of RS materials. This paper should be read in conjunction with the paper by Cameron Petrie et al. in the same issue of *Geosciences*, which explores the Survey of India 1″ to 1-mile map series and outlines methods for using these historical maps for research on historical landscapes and settlement distribution.

Keywords: remote sensing; historical landscapes; landscape archaeology; settlements; colonial studies; river morphology; Indus; floods

---

1. Introduction

The floodplain of the Indus River is one of the largest and most dynamic alluvial landscapes in the world [1,2]. Geomorphological and climatic factors create a landscape characterised by strong seasonality, continuous changes in river course, and active processes of erosion and deposition of alluvial sediments. In that context, human settlement and economic activities present a complex relationship with the fluvial environment [3–5]. The Indus River basin, which includes the Indus River and its main tributaries, has the potential for high agro-pastoral productivity capable of sustaining large populations, including urban centres, and it hosted the first urban settlements in the South Asian Subcontinent, which were established in the third millennium BC. The region has been intensively occupied during different periods, and it is now home to dense clusters of human population and remains one of the most productive agricultural areas in South Asia. Nonetheless, the river and its floodplain are highly active, and have strong potential to be a disturbing force for settlements and the economic dynamics of the area.

In spite of this, the Indus floodplain presents abundant traces of intense and continuous human activity in the past [6,7]. However, erosion and sedimentation processes typical of these changing floodplains have undoubtedly influenced the preservation and interpretation of the archaeological remains of this cultural heritage.

Despite their historical interest, areas that are subject to continuous transformation have been a barrier to Remote Sensing (RS) applications for the study of past landscape dynamics. Indeed, RS was initially applied for archaeological research in areas of long-term soil stability in the Middle East and northern Europe [8] and they have had a high impact in areas of where traces of past are fossilised into the present landscapes [9]. Tropical rainforests and arid areas, where fossilised landscapes can be well preserved have also revealed considerably insights into landscape dynamics in the past thanks to improvements in RS [10–12]. In contrast, active floodplain environments such as that of those of the Indus River Basin that are densely occupied in the present day and also areas of intense agricultural activity have been much less explored due to obvious problems of visibility and alteration. Works developed on those areas have integrated historical data within the RS approach in order to overcome some of those problems [13–15].

In the large Indus region, RS has been extensively used to locate rivers that have disappeared and their relationship to ancient and historical settlements. In that context, RS has been used often as a support of geoarchaeological and archaeological surveys. The Yamuna—Sultej interfluve in northwestern India, and in particular the arid area occupied by the nowadays ephemeral Ghaggar-Hakra river, has been the focus of most of the research, both historically and recently [16–20]. The desert areas of Cholistan and Rajasthan, which lie on the east side of the Indus River basin are one area where RS has strong potential for exploring predominant aridity areas [21,22]. Notably less analysis has been carried out on the central part of the Punjab River System in Pakistan or along the Indus River itself [23,24].

This paper focuses on the analysis of the morphodynamics of a section of the Indus River during a short period of time. The study case presented here was carried out as part of the *WaMStrIn* project, which builds on the *TwoRains* project. Both projects are directed to understand the long-term settlement dynamics in the larger Indus River basin. The Indus River basin is characterised by extremely dynamic fluvial processes, and thus far the work of the *TwoRains* project has focussed on the eastern part of the Indus River basin, which is a large area characterised by historical and ongoing hydrological changes [25–27].

The research presented here aims to address the relationship between river morphodynamics and human settlement within a much more active part of the Indus River basin. The Area of Interest (AOI), around the city of Dera Ghazi Khan (DG Khan), has been selected in order to analyse a human settlement situated in an extremely active river environment. The chronological focus of the study is the nineteenth and twentieth centuries AD, a period for which historical documentation allows the reconstruction of both the history of the settlement and the changes in the river landscapes in considerable detail. The selection of this chronological span makes it possible to explore the approaches to integrating recent historical datasets, and this paper acts as a companion piece in the same issue that explores the potential for historic maps for the detection of archaeological sites [28].

This paper will assess the use of historical documents in the interpretation of RS data and settlement patterns. Firstly, it describes and demonstrates a method for extracting and georeferencing the historical data related to the settlement and river dynamics of the study area in order to obtain a regressive landscape reconstruction. Secondly, it applies different RS techniques and obtain images of the study area surface to compare with the historical landscape. These data are then analysed to address: (1) how well know historical processes are reflected in RS, and; (2) how the historical data can provide valuable information for the analysis of the cultural heritage beyond its specific chronological framework.

2. Study Case: Dera Ghazi Khan and the River Indus

Today, the city of DG Khan is the capital of a homonymous district at the south-west edge of the Pakistani province of Punjab. DG Khan district includes part of the alluvial plains at the west of the Indus River and extends towards the foothills and uplands of the Sulaiman Mountains (Figure 1).

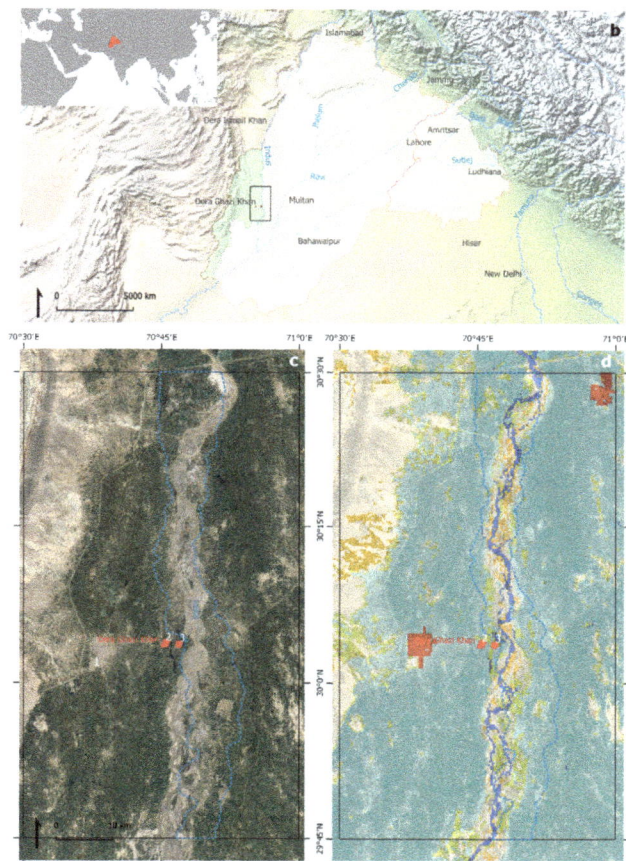

Figure 1. Situation and context of the AOI: (**a,b**) DG Khan district in the context of the moderns Punjab Province (Pakistan) and Punjab State (India); (**c**) actual aspect of the AOI and situation of the DG Khan modern city (1), historical town (3) and nineteenth century British cantonment (2). The dotted line represents the active floodplain of the Indus as it was in 1907; (**d**) actual land-cover classification (GLOBCOVER V. 2.3, 2009) show the area occupied almost entirely by irrigated croplands, except in the riverbed and mountain slopes.

The modern settlement was founded in its current location in 1911 as a consequence of a flood that destroyed the historical town in 1909. The original position of this earlier version of the city was 15 km to the east, and lay within the modern active floodplain of the Indus. The original city was founded in the fifteenth century AD, as one of the regional centres named after prominent members of the Dodai tribe that had recently stablished its control of the west bank of the Indus. DG Khan maintained its role as political centre of the area under different rulers and was incorporated into the British administration in the mid-nineteenth century, in the wake of the Anglo-Sikh wars [29,30]. Under British colonial rule, DG Khan was the capital of the district and a cantonment for colonial troops and civil administration was established on the western side of the old town (Figure 1 num. 2). After the flood of 1909, the newly built city assumed the role of regional administrative and economic centre.

On the west bank of the Indus where DG Khan is situated, there are two interrelated dynamics (Figure 2): (1) the Indus River has a strong seasonal character due to its seasonal floods and continuously moves its main channels across a wide floodplain in a regular basis (Figure 2: a.2–3 and b.2–3); and (2) ephemeral torrents descend from Sulaiman mountains and cross the piedmont zone at their base before joining the Indus River (Figure 2: a.1–2). The water and the sediments deposited by the Indus and hill torrents have been exploited historically for agropastoral purposes in the area, which has an arid climate that in other circumstances would not allow the cultivation of some crops at productive levels.

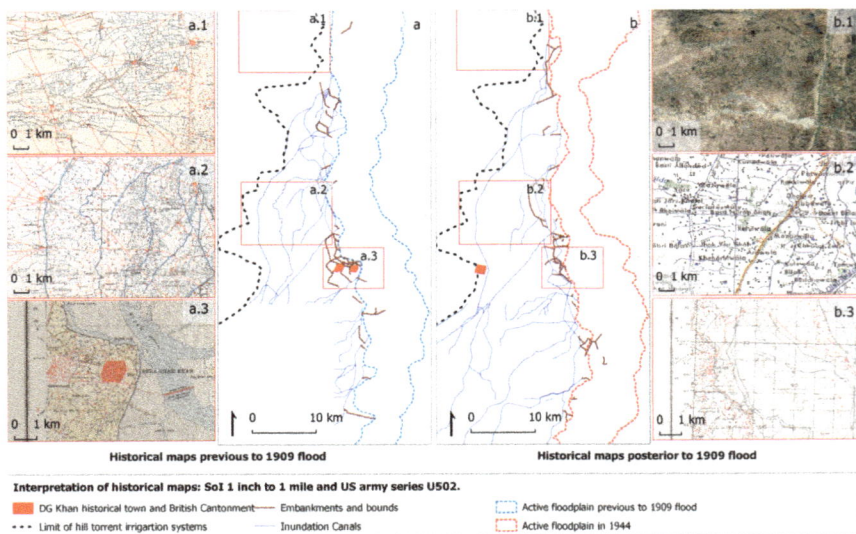

Figure 2. General overview of the AOI before (**a**) and after (**b**) the 1909 flood, according to the information contained in the SoI 1" to 1-mile maps (for more details see Section 3 and Figure 5). The maps contain information of the historical landscape morphology: (**a.1,b.1**) irrigation systems based on hill-torrents; (**a.2,b.2**) inundation canals in the old floodplain; (**a.3,b.3**) active floodplain, earthworks and canals around DG Khan. No maps edited before the flood were available for the South-West sector of the AOI (see Figure 5). SoI maps Image courtesy of UL, Cambridge.

Despite continuous efforts to control the river's water levels, major floods continued to happen on a regular basis in the Indus valley. Almost 9000 causalities and large economic damage were calculated for the period between 1950 and 2011 [31]. One of the more recent major floods in the Indus basin occurred in 2010 (Figure 3). Mirroring what happened 100 years before in DG Khan, much literature [32–36] has addressed its causes (both natural and anthropic) and consequences (in economic terms, health, for future risk planning, and so on), although few of these works uses historical data previous to the 1950's.

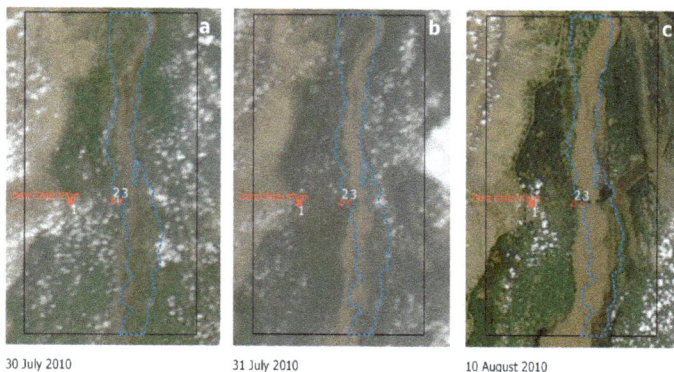

Figure 3. Images of the AOI obtained by the satellite MODIS during the flood event of summer 2010 [37], showing the progression of the water in three moments of the episode in which the clouds allowed to visualize the AOI: 30th (**a**) and 31st (**b**) of July and 10th of August 2010 (**c**). Blue dotted line represents the active floodplain in 1907. Many Satellite-derived products were produced to study this event [38–40]. The images show how the water occupies most of the early twentieth C. floodplain.

3. Materials and Methods

3.1. Historical Sources

The documents produced by the British colonial authorities provide a unique testimony of the flood episode in 1909. In the process of developing colonial rule, the British authorities undertook a huge and systematic collection of data to describe the acquired territories, including the compilation of detailed District Gazetteers and the systematic mapping of the area by the Survey of India [28]. That effort served the practical need of providing information to officials appointed to administer a territory and people about which they often had limited previous knowledge. This process was also linked to the intellectual context of nineteenth century Europe, where there was a general ambition to describe, measure and classify the physical world.

The District Gazetteers are an example of this intended systematic collection of data on particular administrative areas [41]. The 1893–1897 edition of the Dera Ghazi Khan district Gazetteer [42] has been consulted for this research. The changes in the watercourses and their effect on settlements and cultivated lands were a major concern for the DG Khan district authorities. The reports collected in the District Gazetteer contain a detailed account of flood episodes and the engineering works that were developed in response during the nineteenth century. Other documents were produced by administrators, engineers and geologists [43]. The consequences of the 1909 and 1910 floods were described, including some photos, in an article that appeared in 1910 in The Times of India Illustrated Weekly Supplement [44] (Figure 4).

Shortly after the flood, DG Khan was analysed in detail by a former Superintending Engineer of the Punjab Irrigation. The resulting document [45] provides first hand evidence of the works developed between 1897 and 1909, including several sketches, together with a general discussion about the causes and development of the flood.

All this documentation has been analysed in order to investigate different aspects of interest, in particular: (1) references to geomorphological processes; (2) references to dates and character of earlier flood episodes; and (3) descriptions of the engineering works developed in relation to floods. The information obtained has been georeferenced whenever possible.

Figure 4. Image of DG Khan in process of abandonment in 1910 [44].

The documents used here were produced between 1870 and 1912 (Figure 5, Table 1). They describe ongoing processes as they were perceived by representatives of institutions with a direct knowledge of the area. Hence, they provide a unique first-hand description of physical changes in the river environment concerning the episode that is the object of this study. Beyond the direct description of local conditions, the knowledge of the river's morphodynamics reflects the conception and technical limitations of the society in which they have been produced [3,43]. The reconstruction based on these sources has an approximative character in terms of topographic georeferentiation, but many of the described features can be identified in detail through the 1″ to 1-mile maps.

3.2. Historical Map Analysis

A second type of documents are the cartographic material produced by the Survey of India (SoI). The processes involved in the creation of the SoI and the characteristics of the materials produced are presented in another paper presented in this volume [28].

For the work presented here, several editions of the SoI map series were employed. These include those sheets that show the old and new cities of DG Khan and the areas immediately to the south and north (Figure 5). The AOI measures around 80 km N-S and 40 km E-W. The large scale (1″ to 1-mile; 1:63,360) and accuracy of the SoI maps provide a unique source for the study of the area in the years before and after the 1909 flood. These maps were provided (as scanned 600 dpi tif files) by the University Library, University of Cambridge.

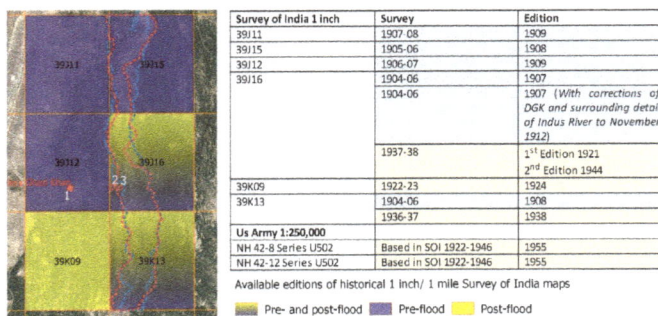

Survey of India 1 inch	Survey	Edition
39J11	1907-08	1909
39J15	1905-06	1908
39J12	1906-07	1909
39J16	1904-06	1907
	1904-06	1907 (With corrections of DGK and surrounding detail of Indus River to November 1912)
	1937-38	1st Edition 1921 2nd Edition 1944
39K09	1922-23	1924
39K13	1904-06	1908
	1936-37	1938
Us Army 1:250,000		
NH 42-8 Series U502	Based in SOI 1922-1946	1955
NH 42-12 Series U502	Based in SOI 1922-1946	1955

Available editions of historical 1 inch/ 1 mile Survey of India maps

■ Pre- and post-flood ■ Pre-flood ■ Post-flood

Figure 5. Editions of the SoI 1″ to 1-mile maps used for this study and its geographical and temporal coverage. The whole area is also covered by two maps from the US Army Map Service, Series U502 (1955).

Table 1. Summary of references to the riverine dynamics during the nineteenth century recorded by the British colonial authorities.

Date	River Dynamics	Other Historical Data
Mid. 15th C. AD		Foundation of DG Khan
1787	Abrupt change in the river	
1812	Flood	
1833	Flood	
1841	Flood	
1842		Sihk Wars. British take control of DG Khan.
1858	Flood swept away cantonment	
1857	Embankment build	
1870's		Original reports compiled in the 1893–1897 Gazetteer.
1878	Embankment overtop damage cantonment	
1882	Embankment breached out puts town and cantonment in risk	
1888–1889	High floods	
1889–91	Building of stone embarkments	
1895	Breach of the stone embankment	
1895–97	Embankment extension Still risk	
1897	Reported problems in buildings and health due high levels of underground water	Compilation of the district Gazetteer
1899	New works increases municipality debt	
1900	Embankment damaged when river cuts its Western banks	
1901	Decision to undertake new works.	
1902	Works interrupted by an early flood	
1904–1906		Surveys for the elaboration of SoI 1" to 1-mile maps
1906	River cut its Western bank	
1907		Edition of the SoI 1" to 1-mile map
1908	Stone embankment collapse	
1909	The Indus enters the old town	
1910	River eroding the remains of the city. City in process of abandonment	Article in The Times of India Illustrated Weekly Supplement
1910/1911	Decision to move the city in its new location	
1912		Publishing of the second edition of "Punjab Rivers and Works"

Georeferenciation of these maps has been carried out following the process described by the authors in a paper published in this same volume [28]. Ground control points (GCPs) were obtained using the high-resolution images available through QGIS world imagery map services. The areas analysed have undergone substantial changes during recent decades, which has resulted in a significant reduction in the number and distribution of reliable GCPs in comparison to other Indus valley areas. Very few reliable GCPs were identified in the area occupied by the Indus river channels. Only 6 to 8 GCPs have been used for each of the maps presented here. Given the low number of GCPs for the area rectification procedures used a first-degree polynomial, which provided medium RMSE value of 27 m. The evaluation of the map accuracy has been possible through comparison with canals and settlements present in modern images. Canals preserved today prove that the maps were very accurate in the representation of those elements. Also, main settlements and roads are drawn accurately, but the accuracy of the positioning of single farms and footpaths on the maps are more difficult to assess.

The earliest series of SoI 1" to 1-mile maps of the DG Khan area were produced between 1907 and 1909 from data collected in 1904–1908, immediately before the 1909 flood. In 1912 a new edition of the maps of the area of DG Khan was produced specifically to document the changes in the river and the effects of the flooding. There was an even later edition produced in 1944 based on survey data of 1937–1938 for the same area. Post-flood editions of the SoI 1" to 1-mile series produced between the 1920s and 1940s, are available for only other two maps. The post-flood image has been completed with the georeferentiation of the 1950's maps produced by the US army, which are based on the Survey of India 1920s–1940s editions (Figure 2: b.2; Figure 4).

3.3. RS Approach

3.3.1. Declassified Data and Satellite Historical Photographs

High-resolution satellite images provided by QGIS 3.2 Imagery Web Services were used to provide a current view of the AOI. A declassified KH4B CORONA image from 1972 has been acquired [46] and georeferenced. Declassified CORONA data have been proved extremely useful in archaeological research [22,47], as the image series provides a unique view of past cultural landscapes before recent landscape changes. Unfortunately, the only scene available at the USGS collections covering our AOI has much of the DG Khan area towards the edge of the scene, and therefore is highly affected by image panoramic distortion [48]. As a result, the rectified image should be considered as informative, but not of sufficient quality for data extraction.

3.3.2. Multispectral Satellite Imagery

Several recently published algorithms [25] have been applied to the Landsat 5 long-term archive to boost the detection of palaeorivers and related geomorphological traits: (1) Normalised Difference Seasonal Vegetation Index (NDSVI), which provides information about the amount of inter-seasonal variability of vegetation and allows the user to select between the following techniques according to the most adequate environmental factors; (2) Seasonal Multi-Temporal Vegetation Indices (SMTVI), which produces an image of long-term vegetation health, which can be related to subsurface features; and (3) multi-temporal seasonal spectral defragmentation techniques, which include Principal Component Analysis (PCA) and Tasselled Cap Transformation (TCT) to allow the exploration of multispectral datasets in different environmental conditions. Details about these algorithms and the code employed to implement them have been previously published [25] (see the Supplementary Material for the version used here). Landsat 5 have been selected because is the longest Earth Observation missions currently available, covering the period between 1984 and 2013.

Another source of information has been the images acquired by MODIS Satellite during the 2010 flood. This satellite monitored the advance of that flood through the Indus River basin on a daily basis, providing an interesting example of flood progression (Figure 3).

3.3.3. Microtopographic Data

Multi-Scale Relief Model (MSRM) ([26], see the Supplementary Material for the version used here) was employed to extract microtopographic information at variable scales using ALOS data (Figure 6). The new 30 m/cell ALOS digital terrain model (DTM) was selected as it offers much-improved topography in comparison to similar freely available global DTMs such as SRTM and ASTER GDEM. This mission is relatively recent (2004–2011), reflecting the current relief of the AOI.

Palaeorivers and associated features such as levees, oxbow lakes and bluff lines even when inactive leave subtle topographic marks that are reflected in DTMs. These micro-topographies are very variable in size, which renders them undetectable by current micro-topographic visualization techniques. MSRM provides a multi-scale rendering of small topographic differences and presents an ideal technique to investigate morphological remains from ancient rivers.

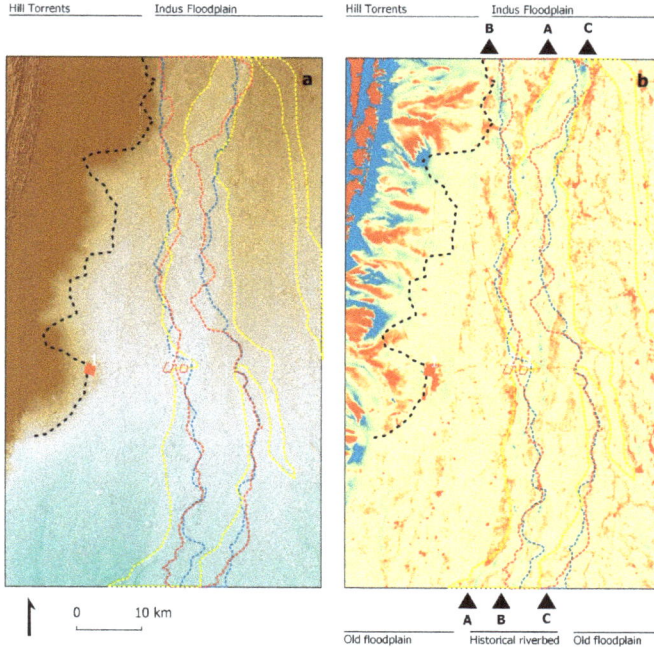

Figure 6. General view of the relief analysis: (**a**) ALOS DEM; (**b**) derived MSRM. Blue represents relative lower areas and red higher areas. Dotted lines represent the limit of hill torrents irrigation systems (black), and the active floodplain in 1907 (blue) and 1944 (red) (see Figure 2). In this area can be identified the actual Indus West levees (A) and historical levees West (B) and East (C). Old floodplains occupy the area between the Suleyman mountain hill-torrent deposits (W) and Indus tributaries (E). Inactive floodplain is irrigated by inundation channels. The micro-topographic analysis show depressions and old river channels occasionally flood. Like the previously mentioned 2010 event (yellow dotted line).

4. Results

4.1. Historical Landscape Reconstruction

The information contained in the historical documents [42,44,45] reveals how the 1909 flood was the culmination of a long process (Table 1). During the nineteenth century, the colonial administration had to deal with a landscape that was undergoing continuous change, with floods seasonally covering part of the administrated territory. They also had to take control of a network of artificial weirs and canals build to regulate floods and use the river water and sediments for agriculture. Despite the presence of these infrastructures, and the multiple improvements undertook during the second half of the nineteenth century (Figures 7 and 8), the historical sources reported at least twelve major flood episodes between 1787 and 1909, which were considered catastrophic or potentially catastrophic for DG Khan. In response the management strategies focused in the building of new embankments and the extension and maintenance of the ones that were already extant.

The Gazetteer acknowledges that the original city of DG Khan was originally built on a river island, with the remains of a channel still visible in the west side of the town where the cantonment was built:"Four hundred years ago, when the city of Dera Ghazi Khan was founded the river basin was travelling eastwards, and the city is said to have been built on an island or chakkar, the creek of the west of which has long ago silted up, though parts of it are traceable in slight depressions" ([42], p. 6). In fact, Charles Masson reportedly crossed the Indus around 3–4 miles from DG Khan in 1826.

An old channel of the river can be spotted in this area (Figure 6C). Between the river and the city, Masson described an " . . . an immense assemblage of date groves and gardens amid which the large, populous and commercial town of Déra Ghází Khân is situated." [49], which indicates that the river floodplain had been inactive for some time in the areas surrounding the town.

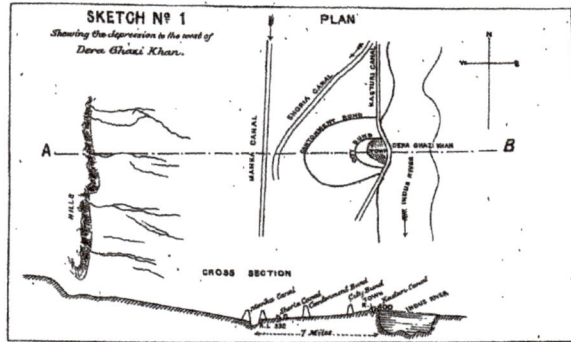

Figure 7. Sketch of the city flood-protection system at the end of the nineteenth century [42].

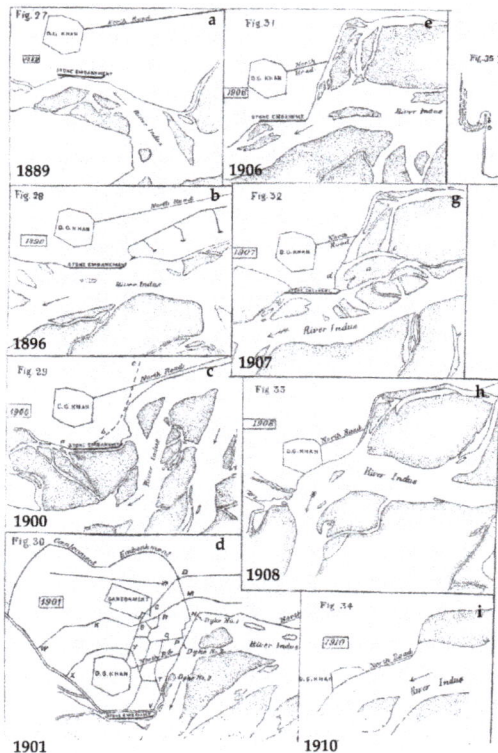

Figure 8. (**a–i**): Sketches showing the movement of the river between 1888 and 1910 [45]. (**d**) shows the protection system planned in 1901, which was already half lost in the 1907 SOI map. Note that the images are orientated West with North at right.

The west paleochannel reported in the Gazetteer was not clearly indicated on the SoI 1″ to 1-mile map produced in 1907 (39/J/16/1907), but it appears in the map edited after the flood (39/J/16/1944). A channel in this area can be spotted in the satellite derivates as a depressed area in the microtopographic image (Figure 6A). It could be argued that the earliest map reflects a mindset whereby the aim of stopping the water from encroaching on the city had been successful, despite the fact that in this map, the stone embankment that had been built in 1888 was already lost (Figure 8). The 1907 map, in fact, depicts a channel turning east precisely where it meets the embankments built north of the town. It is interesting to note that a small body of water is depicted in 1907 precisely in the place where in the 1912 modified map shows that the channel has broken the embankment (Figures 9 and 10a,b). This example is indicative of how maps can produce a static 'frozen in time' picture of a constantly changing landscape.

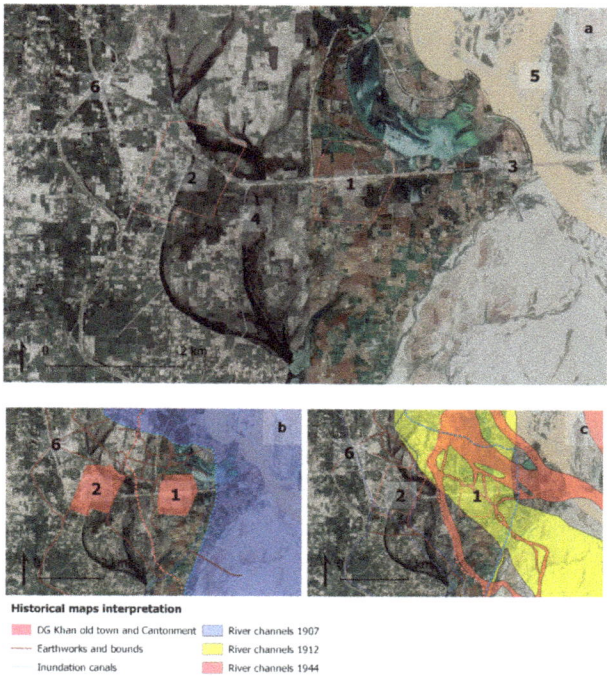

Historical maps interpretation

DG Khan old town and Cantonment		River channels 1907
Earthworks and bounds		River channels 1912
Inundation canals		River channels 1944

Figure 9. (**a**) Detail of the area of the old DG Khan city in a high-resolution satellite image. Below, superposition of the features extracted from the SoI 1″ to 1-mile; (**b**) before the flood (1907 edition) and; (**c**) after (1912 and 1944 editions). The features have been extracted directly from the maps, showing its topographic reliability. That is particularly visible in the right part of the image, less affected by the river movements, where (**6**) part of the earthwork morphology has been preserved by canals and the modern road network. The accuracy of the elements preserved in today's landscape provides a high confidence in the location of disappeared elements. The numbers indicate different areas discussed in detail in the following section: (**1**) the old walled DG Khan; (**2**) the British cantonment; (**3**) the modern N-70 bridge; (**4**) depression between the old city and cantonment and; (**5**) the main Indus channel. In the maps (**b**,**c**) the river bed according the different SoI editions is superimposed.

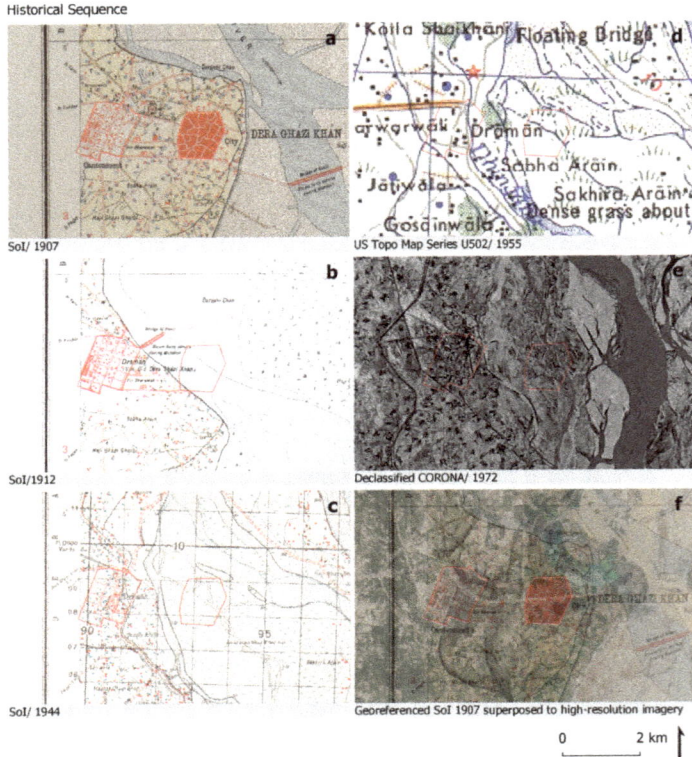

Figure 10. Detail of the area were DG Khan was placed until 1909 in different historical documents: SoI 1″ to 1-mile, editions of (**a**) 1907, (**b**) 1912 and (**c**) 1944; (**d**) US Army Map Service, Series U502 of 1955; (**e**) declassified CORONA of 1972 and (**f**) present day high-resolution satellite image. SoI maps images courtesy of UL, Cambridge.

The works to protect both city and cantonment form the floods are described in the Gazetteer, which includes an engineer's report and sketches of the works (Figure 7), and more details are given in Bellasis work (Figure 8). Both canals and levees are represented in the SoI 1″ to 1-mile maps, which show part of the plans described by Bellasis that were part of the respond to the 1901 flood (compare Figures 8d and 9b and Figure 10a). Those plans were in fact never fully completed and part of the system was already covered by the water when the map was draw [45].

In 1912 a second edition of the 1907 SoI 1″ to 1-mile map was produced specifically to incorporate the changes in the area of DG Khan (Figure 10b). This map shows a channel of the river crossing the area of the city and does not show any traces of the old town, although most of the cantonment is still visible. The map doesn't cover the area where the new city was being built at that time, and the map legend doesn't mention if new surveys had been undertaken in order to re-draw the map or if it was an approximate modification. Furthermore, the map produced in 1944 (Figure 10c) shows the 1912 location of the channel, even though it was dry at that time, and none traces of the old town. The area seems occupied by high grass and other dry channels. That situation is reflected also in the 1950s US version of this map (Figure 10d), and it is also how the area appears in the 1972 CORONA image (Figure 10e). Since then, important works related to road infrastructure, mainly the N-70 road bridge over the Indus, have further modified the area (Figure 9a num. 3 and Figure 10f).

4.2. Remote Sensing Analyses

4.2.1. Dera Ghazi Khan Old Town

Examination of cartographic material produced after 1909 does not show any traces of the old town of DG Khan (Figure 9 num.1 and Figure 10a–d). Satellite images, covering the period from 1972 to the present, show that the area formerly occupied by the old town is now dedicated to agricultural fields, with some roads and channels (Figures 9 and 10e,f)). In fact, the present-day high-resolution images, or historical CORONA shows no remains of the old town or its shape in the landscape. However, part of the cantonment is visible, since some roads and field borders follow the lines defined by the cantonment streets depicted in the SoI maps.

The analysis of long-term multi-spectral images of the area shows that the surface of the old city does not present a unique differentiated signature. The embankment of the N-70 Road bridge and associated infrastructures had a clear impact, however, dividing the area between north and south, with very different dynamics (Figure 9 num.3). In the PCA and TCT analysis (Figure 11c–f), the surface of the old city is not distinguishable from the surrounding area. Only in the SMTVI image, is there a clearly distinctive signature in the north-west quarter of the old city (Figure 11a,g). This signature apparently comes from high values in band 1 (Jan–Feb), as it is the only area in the riverbed where that happens. Unusually high values in band 1 correlate with lower areas in other parts of the AOI, but here there is an exception of this trend, since this area is surrounded by lower areas. On the other bands, the area presents the same signature than that of the areas close to the north side of N70 road embankment (Figure 11c–f). In that sense, it seems that the area is flooded occasionally by water stopped by the embankment.

In the MSRM, the N-70 embankment is also a prominent feature (Figure 11h). Depressions related to channels of the river Indus are visible in both East (Figure 9 num.4) and West (Figure 9 num.5) sides of the old town, enhancing the original image of a river island for the city placement. The lower East area extends until at least the middle of the old town and a lower area is also present on the South-Western corner, leaving the North-Western quarter again as the only observable elevated area (Figure 11h).

4.2.2. River Channels

Today, the old site DG Khan lies within the Indus active floodplain. Its west limit is defined by an elevated bluff line that is clearly distinguishable in the relief, as it shows the MSRM image (Figures 6 and 11h), and can be delineated in the multispectral images as well (Figure 11a–d). This line follows the riverbed as it was draw in early twentieth century maps of the area north of DG Khan old town. South of the old town, the modern bluff line is a few kilometers (2–4 depending on the point) to the west of where it was draw in the early twentieth century maps.

Within the riverbed, the N-70 road bridge has had a major effect on the river movements on this area, which is clearly visible in multi-temporal and multi-spectral image composites. As observed by the first available Landsat 5 images (1984), the bridge and associated embankments were already built (Figure 11). In the Corona image of 1972, this section of the river has a very different aspect, and many elements represented in the 1944 SoI map were still visible (Figure 10).

The superposition of the channels identified in the historical cartography shows that over time almost the entire modern floodplain has been occupied by Indus channels (Figure 9 num.5 and Figure 12). This is particularly telling considering that the SoI maps and the CORONA image that are available only provide information for 4 particular moments, and the longest recent series of remote sensing imagery covers a relatively short period of time (1984–2013 for Landsat 5, that can be complemented by more recent data). In general, the continuous superposition of channels occupies parts of the older ones. Channels identified in multispectral images, both active and non-active, are the product of this processes and have a very recent origin, regardless that they might partially follow the same tracks that channels documented in older maps and photographs had occupied before.

Multi-Temporal Remote Sensing images (From landsat5. 1984-2013)

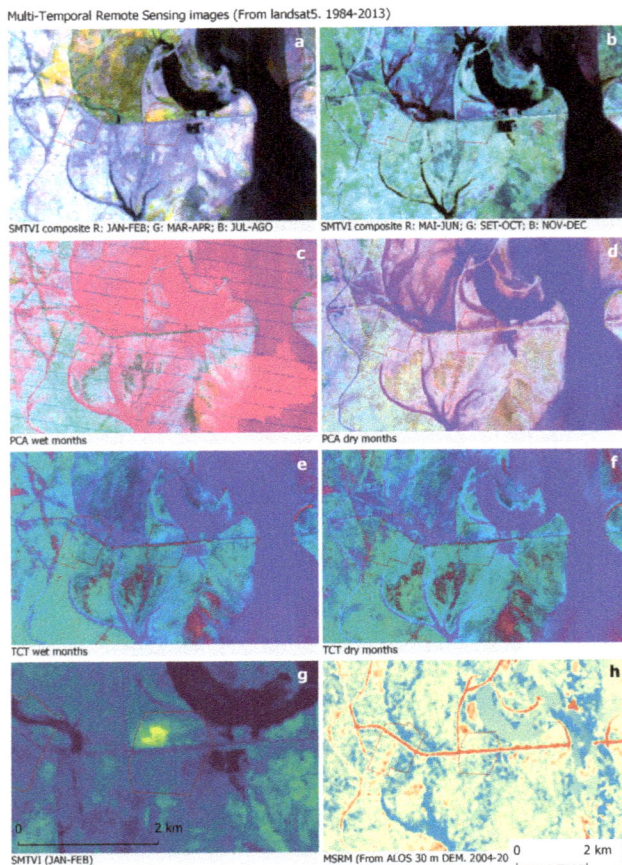

SMTVI composite R: JAN-FEB; G: MAR-APR; B: JUL-AGO

SMTVI composite R: MAI-JUN; G: SET-OCT; B: NOV-DEC

PCA wet months

PCA dry months

TCT wet months

TCT dry months

SMTVI (JAN-FEB)

MSRM (From ALOS 30 m DEM. 2004-20

0   2 km

Figure 11. Detail of the area were DG Khan was situated until 1909 in Multi-Temporal RS products: SMTVI (**a**) wet and (**b**) dry months; PCA (**c**) wet and (**d**) dry months; TCT (**e**) wet and (**f**) dry months; (**g**): Detail of the area occupied by the old DG Khan town, with an anomaly in the Northeastern sector visible in the months of January-February. MSRM (**h**) image of the same area. Blue represents relative lower areas and red higher areas.

Similar circumstances have affected the topography, which broadly reflected in the actual main active channels. In that sense it is interesting to note the depressions East and West of the old DG Khan site, visible in the SoI maps (Figure 10). They were part of the geomorphology of the area documented in the early twentieth century and they have been intermittently active until nowadays. The erosion of the areas formerly occupied by the old town and the cantonment is visible in the presence of lower areas within the limits of the old settlements (Figure 11h).

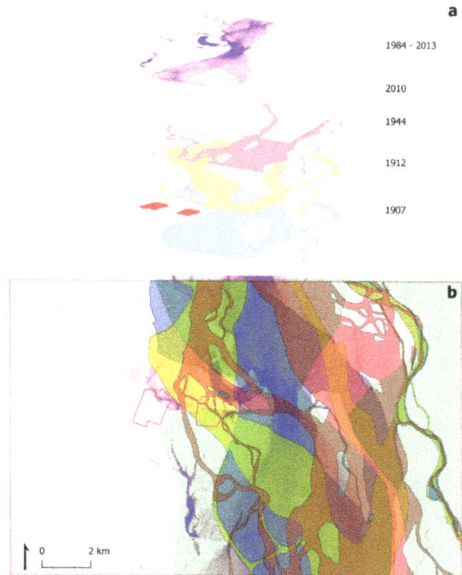

Figure 12. (**a**) active channels of the Indus documented in different historical moments. SoI 1″ to 1-mile editions of 1907, 1912 and 1944. It have been add the extension of the 2010 flood extracted from MODIS (see Figure 3) and the surface water occurrence layer from Global Surface Water Explorer [50,51]; In the image below (**b**) the different documented historical extensions of the Indus riverbed are superimposed in a single map.

5. Discussion and Conclusions

5.1. Integrating Remote Sensing and Recent Historical Data in the Study of Human Settlement and Riverine Dynamics

The regressive analysis of late nineteenth and early twentieth century cartographic and historical sources provide reliable information to place with accuracy distinct landscape features of interest, including: the old DG Khan historical town (Figure 9 num.; Figure 10f), the nineteenth century British cantonment (Figure 9 num. 2), the canal systems (Figure 9 num. 7, build embankments and the road network (Figure 9 num. 6). Moreover, the movement of the Indus channels and other geomorphological features can be partially reconstructed using the same sources (Figures 2, 5 and 9 num. 4–5). Since all of these features have disappeared from the modern landscape, this approach provides a view over the historical landscape that would be impossible to obtain otherwise.

More importantly, for the objectives of this work, this dataset can be compared with Remote Sensing products. The georeferenciation techniques applied here allow to overlay the information contained in the maps over multispectral satellite data-derived products. If we consider the resolution, georeferenciation errors are imperceptible when compared to the 30 m/pixel satellite images used here or even the most common high-resolution satellite imagery (e.g., Google satellite, Bing, and similar services). In addition, the chronological proximity of these maps to the other historical accounts is another highly valuable feature provided by the use of the SoI maps. Since the authors of the analysed reports and the cartographers were describing the same territory, many features described can be identified in the maps, accurately located, and compared with the satellite images. The result is a multi-layer impression that reflects more than 100 years of landscape history and in which the complementarity of the different sources used is boosted.

5.1.1. DG Khan Old Town

In a general overview, the area formerly occupied by the old town (Figure 9 num.1) shows in the RS derivates and historical imagery analysed in this paper a similar aspect than the surrounding riverine environments, with any element that could indicate the presence of a settlement in the past (Figures 10 and 11). A possible exception has been obtained using the enhanced RS approaches. SMTVI and MSRM have uncovered a differentiable signature in the northwestern quarter of the former old city, while the rest of the area has the same aspect of the neighbouring riverbed sectors. The possibility that this signature is related to the old town cannot be dismissed, as for example, the consequence of an area less affected by erosion. Despite that, the possibility that it might also be related to the soil use or the impact of modern infrastructure seems more probable. Therefore, archaeological data from the terrain would be necessary to verify possible hypothesis. This highlights the need to incorporate ground truth in the form of archaeological survey in RS-based archaeological research.

5.1.2. River Channels

The analysis of the river channels complements the view built up by the study of the historical dynamics (Figure 9 num. 4–5). The combined historical sources show how in around 100 years, water has covered almost all the active floodplain in one moment or another (Figure 12). Moreover, channels of the river can move towards parts of the inactive floodplains. The movement of the active floodplain can also be detected in several points of the study area, in particular to the south of DG Khan. RS images reflects essentially the dynamics of the river during the acquisition of the measurements used in the analysis (1984–2013 for Landsat and 2004–2011 for ALOS). Channels detected through RS have punctual coincidences with the historical channels, but this is not necessarily related to the details of the historical channel morphology.

5.1.3. The Value of Historical Sources and Cartographic Material in RS Approaches

Historical documents, in particular the SoI 1" to 1-mile maps, are the only dataset that provide temporal information about the most significant historical landscape features and morphology, which are present in RS data as an accumulation of features that are difficult to differentiate. Some landscape elements are not-visible through analysis derived from images acquired during the last 30 years, and others can only be interpreted with the support of the old maps. The Indus River probably represent an extreme case of strong geomorphological and anthropic dynamics, even in the context of the Indus basin. But it also reflect landscape dynamics that, to a minor degree, are widespread around the areas influenced by this river and major tributaries. Furthermore, those dynamics are at a fundamental level the types of geomorphologic processes that formed the landscapes of the Indus floodplain, and are likely to have continually affected it during the proto-historic and historic phases during which humans have occupied the plain.

Nonetheless, it is important to recognise that the SoI maps have limitations. Although they extend our temporal window, they are limited to providing information about the moment in which the data was acquired, as illustrated by the difference between the editions of the SoI maps. They do not present a continuous view of landscape changes but a set of temporal slices determined by the interests of the agency producing them. In this regard, they might have been and probably were many more geomorphological changes between these editions that we have not been able to reconstruct. Map-based historical landscape analysis is also dependent on the availability, selective criteria and quality of the maps [28]. Finally, it is relatively exceptional for systematic and accurate historical mapping to cover a period that goes before the late nineteenth and twentieth centuries.

Satellite-based sources have proved their high-potential to analyse the landscapes of the Indus and beyond and to extract reliable information on their long-term landscape dynamics [25]. These are an outstanding instrument for exploratory work and to guide fieldwork. The case developed in this paper is an example of how regressive analysis based on recent historical data can enhance RS analysis, by providing information about disappeared features and complementing the interpretation of other sources. This is critical for addressing part of the problematics that highly transformed areas but also identify RS-detected features that have a modern origin. In the case of the Indus, the inclusion of materials like the SoI maps, the district Gazetteers and other similar historical sources need to be systematically integrated in RS workflows [28].

5.2. River Morphodynamics, Settlement Patterns and Archaeological Record in a Long-Term Perspective

The reconstruction of DG Khan's old town and its relationship with the morphodynamics of the Indus river provide data for the study of the history of this area during the last 200 years. However, this example is also significant for the study of past landscape dynamics. Similar geomorphological dynamics, even if influenced by climate change and other factors, can be assumed for the study area throughout the Holocene. Despite the obvious problems involved in extrapolating interpretations extracted from very specific geographical and temporal frameworks, small-scale analyses are an important mechanism for understanding the parameters in play for large-scale dynamics. Elsewhere we have addressed the nature of the map sources and their potential for types of analysis performed here, but in a larger scale for the detection of archaeological sites [28,52]. The specific case of DG Khan presented here allows us to address in more detail the significance of historical morphodynamics as one of the many interpretation filters that can be used when observing the available prehistoric settlement distribution maps of the Indus basin. In the extant site distribution tables and maps [53], large areas of Punjab appear to have been largely unpopulated in terms of archaeological data, at least for the protohistoric periods, but also to some extend the historic periods. Our data underlines that five thousand years of very active river geomorphology could have had an important influence on the number of sites that have been available to document using our current recording methods in combination with the patterns of the known site distributions in Punjab.

River morphodynamics have had in the case of DG Khan a double impact: firstly, in the development of the settlement, starting from the original placement on a river island, and continuing to be a central factor in the urban development, abandonment and final movement of the town to a new place. The second impact is on the preservation of the old town as an archaeological site. The results presented here (Section 4) provide a very interesting picture in which we have three settlements—the historical town, the cantonment and the new town—where we have historical evidence of both synchronic and non-synchronic moments of occupation within a single historical period, but only the new town is visible in the present landscape.

The concealing of a relatively large settlement in a short period of time, not only as an inhabited place but also as a potential archaeological site capable of being physically explored -at least through most RS techniques available-represents a clear example of how the preservation of historical features in the area presents often a random character. Indus archaeologists have long been aware of the potential for the river action to result in a significant loss/occultation of archaeological sites, including big urban sites, which impact both surface analysis and RS. The potential for settlement and/or site loss closely related to riverine dynamics thus necessitates a good understanding of river morphodynamics when studying historic and protohistoric settlement patterns. The challenges must be addressed in two main levels: first to understand the processes of creation and abandonment of the settlements, and; secondly, to understand the processes that created the archaeological record available to archaeological research. The impact of those questions acquires an important role for RS, as it is now often the first step from which archaeological survey projects are developed. We are not only considering here the punctual potential effects that river morphodynamics might have in the abandonment of some sites and the covering of earlier inhabited or cultivated areas, but also its long-term effects on cultural heritage. We might consider DG Khan as an exceptional case, and of course most of the similar settlements documented in the SoI maps were not abandoned during the same period. From an historical perspective, however, the circumstances documented in colonial DG Khan history are processes that have been operating across a wide geographic area and during a long period of time.

An illustrative example is the situation of Mohenjo Daro, a large archaeological site in the lower Indus river basin where substantial works have been carried out on flood defences in recent years in response to the concern about the erosion caused by the river approximation to an archaeological site preserved for 4500 years [6,54].

The analysis of DG Khan flood evidence shows that our understanding of surface river channels, both active and dry, is largely dependent on the dynamics active during the short period of time before the image—whether they be maps or RS—are obtained. Just taking into account the case of 1909 DG Khan we can document channels on the east, west, and actually crossing the area of the city in several moments within a short period of time, that would be impossible to differentiate in historical sedimentary terms. Moreover, the subsoil can potentially contain remains of multiple previous episodes from previous unknown chronologies. In DG Khan, the triangle of surface evidences of palaeochannels, buried alluvial sediments, and proximity to human settlements might be not as straightforward as we might suppose at first glimpse.

5.3. Historical Social Responses to River Morphodynamics

In the specific case of DG Khan, two consecutive main actions were decided and executed in the context of the colonial administration. Both imply the mobilization of intellectual, technical and economic resources controlled by the colonial administrators and had been decisive for the processes studied here: In the first place, since the foundation of the cantonment, and after the flood in 1858, there is a strong effort in the protection of the town, with the building of new embankments and the improvement of those already in place. The turning point was, apparently, the entering of the water in the old town in 1909. The second action was to move the town to a new location, safe from river movements. In consequence, although the end of the historical town was forced by the river geomorphology, it was, at the same time, a planned dismantlement (with the removal of construction and other types of material) by a central power which probably influenced the fact that the old settlement was almost invisible just a few years later. The response of DG Khan inhabitants makes us wonder to what point construction techniques and habitation patterns and practices in the past were influenced by similar river morphodynamics. When studying past settlement in similar landscapes it is, therefore, worth considering a type of habitation characterised by a high degree of mobility. One of the consequences of this mobility can be the multiplication of sites that were only inhabited for relatively shorts periods of time. These can include both short-term semi-nomadic sites and permanent settlements that can be occupied for various generations before their abandonment and re-location caused by river movements and flooding.

It is interesting to note how the new town was created following an orthogonal grid, representing the urbanism promoted by late nineteenth century European rationalism. To address the disturbing effects of the floods, European expertise was imported on landscape management and, on a middle-term perspective reinforced the colonial system, contributing to the ideological justification and increasing the control over the economic organisation, largely depending on the river soil productivity [3,43]. That results in landscapes in which the colonial footprint is increasingly evident over time. That process provides a good example of how internal historical processes -in our case the consolidation of European imperialism- are determinant factors in shaping responses to what we define as natural hazards. The analysis of historical study cases like the one exposed here can contribute as a support of comparative analysis, like the ones developed during the last decade in the Indus valley [55].

Strategies developed by human societies inhabiting Indus alluvial plains adapt the constructed social landscape to the geomorphological conditions and, at the same time, respond to the particular historical socio-political and cultural environment. Monitoring of the river conditions, the building of hydraulic infrastructure, and mobility summarises the basic elements of the management systems documented in the nineteenth century Indus valley [43], and they are not far from current approaches to flood management [31,34,36,40,56]. It is only to be expected that these were part of the landscape management strategies of past societies inhabiting this type of environment.

In first place, different traditional information transmission methods were reported during the nineteenth century [43], and similar methods could have been in use for a long time. The well stablished trade networks documented in the Indus basin since Prehistory [57] could have carried critical information together with the goods documented within the archaeological record. Landscape, hydrological and geomorphological knowledge could have historical significance in a short-term perspective, anticipating floods and undertaking actions to avoid or mitigate damage, but also in a long-term perspective, determining, e.g., where to stablish a new settlement. The role played by traditional knowledge about water management is very difficult to assess without a larger body of ethnographic and historical literature and some works have been developed in that direction the Indus valley [58].

The second set of strategies, hydraulic works, were at the centre of the British colonial system in Punjab [3] and they were object of debate since the beginning [4,43]. The main practical approach included the building of inundation canals and artificial levees to contain and divert the excess of water during flooding episodes. Both have been previously used in the Indus basin. Already in the proto-historical period hydraulic works are known in the urban centres [57,59]. Had either been used in the countryside it is a debated question with no data available [60]. The scale of the works developed during the British colonial rule, however, have no precedents [4]. The previous works were integrated into the new systems or abandoned. As a result, few elements of the systems existent before the late nineteenth century are preserved. The analysis of DG Khan is an example of how channels and embankments are continuously re-adapted and can leave few traces when abandoned. It was also the main strategy adopted since 1858 and until the 1909 flood.

The third group contemplates different forms of mobility. The movement of the whole DG Khan settlement was the decision adopted in 1909–1910, resulting in a new settlement. The possibility of moving entire settlements in a dangerous location to a safer place was not unknown on the Indus region. Historical accounts of early nineteenth century reported the extension of temporary settlement. For example, in relation to the Sindh, Richard Burton wrote in 1849: "most villages could be razed to the ground, transported to the requisite distance, and re-erected in a week, at an expense of probably a couple of rupees per house" [43]. Few years earlier, in 1842, Mountstuart Elphinstone wrote in relation to the Indus river: "the labourers have temporary huts erected, and cultivated" and "and the villages are only temporary, with a few exceptions" [61]. Mobile strategies have been proposed for the interpretation of prehistoric settlement patterns related to irregular water availability in semi-arid areas of Western India [22,60,62]. The setup of temporary agricultural plots and semisedentary husbandry were other well-known traditions in the Indus basin nineteenth century. The role played by semi-nomadic pastoralism in the pre-colonial Sindh [43] and Punjab [3] could be read in terms of that perspective. In contrast, the colonial administration promoted permanent forms of agriculture and settlement [43].

How different societies combined these strategies through time is a significant factor in the processes of formation of the archaeological record and in the interpretation of settlement patterns. The historical moment that defined both the 1909 flood episode and the historical documentation related is characterised by the European Imperialism, in its British version in the case of the Indian subcontinent. The role of the colonial authorities is prominent in the available documentation and the actions and interests of other local and regional actors are much less documented. Special local contributions for the embankments are mentioned in the Times of India Illustrated [44] and the active destruction of new embankments by local farmers is reported in other areas of British India and there were claims that it happened as well in DG Khan [43,45], but the complexity of the local social and economic interests and conflicts are largely ignored in the documentation or presented exclusively from the British colonial authorities point of view. There is already a significant research on how strategies related to hydrological dynamics shaped the colonial Punjabi landscape between 1842 and

1947, its role in the larger context of the Indian subcontinent and its impact until the present [3,4,43]. Much less research is available for previous periods, a necessary step to understand the long-term dynamics that shaped the Indus basin landscape.

Supplementary Materials: For access the code used for the analysis of Multispectral satellite imagery follow this link: https://github.com/wamstrin/Orengo_Petrie_2017_RS/blob/patch-1/version_Garcia_etal_geosciences. For access the code used to create the Multi-Scale Relief Model (MSRM) follow this link: https://github.com/wamstrin/Orengo_Petrie_2018_MSRM/blob/patch-1/version_Garcia_etal_geosciences.

Author Contributions: Conceptualization, A.G.; methodology, A.G. & H.A.O.; software, A.G., H.A.O. & F.C.C.; formal analysis, A.G.; investigation, A.G.; resources, A.G., H.A.O., F.C.C., A.S.G., C.A.P.; data curation, A.G.; writing—original draft preparation, A.G.; writing—review and editing, A.G., H.A.O., F.C.C., A.S.G. & C.A.P.; visualization, A.G.; supervision, H.A.O., C.A.P.; project administration, A.G.; funding acquisition, A.G.

Funding: *WaMStrIn* project (A.G.) has received funding from the European Union's Horizon 2020 research and innovation programme under the Marie Skłodowska-Curie grant agreement No. 746446. The *TwoRains* project (H.A.O., A.S.G. & C.A.P.) has been primarily funded by the European Research Council under the European Union's Horizon 2020 research and innovation program (grant agreement no. 648609), but has also received support from DST/UKIERI, the British Academy and the McDonald Institute for Archaeological Research. The *Marginscapes* project (F.C.C.) has been funded by the European Union's Horizon 2020 Research and Innovation programme under the Marie Sklodowska-Curie grant agreement No. 794711. HAO work has been funded by the Ramón y Cajal program, Spanish Ministry of Science, Innovation and Universities.

Acknowledgments: The authors would like to thank the heads of the Department of AIHC and Archaeology at Banaras Hindu University and the Department of Archaeology at the University of Cambridge who have provided their support to the *TwoRains*, *WaMStrIn* and *Marginscapes* projects. We would especially like to thank the staff of the Map Room and Imaging Services at the University Library at the University of Cambridge for providing access to and high-resolution copies of the Survey of India 1" to 1-mile maps in their collection.

Conflicts of Interest: The authors declare no conflict of interest.

References

1. Meadows, A.; Meadows, P. (Eds.) *The Indus River: Biodiversity, Resources, Humankind*; OUP Pakistan: Karachi, Pakistan, 1999; ISBN 978-0-19-577905-9.

2. Dong, S.; Bandyopadhyay, J.; Chaturvedi, S. *Environmental Sustainability from the Himalayas to the Oceans: Struggles and Innovations in China and India*; Springer: Cham, Switzerland, 2017.

3. Agnihotri, I. Ecology, land use and colonisation: The canal colonies of Punjab. *Indian Econ. Soc. Hist. Rev.* **1996**, *33*, 37–58. [CrossRef]

4. Hill, C.V. *River of Sorrow: Environment and Social Control in Riparian North India, 1770–1994/Christopher V. Hill*; Association for Asian Studies: Ann Arbor, MI, USA, 1997.

5. Schuldenrein, J.; Wright, R.P.; Mughal, M.R.; Khan, M.A. Landscapes, soils, and mound histories of the Upper Indus Valley, Pakistan: New insights on the Holocene environments near ancient Harappa. *J. Archaeol. Sci.* **2004**, *31*, 777–797. [CrossRef]

6. Jansen, M.; Mulloy, M.; Urban, G. *Forgotten Cities on the Indus: Early Civilization in Pakistan from the 8th to the 2nd Millennium BC*; Verlag Philipp von Zabern: Mainz, Germany, 1991.

7. Mughal, M.R. A Preliminary Review of Archaeological Surveys in Punjab and Sindh: 1993–95. *South Asian Stud.* **1997**, *13*, 241–249. [CrossRef]

8. Crawford, O.G.S. A Century of Air-photography. *Antiquity* **1954**, *28*, 206–210. [CrossRef]

9. Orengo, H.A.; Knappett, C. Toward a Definition of Minoan Agro-Pastoral Landscapes: Results of the Survey at Palaikastro (Crete). *Am. J. Archaeol.* **2018**, *122*, 479. [CrossRef]

10. Menze, B.H.; Ur, J.A.; Sherratt, A.G. Detection of Ancient Settlement Mounds: Archaeological Survey Based on the SRTM Terrain Model. *Photogramm. Eng. Remote Sens.* **2006**, *72*, 321–327. [CrossRef]

11. Chase, A.F.; Chase, D.Z.; Weishampel, J.F.; Drake, J.B.; Shrestha, R.L.; Slatton, K.C.; Awe, J.J.; Carter, W.E. Airborne LiDAR, archaeology, and the ancient Maya landscape at Caracol, Belize. *J. Archaeol. Sci.* **2011**, *38*, 387–398. [CrossRef]

12. Evans, D.H.; Fletcher, R.J.; Pottier, C.; Chevance, J.-B.; Soutif, D.; Tan, B.S.; Im, S.; Ea, D.; Tin, T.; Kim, S.; et al. Uncovering archaeological landscapes at Angkor using lidar. *Proc. Natl. Acad. Sci. USA* **2013**, *110*, 12595–12600. [CrossRef]

13. Pournelle, J.R. *Marshland of Cities: Deltaic Landscapes and the Evolution of Early Mesopotamian Civilization*; University of California: San Diego, CA, USA, 2003.

14. Hritz, C. Tracing Settlement Patterns and Channel Systems in Southern Mesopotamia Using Remote Sensing. *J. Field Archaeol.* **2010**, *35*, 184–203. [CrossRef]

15. Orengo, H.A.; Krahtopoulou, A.; Garcia-Molsosa, A.; Palaiochoritis, K.; Stamati, A. Photogrammetric re-discovery of the hidden long-term landscapes of western Thessaly, central Greece. *J. Archaeol. Sci.* **2015**, *64*, 100–109. [CrossRef]

16. Yashpal, S.B.; Sood, R.K.; Agarwal, D.P. Remote sensing of the lost Saraswati river. *Proce. Indian Acad. Sci. (Earth Planet. Sci.)* **1980**, *89*, 317–337.

17. Ramasamy, S.M.; Bakliwal, P.C.; Verma, R.P. Remote sensing and river migration in Western India. *Int. J. Remote Sens.* **1991**, *12*, 2597–2609. [CrossRef]

18. Gupta, A.K.; Sharma, J.R.; Sreenivasan, G. Using satellite imagery to reveal the course of an extinct river below the Thar Desert in the Indo-Pak region. *Int. J. Remote Sens.* **2011**, *32*, 5197–5216. [CrossRef]

19. Rajani, M.B.; Rajawat, A.S. Potential of satellite based sensors for studying distribution of archaeological sites along palaeo channels: Harappan sites a case study. *J. Archaeol. Sci.* **2011**, *38*, 2010–2016. [CrossRef]

20. Dave, A.K.; Courty, M.-A.; Fitzsimmons, K.E.; Singhvi, A.K. Revisiting the contemporaneity of a mighty river and the Harappans: Archaeological, stratigraphic and chronometric constraints. *Quat. Geochronol.* **2019**, *49*, 230–235. [CrossRef]

21. Balbo, A.L.; Rondelli, B.; Cecília Conesa, F.; Lancelotti, C.; Madella, M.; Ajithprasad, P. Contributions of geoarchaeology and remote sensing to the study of Holocene hunter–gatherer and agro-pastoral groups in arid margins: The case of North Gujarat (Northwest India). *Quat. Int.* **2013**, *308–309*, 53–65. [CrossRef]

22. Conesa, F.C.; Madella, M.; Galiatsatos, N.; Balbo, A.L.; Rajesh, S.V.; Ajithprasad, P. CORONA Photographs in Monsoonal Semi-arid Environments: Addressing Archaeological Surveys and Historic Landscape Dynamics over North Gujarat, India. *Archaeol. Prospect.* **2015**, *22*, 75–90. [CrossRef]

23. Giosan, L.; Clift, P.D.; Macklin, M.G.; Fuller, D.Q.; Constantinescu, S.; Durcan, J.A.; Stevens, T.; Duller, G.A.T.; Tabrez, A.R.; Gangal, K.; et al. Fluvial landscapes of the Harappan civilization. *Proc. Natl. Acad. Sci. USA* **2012**, *109*, E1688–E1694. [CrossRef]

24. Wright, R.; Hritz, C. Satellite Remote Sensing Imagery: New Evidence for Sites and Ecologies in the Upper Indus. In *South Asian Archaeology 2007*; BAR International Series: Oxford, UK, 2013; pp. 315–321.

25. Orengo, H.A.; Petrie, C.A. Large-Scale, Multi-Temporal Remote Sensing of Palaeo-River Networks: A Case Study from Northwest India and its Implications for the Indus Civilisation. *Remote Sens.* **2017**, *9*, 735. [CrossRef]

26. Orengo, H.A.; Petrie, C.A. Multi-scale relief model (MSRM): A new algorithm for the visualization of subtle topographic change of variable size in digital elevation models. *Earth Surf. Process. Landf.* **2018**, *43*, 1361–1369. [CrossRef]

27. Singh, R.N.; Green, A.S.; Ranjan, L.M.; Alam, A.; Petrie, C.A. Between the Hinterlands: Preliminary Results from the TwoRains Survey in Northwest India 2017. *Man Environ.* **2018**, in press.

28. Petrie, C.A.; Orengo, H.A.; Green, A.S.; Walker, J.S.; Garcia, A.; Conesa, F.C.; Singh, R.N.; Knox, J.R. Mapping archaeology by mapping an empire: Using historical maps to reconstruct ancient settlement landscapes in modern India & Pakistan. *Geosciences* **2018**, in press.

29. Singh, M. The British Indian Empire: Military Geography and Emergence of New Military Landscapes in Punjab. *Shabd–Braham Int. Res. J. Indian Lang.* **2017**, *5*, 32–35.

30. Akhtar, S. The Historical and Archeological Significance Dera Ghazi Khan District through Ages. *Archaeol. Anthropol. Open Access* **2018**, *2*, 4. [CrossRef]

31. Akhtar, A. *Indus Basin Floods: Mechanisms, Impacts, and Management*; Asian Development Bank: Mandaluyong City, Philippines, 2013; p. 67.

32. Webster, P.J.; Toma, V.E.; Kim, H.-M. Were the 2010 Pakistan floods predictable? *Geophys. Res. Lett.* **2011**, *38*. [CrossRef]

33. Gaurav, K.; Sinha, R.; Panda, P.K. The Indus flood of 2010 in Pakistan: A perspective analysis using remote sensing data. *Nat. Hazards* **2011**, *59*, 1815. [CrossRef]

34. Hashmi, H.N.; Siddiqui, Q.T.M.; Ghumman, A.R.; Kamal, M.A. A critical analysis of 2010 floods in Pakistan. *Afr. J. Agric. Res.* **2012**, *7*, 1054–1067.

35. Kirsch, T.D.; Wadhwani, C.; Sauer, L.; Doocy, S.; Catlett, C. Impact of the 2010 Pakistan Floods on Rural and Urban Populations at Six Months. *PLoS Curr.* **2012**, *4*. [CrossRef]

36. Tariq, M.A.U.R.; van de Giesen, N. Floods and flood management in Pakistan. *Phys. Chem. Earth Parts A/B/C* **2012**, *47–48*, 11–20. [CrossRef]

37. Daily Images of the 2010 Flood Episodes Are Available in NASA Worldview Web Service. Available online: https://worldview.earthdata.nasa.gov/?p=geographic&l=MODIS_Terra_CorrectedReflectance_TrueColor,Graticule(hidden),Reference_Labels(hidden),Coastlines(hidden)&t=2018-11-14-T00%3A00%3A00Z&z=3&v=51.000945871291805,5.956715959569404,110.8720396212918,39.14421595956941 (accessed on 14 November 2018).

38. See, e.g., The Cartographic Material about Pakistan Collected by the UN Program UNOSAT. Available online: http://www.unitar.org/unosat/maps/PAK (accessed on 14 November 2018).

39. Heavy Rains and Dry Lands Don't Mix: Reflections on the 2010 Pakistan Flood. Available online: https://earthobservatory.nasa.gov/Features/PakistanFloods (accessed on 14 November 2018).

40. Amarnath, G.; Rajah, A. An evaluation of flood inundation mapping from MODIS and ALOS satellites for Pakistan. *Geomat. Nat. Hazards Risk* **2016**, *7*, 1526–1537. [CrossRef]

41. Scholberg, H. *The District Gazetteers of British India. A Bibliography*; Bibliotheca Asiatica; [Poststr. 4]; Inter Documentation: Zug, Switzerland, 1970; Volume 3.

42. *Gazetter of the Dera Ghazi Khan District, 1893–1897*; Reprint; Sang-e-Meel Publications: Lahore, Pakistan, 1990.

43. Weil, B. The Rivers Come: Colonial Flood Control and Knowledge Systems in the Indus Basin, 1840s–1930s. *Environ. History* **2006**, *12*, 3–29. [CrossRef]

44. Times of India. *The Passing of Dera Ghazi Khan*; Times of India: Bombay, India, 1910.

45. Bellasis, E.S. *Punjab Rivers and Works*; E. & F.N. Spon, Ltd.: London, UK, 1912.

46. Declassified Satellite Imagery—1 | The Long Term Archive. Available online: https://lta.cr.usgs.gov/declass_1 (accessed on 14 November 2018).

47. Casana, J.; Cothren, J. Stereo analysis, DEM extraction and orthorectification of CORONA satellite imagery: Archaeological applications from the Near East. *Antiquity* **2008**, *82*, 732–749. [CrossRef]

48. Scollar, I.; Galiatsatos, N.; Mugnier, C. Mapping from CORONA: Geometric Distortion in KH4 Images. *Photogramm. Eng. Remote Sens.* **2016**, *82*, 7–13. [CrossRef]

49. Masson, C. *Narrative of Various Journeys in Balochistan, Afghanistan, and the Panjab: Including a Residence in Those Countries from 1826 to 1838/Charles Masson*; Richard Bentley: London, UK, 1842.

50. Global Surface Water Explorer. Available online: https://global-surface-water.appspot.com/ (accessed on 15 November 2018).

51. Pekel, J.-F.; Cottam, A.; Gorelick, N.; Belward, A.S. High-resolution mapping of global surface water and its long-term changes. *Nature* **2016**, *540*, 418–422. [CrossRef] [PubMed]

52. Green, A.S.; Singh, R.N.; Alam, A.; Garcia, A.; Greene, L.M.; Conesa, F.C.; Orengo, H.A.; Ranjan, A.; Petrie, C.A. Re-discovering dynamic ancient landscapes: Archaeological survey of features from historical maps in northwest India and their implications for the large-scale distribution of settlements throughout South Asia. *Remote Sens.* **2018**. in preparation.

53. Possehl, G.L. *Indus Age: The Beginnings*; Oxford & IBH Publihing Co.: New Delhi, India, 1999; ISBN 81-204-1296-6.

54. Centre, U.W.H. Archaeological Ruins at Moenjodaro. Available online: https://whc.unesco.org/en/list/138/ (accessed on 14 November 2018).

55. Miller, H.M.-L. Surplus in the indus Civilization. In *Surplus: The Politics of Production and the Strategies of Everyday Life*; University Press of Colorado: Boulder, CO, USA, 2015; pp. 97–120.

56. Munir, B.A.; Iqbal, J. Flash flood water management practices in Dera Ghazi Khan City (Pakistan): A remote sensing and GIS prospective. *Nat. Hazards* **2016**, *81*, 1303–1321. [CrossRef]

57. Wright, R.P. *The Ancient Indus: Urbanism, Economy, and Society, Case Studies in Early Societies*; Cambridge University Press: Cambridge, UK, 2010.

58. Miller, H.M.-L. Water supply, labor requirements, and land ownership in Indus floodplain agricultural systems. In *Agricultural Strategies*; The Cotsen Institute of Archaeology Press: Los Angeles, CA, USA, 2006; pp. 92–128.

59. Jansen, M. Water supply and sewage disposal at Mohenjo-Daro. *World Archaeol.* **1989**, *21*, 177–192. [CrossRef]

60. Petrie, C.A.; Singh, R.N.; Bates, J.; Dixit, Y.; French, C.A.I.; Hodell, D.A.; Jones, P.J.; Lancelotti, C.; Lynam, F.; Neogi, S.; et al. Adaptation to Variable Environments, Resilience to Climate Change: Investigating *Land, Water and Settlement* in Indus Northwest India. *Curr. Anthropol.* **2017**, *58*, 1–30. [CrossRef]
61. Elphinstone, M. *An Account of the Kingdom of Caubul, and Its Dependencies, in Persia, Tartary, and India (1842)*; Richard Bentley: London, UK, 1842.
62. Conesa, F.C.; Devanthéry, N.; Balbo, A.L.; Madella, M.; Monserrat, O. Use of Satellite SAR for Understanding Long-Term Human Occupation Dynamics in the Monsoonal Semi-Arid Plains of North Gujarat, India. *Remote Sens.* **2014**, *6*, 11420–11443. [CrossRef]

geosciences

MDPI

Article

Towards National Archaeological Mapping. Assessing Source Data and Methodology—A Case Study from Scotland

Łukasz Banaszek *, Dave C. Cowley and Mike Middleton

Historic Environment Scotland, John Sinclair House, 16 Bernard Terrace, Edinburgh EH8 9NX, UK;
dave.cowley@hes.scot (D.C.C.); mike.middleton@hes.scot (M.M.)
* Correspondence: lukasz.banaszek@hes.scot; Tel.: +44-131-651-6804

Received: 6 July 2018; Accepted: 24 July 2018; Published: 26 July 2018

Abstract: While the National Record of the Historic Environment (NRHE) in Scotland contains valuable information on more than 170,000 archaeological monuments, it is clear that this dataset is conditioned by the disposition of past survey and changing parameters of data collection strategies over many decades. This highlights the importance of creating systematic datasets, in which the standards to which they were created are explicit, and against which the reliability of our knowledge of the material remains of the past can be assessed. This paper describes issues of data structure and reliability, then discussing the methodologies under development for expediting the progress of national-scale mapping with specific reference to the Isle of Arran. Preliminary outcomes of a recent archaeological mapping project of the island, which has been used to develop protocols for rapid large area mapping, are outlined. The primary sources for the survey were airborne laser scanning derivatives and orthophotographs, supplemented by field observation, and the project has more than doubled the number of known monuments of Arran. The survey procedures are described, followed by a discussion of the utility of 'general purpose' remote sensed datasets, focusing on the assessment of strengths and weaknesses for rapid mapping of large areas.

Keywords: airborne laser scanning; orthophotographs; archaeological survey; field reconnaissance; Arran; national archaeological mapping programme

1. Introduction

Approaches to archaeological survey and mapping of the historic environment in Scotland range from broad-brush characterisation, such as Historic Land-use Assessment (HLA) [1,2], to exploration through archaeological survey of landscapes and sites in varying degrees of detail [3–5]. The former has achieved national coverage, providing generalised mapping of Scotland's historic environment that is useful for strategic purposes such as feeding into national planning policy and management [6]. Archaeological survey has provided systematic and detailed information, which is valuable for site and landscape management and understanding of Scotland's past, for relatively small areas of the country. For example, field survey projects by the former Royal Commission on the Ancient and Historical Monuments of Scotland (RCAHMS) since the mid-1980s, and currently run by Historic Environment Scotland (HES), have covered about 10% of Scotland's landmass. These surveys, which mapped the visible topographic archaeological remains at various scales may add in these areas anything between 50% and over 200% of the previously known inventory of sites to the National Record of the Historic Environment (NRHE—formerly known as the National Monuments Record of Scotland (NMRS)). This shows that the contents of the NRHE are unrepresentative when considered at regional and national scales, and that the majority of the archaeological sites and monuments that survive in Scotland's landscape are not on record (Figure 1). RCAHMS/HES' established approaches to archaeological survey cannot realistically expect to achieve national levels of systematic data collection without either a massive increase in human resourcing or conducting such survey over very long periods of time, measured in centuries. This means that the prospect of creating systematic national coverage of archaeological information at a level that can inform site and landscape-scale management and understanding is a very distant prospect, leaving us with the knowledge that large numbers of sites and large areas of archaeological landscape remain unrecorded.

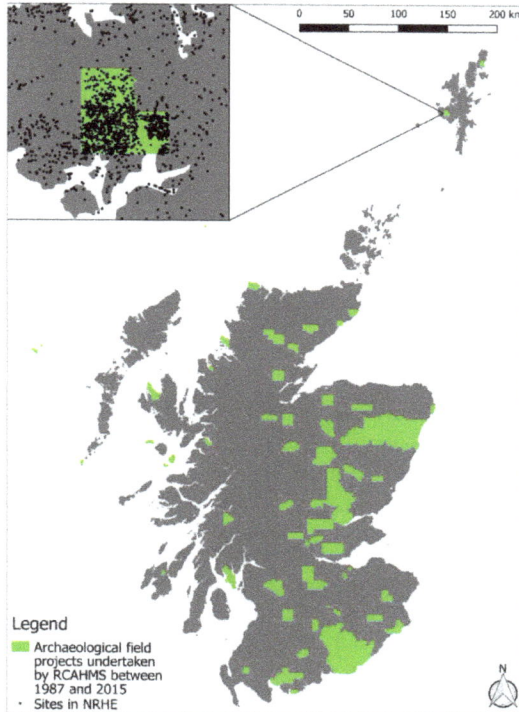

Figure 1. The areas of systematic, extensive field survey to modern standards undertaken over the last 30 years have covered some 10% of Scotland's 80,077 km² land mass, drawing on a variable human resource of anything up to eight field workers. This highlights the lack of scalability in traditional approaches to large area mapping, which rely heavily on human resources and field visits. The significant addition to the NRHE that systematic area survey can make is illustrated by West Mainland, Shetland (top left), where the discrete area investigated in 2010 is clearly visible in the increased density of site locations. © Historic Environment Scotland.

For HES, as the lead public body for the historic environment in Scotland, the knowledge that the existing data available in the NRHE is biased, and only very generally representative of what survives in the landscape, is a challenge. Knowing that established approaches to field survey are not scalable, a response, supported by the increasing availability of remotely sensed datasets, has been proposed. Using the Isle of Arran as a case study, this paper presents a methodology under development in which remote sensing data is the primary source for archaeological prospection of large areas. This approach aims to make best use of the resources available in Scotland, responding to variability in local environmental conditions and addressing the challenges of developing survey practice.

The paper begins by outlining the history and character of Scotland's NRHE, within a context of other national mapping programmes in Scotland and beyond (Section 2). Thereafter, we explain why Arran was chosen as an outdoor laboratory for scoping our approach to rapid large area mapping (Section 3). This is followed by discussion of the remote sensed datasets that were used, and the desk-based and field survey methodologies (Section 4). Preliminary results from the Arran survey are presented (Section 5), providing context for elaboration of initial survey outcomes, in particular addressing the reliability of desk-based interpretation and the roles of targeted field observation (Section 6). The paper concludes with a discussion of the utility of the available remote sensing datasets and outlines some next steps in scoping our approach to national archaeological mapping (Section 7).

2. A National Archaeological Record—Setting the Scene

Reliable and systematic information about the distribution and character of the archaeological record are amongst the principal foundations of effective heritage management and sound research. However, critical analysis of heritage datasets will usually quickly reveal how unrepresentative they are, constituted as they may be from multiple sources. This issue is the central objective that the development of national archaeological mapping protocols seeks to address—by leveraging the proliferation of remote sensed data and devising appropriate survey methodologies to generate systematic data. In Scotland, this means exploring the potential to address the current limitations of the NRHE through systematic archaeological survey, at a national scale, and in a timescale of years rather than many decades or even centuries.

2.1. The National Record of the Historic Environment—Origins and Character

Scotland's NRHE, delivered online through an interface known as Canmore [7], is many things. It is an index to the Historic Environment Scotland archive—a national collection of material relating to archaeology, buildings, industrial and maritime heritage. It is a catalogue of images relating primarily, but not exclusively, to Scotland and it is an inventory of recorded historic monuments, archaeological fieldwork and architectural and aerial surveys [8–10]. It is important to recognise that it is not a single, comprehensive survey, but has grown organically with roots linking it to numerous sources of information This work spans over 200 years of investigation and collection and includes the work of map makers who wished to depict archaeological landmarks, inventorying programmes, and emergency surveys in response to land use change, amongst others. Each had different motivations and specifications for what they included, and these continue to have a profound impact on the character of the incomplete and patchy information that makes up the NRHE.

Early modern surveyors, such as Timothy Pont in the late 16th century [11], were some of the first to record archaeological monuments in Scotland on an ad hoc basis. William Roy, the military cartographer, took a more systematic approach, publishing his "Military Antiquities of the Romans in North Britain", in 1793 [12]. Between 1843 and 1882 the Ordnance Survey (OS) carried out the first nationwide survey of Scotland [13]. This included many of the more prominent archaeological monuments, but was not necessarily systematic, drawing on the knowledge of local informants. The depiction of archaeological sites on OS maps continued on an ad hoc basis with subsequent editions until the 1920s when the OS appointed O.G.S. Crawford as their first archaeology officer, a position he grew and developed into what became the Ordnance Survey Archaeology Division [14]. This body was responsible for documenting the ancient monuments that were included on the OS maps in a card index [15], with field checking and inspection of aerial photographs in support. When the Ordnance Survey embarked on a Scotland wide map revision, this included the OS Archaeology Division carrying out a nationwide programme of field survey between 1947 and 1983 [14].

The establishment of the Society of Antiquaries of Scotland in 1780 provided another focus for growing antiquarian work, including the results of field survey work [16]. This growing interest in ancient monuments became a driver for State funded survey, with the establishment in 1908 of RCAHMS to, " ... make an inventory of the ancient and historical monuments ... " [9]. Work on a national inventory began by creating a series of regional inventories, beginning with the publication of the County of Berwick in 1909 [17], in a programme that was stopped in the mid-1980s, and replaced with an area-based approach to field survey.

In 1983 the OS Archaeology Division was disbanded, and its functions passed to other organisations. At the same time the antecedent of the NRHE was created, with at its core, the OS record cards [18]. This record continues to be a primary repository for archaeological survey and other information derived from a range of sources. Significant contributors include the Council for Scottish Archaeology (CSA) who since 1947 have published an annual list of fieldwork and discoveries [19]. This, along with an online recording environment known as OASIS, provide the public, academics and commercial organisations with a mechanism by which to contribute new information to Canmore [20].

The archaeological survey work of RCAHMS/HES has developed over the years. Field recording has moved away from creating inventories to recording smaller areas. Field survey has been complemented by the addition of an Aerial Survey programme initiated in 1975 [9,21]. Desk-based projects, such as the First Edition Survey Project (FESP) which assimilated evidence of unroofed buildings and other remains from the mid-19th century maps into the NRHE [22], have also contributed (Figures 2 and 3). And, as might be expected, the definitions of what comprises an archaeological site have changed, with the progressive routine inclusion of 19th century remains and military archaeology of 20th century date [23–25].

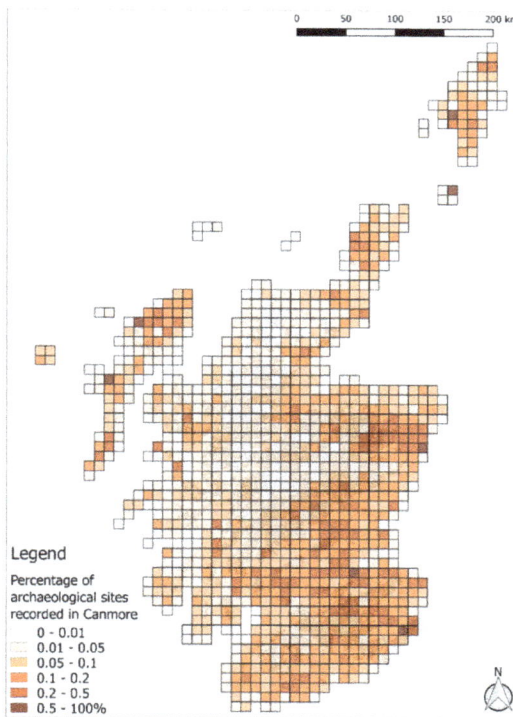

Figure 2. Map, by 10 km grid square, showing the density of archaeological records with an accuracy recorded as being within 10m. The east and southern lowlands are noticeable for having most records while the central and western highlands and islands have a noticeably lower density. Areas of intensive survey are apparent as areas of darker tones. © Historic Environment Scotland (Source: NRHE/Canmore).

It should be clear from these examples of the many sources of information which make up the NRHE that the Record's holdings do not represent a comprehensive survey of Scotland. It is an inventory of what has been recorded. It reflects the interests and recording policies of those that created it, with clear bias in contents as a result [26]. Thus, areas of systematic survey can be identified (Figure 2), with areas that have attracted less attention also evident, such as the northwest Highlands. While the latter may be in part due to the terrain in this area, it does also reflect a lower intensity of survey activity. A bias towards the lowland areas of Scotland is reinforced by the aerial reconnaissance programme which has long maintained a focus on lowland agricultural landscapes, most conducive to cropmarking. Most biases in the NRHE are subtle and may reflect the aims of the individuals involved in the recording, or the focus of a specific project (Figure 3).

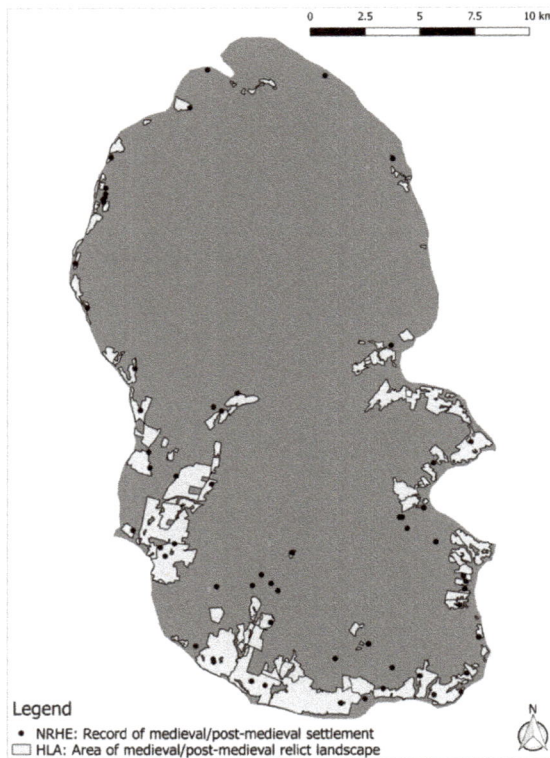

Figure 3. The NRHE and HLA record of medieval/post-medieval settlement and land use, as it was before the recent survey, provide two different views of the extent of known remains of these dates on Arran, both with their limitations depending on survey specification and source data. The distribution of township records is largely due to the First Edition Survey Project [22], which identified unroofed buildings on the earliest Ordnance Survey (OS) maps, but where decisions on what to map and what to ignore depended entirely on the work of the 19th century OS surveyors. They operated within specifications that included a minimum height requirement (0.3 m) before a feature qualified for inclusion on the map. These decisions still have an impact on the nature of the NRHE over 150 years later. The polygons from the HLA document crofting townships, medieval and post-medieval settlement and agriculture as recognised from a national land-use characterisation programme, primarily from aerial photographs and only for areas greater than one hectare. © Historic Environment Scotland (Sources: NRHE/Canmore and HLA).

2.2. Scoping a National Archaeological Mapping Programme

When the national programme of County Inventories for Scotland was initiated in 1908 there was an aspiration to cover the entire country, though it was quickly recognised that this was not readily achievable [9]. Since then, broad brush characterisation and desk-based assessments like FESP and HLA have achieved national coverage at a resolution of 1:10,000 and 1:25,000, respectively. However, aspirations to achieve systematic national archaeological mapping have been a distant hope. Indeed, resourcing constraints and survey policy have rather focussed on discrete area surveys. Thus, nothing has taken place in Scotland like the National Mapping Programme (NMP) developed in England [27–29] to identify, map, record, and better understand archaeological sites and landscapes through effective use of remotely sensed data. This programme is being reviewed at the time of writing, but it covered just over half of England's land mass, documenting over 120,000 newly discovered archaeological sites over several decades with significant human resourcing. This is due, in part, to its aspiration to basic line work and polygonised depiction of the visible archaeological features. A rather different approach has been taken in Baden-Württemberg, Germany, where rapid mapping based on airborne laser scanning (ALS) derivatives by a single operator has generated large area coverage and some 600,000 monuments over less than a decade [30]. These are amongst a number of approaches to large area or national mapping, and have most relevance to a Scottish context.

These examples of very large area mapping have a bearing on considerations of how to frame an approach to rapid national archaeological mapping in Scotland. The need to balance area coverage and rapidity of work with the level of detail recorded has been identified as a central issue. Indeed, a tendency to record in increasing degrees of detail can be identified in some of the survey projects shown on Figure 1. The second-named author (DCC) started working in field survey in the late 1980s in a project aiming to sample areas at general threat from afforestation. This Afforestable Land Survey initially used basic survey techniques to map coordinates and site areas quickly, supplemented by text descriptions. In the late 1980s an electronic theodolite was acquired, representing a big improvement in the metrical accuracy that could be achieved in site location. Initially, rapid mapping methods were maintained, using one- and two-point codes to map rectangular and round buildings at 1:10,000 map scale (Figure 4). However, in fairly short order, the team moved to more detailed recording, for example mapping all four corners of a building, and its partitions and other internal features. Therefore, what had started as recording a single centre point, or two points at either end of a building that took one minute to record, became a five-minute exercise collecting eight or more survey points. This, in microcosm, illustrates a tendency for archaeological recording to be drawn into detail, with consequent implications for the resource required, and a drift in the survey specification from rapid mapping to more detailed considered treatment. While this had a benefit in detail of record, the casualty was the rate of area coverage.

Keeping in mind the tendency for a drift to detail, the specification for the development of a modern national archaeological mapping emphasises speed and economy over detail, recognising that it is one approach in a nested set of levels of survey.

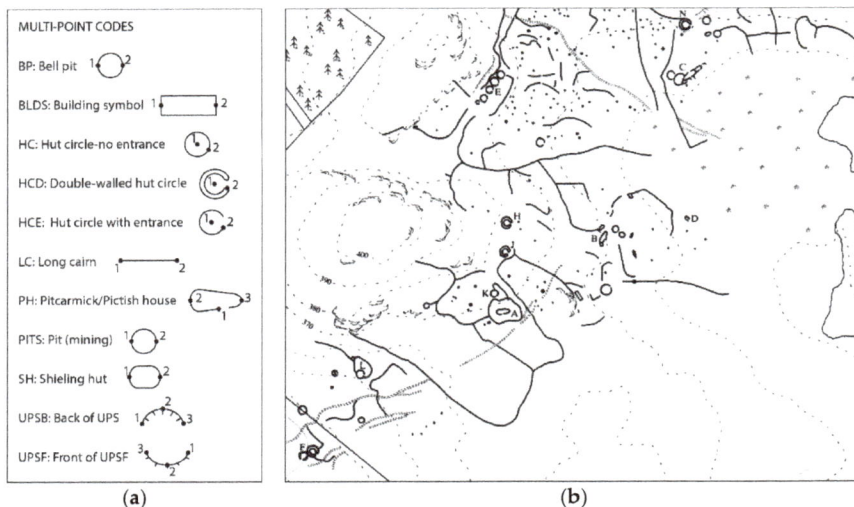

Figure 4. The use of simple codes for landscape mapping at 1:10,000 scale was designed to help expedite rapid coverage, as illustrated by an extract from a field reference sheet (**a**). However, a clear tendency can be identified towards mapping in more detail, with a shift in survey specification, sometimes underpinned by an implicit assumption that more detail is automatically 'better' illustrated (**b**) by an extract from mapping in north-east Perthshire [31]). © Historic Environment Scotland.

3. Case Study Area: 'Scotland in Miniature'

The requirement to develop an approach to national archaeological mapping, as one strand of HES' survey programmes, is underpinned by a belief that the NRHE should be as comprehensive as possible to inform understanding and management; including within the planning system and for designation. The increasing availability of high-resolution topographic (e.g., ALS) and optical (e.g., colour orthophotographs) data provides an opportunity to explore methods of area archaeological survey that can achieve rates of coverage considerably in excess of existing approaches. Economy and speed are key considerations if national coverage archaeological survey is to be achieved, and this foregrounds the efficient use of remote sensing (RS) data. Moreover, the RS data are 'general purpose' ALS and orthophotographs, that are freely or cheaply available, supplemented by oblique aerial photographs, historic aerial photographs and field observation. The island of Arran has been used as a laboratory to develop protocols for a National Archaeological Mapping Programme (NAMP), where the use of remote sensing data has been prioritised in the methodology, using ALS data as the primary source, with orthophotographs, field reconnaissance and field observation in supporting roles to address particular issues, such as poor RS data quality, dubiety in interpretation, or monument types that will not register in a digital terrain model (e.g., very small structures).

Arran, which is often referred to as 'Scotland in Miniature' because of its range of topographic and land use types, extends to some 432 km^2 and lies in the Firth of Clyde, on the west coast of Scotland. It has been selected as a laboratory for developing and testing NAMP approaches for several reasons. Firstly, it has been entirely covered with publicly available ALS data, orthophotographs, and historic aerial images (cf. Section 2.1). Secondly, the island represents a broad range of topographic, geological and land use types (Figure 5). Thirdly, there is a great diversity of archaeological remains on the island, which mirror the evidence across Scotland, ranging from the Neolithic to the Cold War. Fourthly, given the available resources, the size of the island was ideal to undertake a survey in a relatively short time. Finally, Arran is easily accessible from mainland, simplifying fieldwork logistics.

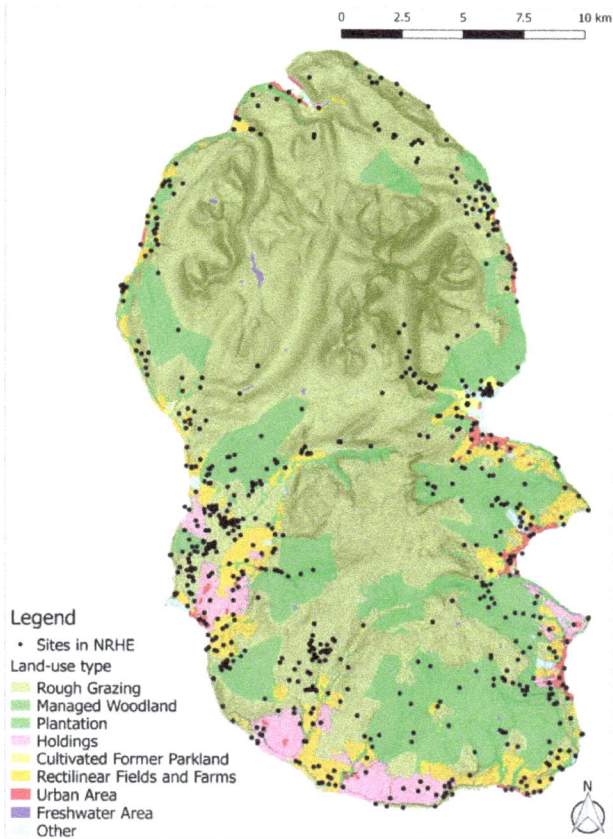

Figure 5. The distribution of the archaeological sites known on Arran before the recent HES survey, shown against a land use character map superimposed on a terrain model. © Historic Environment Scotland (Sources: NRHE, HLA and airborne laser scanning (ALS)).

4. Materials and Methods

As digital topographic datasets have become more readily available in Scotland, there has been a growing recognition of their value for landscape survey. Preliminary assessment of the Arran ALS data in July 2017 indicated its effectiveness in capturing many archaeological features and also the textures of the landscape [32]. This is in marked contrast to the available orthophotographs where summer vegetation (e.g., bracken) obscures many areas and lighting conditions in winter imagery are variable in the extent to which they enhance archaeological remains, though they provided other valuable information. Following preliminary assessment in 2017, the work was structured around a desk-based mapping phase (Section 4.2) in late 2017 and early 2018, followed by a field survey phase in spring 2018 (Section 4.3).

4.1. Data Sources

Keeping in mind that the methodology under development requires to be scaled up to a national programme, the survey of Arran made use of general purpose RS datasets [33]. This approach has significant implications, primarily accepting that the ALS data specification was not optimised for archaeological purposes. However, while recognising the importance of controlling data acquisition parameters in ideal situations [33–35], since the core objective is to achieve national coverage, using freely available data is very important. This is because it is highly unlikely that commissioning bespoke ALS survey of the entire country for archaeological purposes would ever be affordable.

General purpose ALS data have been already collected for some of Scotland, with full coverage planned in the future. To date, two data collections of Scottish Public Sector ALS (Figure 6) have been commissioned by Scottish Government, Scottish Environment Protection Agency, Scottish Water (phase I only), sportscotland (phase II only), and 13 local authorities (phase II only) collectively. In the second phase, the data was acquired between November 2012 and April 2014 for 66 areas across Scotland, including Arran, covering a total of 3,516 km^2. The commissioned point density was a minimum of 1 pts/m^2, and approximately 2 pts/m^2 on average. However, the data collected for Arran is notably higher quality, with an overall point density of 4.67 pts/m^2, with last echoes at 3.86 pts/m^2, and average ground point density of 2.95 pts/m^2.

Modern orthophotographs were also used in the project, supplied to HES as part of a Service Level Agreement, also administered through the Scottish Government. This comprises tiled imagery, at 25 cm ground sampling distance, derived from two surveys in winter 2008 and June 2011. In addition, aerial photographs taken from light aircraft and those from the aerial photographic archives of HES have been used as supporting information, with images dating back to the 1940s.

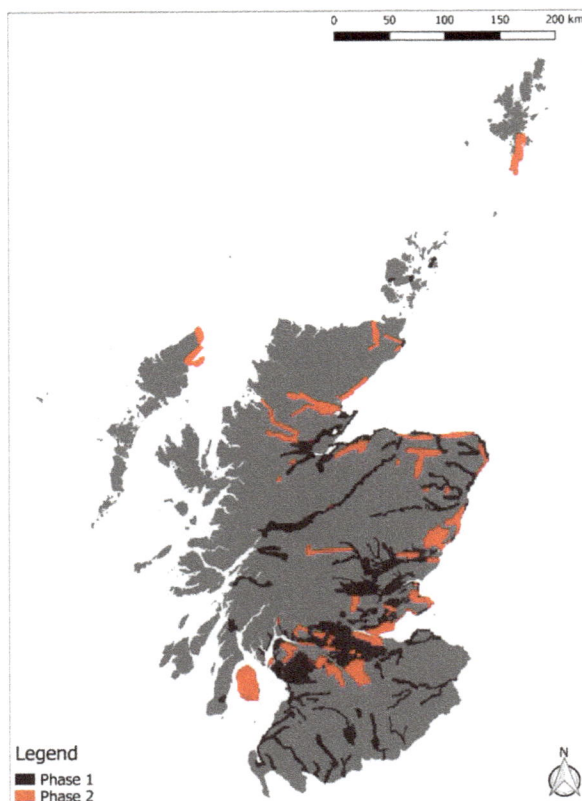

Figure 6. Scottish Public Sector ALS coverage as of July 2018. Arran is the biggest island fully covered by ALS data. © Historic Environment Scotland.

4.2. Desk-Based Survey Methodology

The ALS data was processed to produce a digital terrain model (DTM) at 0.5 m spatial resolution, based on ALS point clouds of 0.58 m ground point spacing. Next, several DTM) visualisations were generated using the Relief Visualisation Toolbox [36,37], and uploaded to the ArcGIS environment, where all digital datasets were analysed and interpreted.

Before the survey the NRHE contained 774 archaeological records on Arran, including stray finds (i.e., axeheads, flint tools, pottery), earthworks and standing remains (i.e., castles, buildings, huts, cairns), and a range of other monument types (historic drainage, field systems, quarries, standing stones, tracks, etc.). Stray finds were excluded from further processing and during preparatory work the remaining 671 site locations in the NRHE had site areas polygons produced with reference to paper record maps, orthophotographs and the ALS data. As a result, the locations and extents of known archaeological features were determined and informed later phases of the project.

Thereafter, in preparation for a programme of field survey in Spring 2018, eight interpreters with field survey experience undertook rapid interpretative mapping on the basis of ALS derivatives (Table 1), orthophotographs, and supporting information such as 19th century maps. The interpreters had different backgrounds and levels of experience in dealing with aerial imagery and/or remote sensing products. Within a defined timeframe (3 days), an average coverage of 90 km^2 per person was achieved. All areas were covered by at least two observers, and an experienced interpreter examined the entire island and undertook some validation of initial outcomes. Apart from demarcating the extent of the archaeological features identified during the desk-based work, the interpreters were asked to provide basic classifications of site type and to assess how confident they felt about their observations. Three different levels of confidence were established (Figure 7). Features clearly visible on the ALS and/or the orthophotographs and easily classified were assigned the highest confidence. Less certain interpretation was marked with a medium level of confidence, while ambiguous features that nevertheless appeared to be of potential interest, were identified with the lowest confidence level. The initials of interpreters were documented in each record to support subsequent analysis of results.

Table 1. Three sets of DTM visualisations, developed by Žiga Kokalj (1, 2) and, Łukasz Banaszek (3), were used as the primary source for the desk-based interpretation.

Stack No.	Visualisations Stacked	Transparency (%)
1	Sky-view Factor (Radius–5 m; Directions–16)	70
	Slope	80
	Multiple Hill-shade (Sun elevation 35 deg.; Red–band 14, Green–band 2, Blue–band 4)	0
2	Sky-view Factor (Radius–5 m; Directions–16)	25
	Openness-Positive (Radius–5 m; Directions–16)	50
	Slope	50
	Hill-shade (Sun elevation 35 deg.; Sun azimuth–315 deg.)	0
3	DTM colour values	80
	Local Dominance M10-20 DI1 A15 OH1.7	30
	Hill-shade (Sun elevation 35 deg.; Sun azimuth–315 deg.)	0

Figure 7. An example of the results of desk-based interpretation of ALS derivatives (1–the highest level of confidence) superimposed over a multi-direction hill-shade visualisation. Note the differences between the outputs from three different interpreters (A, B, C). While some archaeological features were identified by a single person, others were recognized by many. In the latter case, a single monument might be dealt with differently by different interpreters including: (a) with a varied level of confidence; (b) with a variable extent; (c) by including nearby features and thus creating large polygons or by drawing multiple smaller outlines; and (d), in some cases where the same archaeological feature was identified with a similar extent by two or more observers, with different classifications. Thus, a hut circle, as interpreted by one person, could be a shieling according to another. Additionally, while some interpreters used a detailed description, others described the same features using more general terms. This provides an insight into the variability across different operators in interpreting the same datasets. © Historic Environment Scotland.

4.3. Field Survey Methodology

The fieldwork phase of the project was conducted over six weeks in Spring 2018 by three teams each of 2–3 members, amounting to a total field input of about 150 person days. The teams, mainly made up of staff that were involved in the desk-based mapping phase, were supplied with personal GPS units and pen computers with a fully operational ArcGIS. Thus, whilst surveying, team members had access to the RS data and layers of current NRHE records as well as the outcomes of desk-based assessment. Field survey was directed towards visiting and assessing the identified low confidence level targets. Polygons marked with the highest level of confidence (level 1) were not investigated unless these were in the vicinity of level 2 and 3 targets and could be easily accessed or where there was another factor, e.g., obvious misinterpretation. In the same manner, existing NRHE records were not visited, since the project aimed to maximise the returns of previously unknown sites. Basic attribute data fields were used to document the identities of those who undertook the field observation and basic notes of observations. All field workers documented routes taken during field investigation using personal GPS units. The fieldwork teams were also encouraged to explore the landscape, especially in gaps in desk-based identifications and areas where the ALS might not be as effective (e.g., areas of woodland), to establish what additional sites might be identified.

5. Results

The desk-based mapping identified about 2100 targets (Figure 8), of which more than half (ca. 59%) were given a high level of confidence in interpretation, with nearly 27% given a medium level of confidence and about 14% with the lowest level of confidence. During the field survey weeks nearly 94% of the low confidence identifications and about 10% of the highest confidence targets were visited and evaluated. However, the above figure does not represent the number of newly discovered archaeological monuments on Arran. Whereas some of the identified features were recognised by more than one interpreter, resulting in multiple polygons (Figure 8), other targets were dismissed during the field reconnaissance and observation phase and classified as natural objects or identified as current land use.

Figure 8. The distribution of archaeological sites identified during the project, shown against features known before the survey and fieldwork tracks superimposed on a terrain model. This map does not show large areas of dense coniferous plantations, within which reconnaissance was limited to gaps in the forest, while many of the mountainous areas in the north were judged to be beyond the likely limit of surviving archaeological remains. Glen Rosa, the L-shaped trough valley in the northeast, was covered without a personal GPS unit and thus does not show any track. © Historic Environment Scotland (Sources: NRHE and ALS).

In addition, more than 300 sites were recorded during field reconnaissance only. Amongst these were: (a) small monuments, which could not be recognised in the ALS derivatives due to poor survey point distribution, (b) structures indistinguishable within boulder fields and scree, (c) sites in dense coniferous woodland, where ALS penetration capacity was limited, and (d), monuments visible in the ALS visualisations, but which were not noticed during desk-based mapping.

The recording of most walking by personal GPS units demonstrate extensive patterns of walking (Figure 8). It is clear that most of the island where targets had been identified during desk-based mapping were covered. This excludes areas in the centre and southeast where nesting birds created exclusion zones (to be visited at a later date), and high-altitude areas in the north, where potential targets identified during desk-based mapping were judged with the benefit of field experience as likely to represent geological features rather than archaeological monuments.

6. Discussion: What the Laser Saw, What the Lens Saw, and What the Surveyor Saw

The methodology outlined above represents a shift from the predominant reliance of HES area archaeological survey on field survey and ground observation, informed and supported by available aerial photographs. Rather, the approach foregrounds remote sensing data, in this instance, ALS, as the primary source for identification of earthworks and other archaeological remains, with targeted field reconnaissance and observation to explore specific issues of identification and weaknesses in the ALS data.

This methodology is not without risks and challenges. By limiting time spent in the field, and not visiting the majority of identified sites, the approach requires the development of experience and trust in interpreting RS data. This is built on field observation skills and knowledge, but nonetheless requires additional training in reading ALS derived visualisations. It is clear that there are differences between the products of the individuals undertaking the desk-based interpretations, an issue that will need to be addressed to produce coherent standardised results if the approach is to be scaled to a national level. This will include the need to develop clear protocols for defining site extents and classifications and for collecting metadata, for example, on which visualisations were used. Additionally, archaeological sites which were identified only during fieldwork, yet were clearly evident in ALS derivatives, demonstrate that improving skills in interpretative desk-based mapping is required to efficiently use data. Nevertheless, in many cases, the relatively high-resolution Arran data was not sufficient to visualise small archaeological features, such as huts and kilns, that were identified during field reconnaissance. In other areas, with sparse ALS ground points, site classifications from desk-based mapping were not correct and thus modified in the field. These are among the issues which demonstrate the importance of the analysis of data characteristics and survey methodology to inform the understanding of biases in the outputs of such survey [38]. Indeed, given the variability from area to area in factors such as data quality, the character of archaeological remains and vegetation cover, local assessments of data characteristics, including point density, are vital to analysing the levels of confidence of the survey outputs.

For instance, poor ALS penetration of coniferous plantations significantly limited the productiveness of the desk-based interpretation in such areas. For this reason, targeted field reconnaissance was undertaken in accessible areas of plantations (e.g., gaps in the trees along water courses, thin plantings). In particular, prospective ground reconnaissance was undertaken along small valleys in the knowledge that medieval and post-medieval shieling huts might be expected in such locations. This proved to be the case, with the discovery of groups of huts which were entirely invisible in the ALS data. In addition, other archaeological monuments were recorded solely through field observation in open land with good point densities. For example, sites lying within boulder fields and scree were often not identified during the desk-based mapping because of the 'noise' created by the random distribution of boulders and survey points which obscured the archaeological structures. These examples clearly demonstrate that field observation is necessary to examine areas where the ALS data is identifiably poor, or where land use or topographic factors may conspire to obscure features of interest. The identification of these factors at a landscape scale is key to allocating resources in a cost-effective way; for example, recognising that large areas of open land are very effectively documented in the ALS derivatives, while in other discrete areas where the ALS is less effective, targeted field observation can be deployed to very good effect.

In many cases where the ALS data proved less effective, alternative remotely sensed data sources also proved limited in their utility. For the extensive coniferous plantations established in the second half of the 20th century the recent orthophotographs only document the extent of the trees. However, there is coverage of these areas on historic vertical aerial photographs taken in the 1940s. While the systematic examination of this imagery for Arran has not yet been undertaken, preliminary assessment demonstrates that they have value in providing archaeological information for areas that are otherwise largely a closed book. Utilising such imagery is not without its challenges, as, for instance, the photographs need to be georeferenced and variable scales and quality of imagery are present.

This variability in potential utility is also a characteristic of the modern orthophotographs, where land use, vegetation cover, the date of photography and the conditions at the time affect the visibility of archaeological features. While the available orthophotographs for Arran provided additional information, for example in documenting vegetation, that was valuable during the desk-based interpretation, it is clear that ALS derived visualisations offer better visibility of the earthworks and other types of monuments. Like the ALS, these orthophotographs are general purpose in that they were not collected with archaeological imperatives in mind. They vary in their utility for archaeological survey, depending on the conditions under which they were taken, and for Arran they certainly provided useful corroborative information in many cases. Indeed, the aerial imagery was used to build confidence in interpretation of remains that were not clearly observable in the ALS derivatives and benefited from this additional source of information. However, it is worth noting that no archaeological topographic remains were identified solely from the orthophotographs. Furthermore, a rapid assessment of the varying visibility of remains between the orthophotographs and the ALS derivatives for Arran demonstrates that the ALS was overwhelmingly more effective at capturing the topographic archaeological remains (i.e., disregarding areas where vegetation proxies are the main source of archaeological information). This statement should, of course, be qualified as specifically relating to Arran and those particular orthophotographs, recognising also that the extent to which it applies elsewhere will vary from region to region. It would be wrong to downplay the importance of the orthophotographs for archaeological survey, though it is worth restating that the ALS for Arran was an overwhelmingly superior source of archaeological information within the scope of the survey described here.

7. Conclusions: On the Utility of General Purpose Remote Sensed Data for Archaeological Survey

Scoping a National Archaeological Mapping Programme for Scotland requires identification of appropriate expertise and resourcing, and defining protocols and time-scales. The Arran survey described here has used the island and the ALS-derived record of its topography as a laboratory to explore these issues. The survey has demonstrated the strengths and weaknesses of the available general purpose remote sensed data. It has also highlighted the variability of desk-based interpretation between individuals, as well as the necessity for targeted field observation in areas with poor data coverage and where background noise obscures the visibility of archaeological features in the ALS derived visualisations. In general, the survey has established that relying primarily on ALS data, where it is available, for archaeological survey is an effective approach that will be applicable in a broader context, and when suitable ALS data is collected for the entire country, to a national level. Skills in interpretation of remote sensed data will need to be practised in tandem with explicit protocols to ensure consistent outputs from desk-based mapping and effective deployment of field observation. Nevertheless, this is not to downplay the important supporting roles of other sources of information, such as historic aerial photographs that document landscapes which have since been heavily modified. For these areas such sources are unique. It is, however, clear that the survey of Arran has provided a massive improvement in the knowledge base for the island, and that it is the ALS derivatives that have provided the majority of this information dividend.

While this paper has focused on the development of protocols for desk-based mapping and field work in pursuit of large-area mapping, a further element of the project in support of national scaled archaeological mapping has been a proof of concept to test the contribution from developments in computer vision and deep learning. This is a work in progress in collaboration with the Norwegian Computer Center to test the utility of Deep Learning and Neural Networks to provide heavily automated identifications of archaeological remains. This is an approach that offers great potential dividends in increasing rates of coverage far beyond what the essentially manual approaches to mapping described above can achieve [39–41]. The outcomes of the Arran proof of concept are encouraging, with good performance for the morphologically distinctive remains of prehistoric round houses. However, the more challenging variable remains of small field clearance cairns and huts illustrate the challenges of identifying morphologically diverse groups of monuments.

The survey of Arran is a first step in the definition of an approach to national archaeological mapping, and the protocols developed during this work will be further tested on different landscapes and data. These will include assessment of areas with high ALS data point density (e.g., 25 pts/m^2), as well as much lower point densities (e.g., 1 pt/m^2). These assessments will be designed to further explore the relationships between source data quality and the range of data sources, including orthophotographs, to better understand the variability of the archaeological survey outputs, and the impact that has on the reliability of the NRHE for those areas.

Author Contributions: Conceptualization, D.C.C.; Formal analysis, Ł.B., D.C.C. and M.M.; Investigation, D.C.C., Ł.B.; Methodology, D.C.C., Ł.B. and M.M.; Project administration, D.C.C.; Resources, Ł.B. and M.M.; Visualization, Ł. B.; Writing—original draft, Ł.B., D.C.C. and M.M.; Writing—review & editing, Ł.B. and D.C.C.

Acknowledgments: Many people have contributed to the scoping of the Arran survey, ALS data processing, the desk-based mapping stages and fieldwork, namely Ł.B., Georgina Brown, Andreas Buchholz, D.C.C., Piers Dixon, Angela Gannon, George Geddes, Alex Hale, Luke Hooper, Žiga Kokalj, Adara López-López, Alison McCaig., Peter McKeague, M.M., John Sherriff, Robin Turner and Sasya Zeefat. The ALS point cloud is licensed under the Non-Commercial Government Licence v2.0, copyright Scottish Government and SEPA (2014). We are grateful to Piers Dixon and Robin Turner for their comments on the paper, and to the reviewers for their input to improving it.

Conflicts of Interest: The authors declare no conflict of interest.

References

1. Watson, F.; Dixon, P. *A History of Scotland's Landscapes*, 1st ed.; Historic Environment Scotland: Edinburgh, UK, 2018; ISBN 978-1902419930.

2. Millican, K.; Dixon, P.; Macinnes, L.; Middleton, M. Mapping the Historic Landscape: Historic Land-Use Assessment in Scotland. *Landscapes* **2017**, *18*, 71–87. [CrossRef]

3. RCAHMS. *'Well Shelterd & Watered': Menstrie Glen, A Farming Landscape Near Stirling*, 1st ed.; RCAHMS: Edinburgh, UK, 2001; ISBN 978-1902419251.

4. RCAHMS. *In the Shadow of Bennachie: A Field Archaeology of Donside, Aberdeenshire*, 1st ed.; RCAHMS: Edinburgh, UK, 2009; ISBN 978-1902419619.

5. Cavers, G.; Barber, J.; Ritchie, M. The survey and analysis of brochs. *Proc. Soc. Antiq. Scotl.* **2015**, *145*, 153–176. [CrossRef]

6. Landscape Character Assessment. Available online: https://www.nature.scot/professional-advice/landscape-change/landscape-character-assessment (accessed on 19 July 2018).

7. Canmore. The National Record of the Historic Environment. Available online: https://canmore.org.uk/ (accessed on 25 June 2018).

8. McKeague, P.; Thomas, D. Evolution of national heritage inventories for Scotland and Wales. *J. Cult. Herit. Manag. Sustain. Dev.* **2016**, *6*, 113–127. [CrossRef]

9. Dunbar, J.G. The Royal Commission on the Ancient and Historical Monuments of Scotland: The First Eighty Years. *Trans. Anc. Monum. Soc.* **1992**, *36*, 1–58.

10. McKeague, P.; Cowley, D.C. From paper to digital, and point to polygon—The application of GIS in a national body of survey and record. *Int. J. Herit. Digit. Era* **2013**, *2*, 677–694. [CrossRef]

11. Cunningham, I. *The Nation Survey'd: Timothy Pont's Maps of Scotland*, 1st ed.; John Donald: Edinburgh, UK, 2006; ISBN 978-0859766807.

12. William Roy—Military Antiquities of the Romans in North Britain. 1793. Available online: https://maps.nls.uk/roy/antiquities/romans.html (accessed on 25 June 2018).

13. Ordnance Survey Maps—Six-Inch 1st Edition, Scotland, 1843–1882. Available online: https://maps.nls.uk/os/6inch/ (accessed on 25 June 2018).

14. Bowden, M.; MacKay, D. Archaeology and the Ordnance Survey Revisited: Field Investigations by the Ordnance Survey Archaeology Division 1947–1983. *North. Archaeol.* **1999**, *17*, 1–14.

15. Davidson, J.L.; Cowley, D.C.; Barneveld, J.; Ferguson, L.M. Archaeological Mapping in the North of Scotland. In *'We Were Always Chasing Time' Papers Presented to Keith Blood*; Frodsham, P., Topping, P., Cowley, D., Eds.; Northumberland Archaeological Group: Newcastle, UK, 1999; pp. 15–21.

16. The Society of Antiquaries of Scotland. Available online: https://www.socantscot.org/ (accessed on 3 July 2018).

17. First Report and Inventory of Monuments and Constructions in the County of Berwick. Available online: http://canmore-pdf.rcahms.gov.uk/wp/00/WP003901.pdf (accessed on 25 June 2018).

18. Ordnance Survey Archaeology Division Revision Programme. Available online: https://canmore.org.uk/project/1014509 (accessed on 25 June 2018).

19. Discovery and Excavation in Scotland. Available online: https://archaeologyscotland.org.uk/join-us/discovery-and-excavation-scotland (accessed on 3 July 2018).

20. OASIS Scotland. Available online: https://oasis.ac.uk/pages/wiki/Scotland (accessed on 25 June 2018).

21. Cowley, D.C. Creating the cropmark archaeological record in East Lothian, southeast Scotland. In *Prehistory without Borders: Prehistoric Archaeology of the Tyne-Forth Region*, 1st ed.; Crellin, R., Fowler, C., Tipping, R., Eds.; Oxbow: Oxford, UK, 2016; Volume 1, pp. 59–70, ISBN 978-1785701993.

22. Dixon, P. *But the Walls Remained: A Survey of Unroofed Rural Settlement Depicted on the First Edition of the Ordnance Survey 6-inch Map of Scotland*, 1st ed.; RCAHMS: Edinburgh, UK, 2002; ISBN 978-1902419275.

23. Stell, G. *Orkney at War: Defending Scapa Flow*, 1st ed.; Orcadian Limited: Kirkwall, UK, 2011; ISBN 978-1902957487.

24. Barclay, G. *If Hitler Comes: Preparing for Invasion: Scotland 1940*, 1st ed.; Birlinn: Edinburgh, UK, 2013; ISBN 978-1843410621.

25. Kilpatrick, A. World War I Remains in Scotland: Aerial Photography as Heritage. In *Conflict Landscapes and Archaeology from Above*, 1st ed.; Stichelbaut, B., Cowley, D.C., Eds.; Routledge: Abingdon, UK, 2016; Volume 1, pp. 59–72, ISBN 1472464389.

26. Cowley, D.C. What do the patterns mean? Archaeological distributions and bias in survey data. In *Digital Methods and Remote Sensing in Archaeology—Archaeology in the Age of Sensing*, 1st ed.; Campana, S., Forte, M., Eds.; Springer: New York, NY, USA, 2016; Volume 1, pp. 147–170, ISBN 3319406566.

27. A Strategy for the National Mapping Programme. Available online: http://www.english-heritage.org.uk/content/imported-docs/a-e/astrategyforthenationalmappingprogramme2009.pdf (accessed on 28 March 2010).

28. Horne, P.D. The English Heritage National Mapping Programme. In *Remote Sensing for Archaeological Heritage Management*, 1st ed.; Cowley, D.C., Ed.; Archaeolingua: Budapest, Hungary, 2011; Volume 1, pp. 143–151, ISBN 9789639911208.

29. Winton, H.; Horne, P. National archives for national survey programmes: NMP and the English Heritage aerial photograph collection. In *Landscapes Through the Lens: Aerial Photographs and the Historic Environment*, 1st ed.; Cowley, D.C., Standring, R., Abicht, M., Eds.; Oxbow: Oxford, UK, 2010; Volume 1, pp. 7–18, ISBN 1842179810.

30. Hesse, R. The changing picture of archaeological landscapes: lidar prospection over very large areas as part of a cultural heritage strategy. In *Interpreting Archaeological Topography. 3D Data, Visualisation and Observation*, 1st ed.; Opitz, R.S., Cowley, D.C., Eds.; Oxbow: Oxford, UK, 2013; pp. 171–183, ISBN 978-1842175163.

31. RCAHMS. *North East Perth: An Archaeological Landscape*, 1st ed.; HMSO: Edinburgh, UK, 1990; ISBN 0114934460.

32. Cowley, D.C.; López-López, A. Developing an approach to national mapping—Preliminary work on Scotland in miniature. *AARGNews* **2017**, *55*, 19–25.

33. Doneus, M.; Briese, C. Full-waveform airborne laser scanning as a tool for archaeological reconnaissance. In *From Space to Place, Proceedings of the 2nd International Workshop on Remote Sensing in Archaeology, CNR, Rome, Italy, 4–7 December 2006*; Campana, S., Forte, M., Eds.; Archaeopress: Oxford, UK, 2006; pp. 99–106.

34. Doneus, M.; Briese, C. Airborne Laser Scanning in forested areas—Potential and limitations of an archaeological prospection technique. In *Remote Sensing for Archaeological Heritage Management, Proceedings of the 11th EAC Heritage Management Symposium, Reykjavik, Iceland, 25–27 March 2010*; Cowley, D.C., Ed.; Europae Archaeologiae Consilium: Brussels, Belgium, 2011; pp. 59–76.

35. Opitz, R.S.; Cowley, D.C. Interpreting archaeological topography: Lasers, 3D data observation, visualisation and applications. In *Interpreting Archaeological Topography. 3D Data, Visualisation and Observation*, 1st ed.; Opitz, R.S., Cowley, D.C., Eds.; Oxbow: Oxford, UK, 2013; pp. 1–13, ISBN 978-1842175163.

36. Kokalj, Ž.; Zakšek, K.; Oštir, K. Application of Sky-View Factor for the Visualization of Historic Landscape Features in Lidar-Derived Relief Models. *Antiquity* **2011**, *85*, 263–273. [CrossRef]

37. Zakšek, K.; Oštir, K.; Kokalj, Ž. Sky-View Factor as a Relief Visualization Technique. *Remote Sens.* **2011**, *3*, 398–415. [CrossRef]

38. Cowley, D.C.; Banaszek, Ł. Towards accountability and confidence in archaeological prospection and remote sensing archaeology. *Remote Sens.* **2018**, in press.

39. Cowley, D.; De Laet, V.; Bennett, R. Auto-extraction techniques and cultural heritage databases—Assessing the need, evaluating applicability and looking to the future. In *Archaeological Prospection, Proceedings of the 10th International Conference on Archaeological Prospection, Vienna, Austria, 29 May–2 June 2013*; Neubauer, W., Trinks, I., Salisbury, R.S., Einwögerer, C., Eds.; Austrian Academy of Sciences: Vienna, Austria, 2013; pp. 406–408.

40. Bennett, R.; Cowley, D.C.; De Laet, V. The data explosion: tackling the taboo of automatic feature recognition in the use of airborne survey data for historic environment applications. *Antiquity* **2014**, *88*, 896–905. [CrossRef]

41. Traviglia, A.; Cowley, D.C.; Lambers, K. Finding common ground: Human and computer vision in archaeological prospection. *AARGnews* **2016**, *53*, 11–24.

geosciences

MDPI

Article

Sumerian Pottery Technology Studied Through Neutron Diffraction and Chemometrics at Abu Tbeirah (Iraq)

Giulia Festa [1,*], Carla Andreani [1,2], Franco D'Agostino [3], Vanessa Forte [4], Matteo Nardini [5], Antonella Scherillo [6], Claudia Scatigno [2], Roberto Senesi [1,2] and Licia Romano [3]

[1] Centro Fermi - Museo Storico della Fisica e Centro Studi e Ricerche "Enrico Fermi", Piazza del Viminale 1, 00184 Rome, Italy; carla.andreani@uniroma2.it (C.A.); roberto.senesi@uniroma2.it (R.S.)
[2] Physics Department & NAST Centre, Università degli Studi di Roma "Tor Vergata", Via della Ricerca Scientifica 1, 00133 Rome, Italy; claudia.scatigno@uniroma2.it
[3] Dept. Instiute of Oriental Studies, Sapienza Università di Roma, Piazzale Aldo Moro 5, 00185 Rome, Italy; franco.dagostino@gmail.com (F.D.); licia.romano@gmail.com (L.R.)
[4] McDonald Institute for Archaeological Research, University of Cambridge, Downing street, Cambridge CB2 3EF, United Kingdom; vf261@cam.ac.uk
[5] Institute of Physics, Fondazione Policlinico A. Gemelli IRCCS, Università Cattolica Sacro Cuore, Largo Francesco Vito 1, 00168 Roma, Italy; matteo.nardini01@icatt.it
[6] Science and Technology Facilities Council (STFC), Rutherford Appleton Laboratory, ISIS Facility, Harwell OX11 0QX, United Kingdom; antonella.scherillo@stfc.ac.uk
* Correspondence: giulia.festa@centrofermi.it

Received: 24 October 2018; Accepted: 28 January 2019; Published: 31 January 2019

Abstract: Pottery is the most common material found in archaeological excavations and is used as the main tool for chronological dating. Due to the geopolitical instability of the Middle East during the latter part of the last century until recent years, Sumerian pottery studies and analyses are limited. The resumption of archaeological excavations in Iraq during the last ten years allows the acquisition of new information and the study of archaeological material excavated through modern stratigraphic methodologies. This paper presents the results of the non-destructive analyses of Abu Tbeirah Sumerian pottery (Iraq) from the 3rd millennium BC and is aimed at analysing the crystallographic composition of ceramic material, therefore contributing to fill a gap in the knowledge of early Southern Mesopotamian pottery production, shedding new light on details of ancient technology and manufacturing techniques. Among the wide range of analytical techniques available, neutron-based ones have been chosen, obtaining detailed analyses in a non-destructive manner. Non-destructive and non-invasive neutron diffraction (ND) was applied in combination with chemometrics such as Principal Component Analysis (PCA) and Cluster Analysis (CA). ND confirms a general uniformity of the raw materials and a local Mesopotamian provenance through a comparison with modern local clay. Moreover, secondary minerals and their marker-temperature formation imply two different ranges of firing temperature that never exceeded 1000 °C, a temperature easily reachable through pit-firing techniques.

Keywords: Sumerian pottery; neutron techniques; neutron diffraction; chemometric analysis

1. Introduction

Pottery is a human-made material that includes porous, unvitrified clay bodies fired at a wide range of temperatures from 800–900 °C up to 110–1200 °C [1]. The primary raw materials are clays, water and temper [1]. Clay is a sedimentary rock composed mainly of silicates (silicon, aluminium and oxygen, calcium, magnesium, sodium, potassium, iron, manganese and titanium). Temper is a material that includes organic substances such as dung, straw and hay, or inorganic elements like shells, sand, calcite, and sandstone, etc. Temper is added by the potter to modify the clay paste in order to allow, for example, the evaporation of water contained in the amalgam, minimizing the contraction of the clay during firing, at the same time helping the vitrification process and providing plasticity [2]. Colour and composition are dependent on manufacturing processes (kiln environment and temperature of the firing process) as well as on the nature of raw materials and amount of temper added as inclusions (such as straw, quartz, carbonates, ground fired clays, etc.) [3,4]. During firing, the original mineral structure changes and the clay becomes permanently hardened. The loss of water (adsorbed and combined water) and volatile materials in the clay mineral structure (such as organic components, some impurities and inclusions) contribute to the changing of the clay mineral itself (recrystallization and formation of new silicates), which lead to the formation of new minerals (calcium silicates or ferrosilicates), characteristic of high-temperature firings [1]. Above 900–1000 °C, additional changes, such as sintering and vitrification, are promoted by the presence of impurities and inclusions, producing a glassy, nonporous ceramic body [1]. In the case of clays with carbonate fraction, two main thermic processes are involved: clay dihydroxylation—loss of hydroxyls (ca. 400–600 °C) and decarbonation of calcareous materials (750–850 °C). It is widely accepted that quartz and feldspar do not undergo significant variations at low temperatures, except for "alpha and beta" transition at ~573 °C [5]. These changes in the original clay mineral structure can be considered an indication of the temperature reached by the clay object during firing [6]. In particular, illite/muscovite, minerals naturally present in Abu Tbeirah's clays are stable during the firing process up to 900 °C [6]. Clay minerals such as illite/muscovite are dehydroxylated close to 950° [6]. At this temperature, new calcium minerals, such as diopside ($CaMgSi_2O_6$), are produced from calcite and dolomite reacting with other components [6]. Human modification of the clay body should be also taken into account in considering illite/muscovite and diopside as indicators of the firing temperature reached by the clay object. In particular, during slipping, a thin superficial layer of a lighter and fluid suspension of the same clay is applied before firing over the vase [7,8]; this procedure brings the finest components of the clay to the surface [9]. This external layer easily reaches high temperatures [1]; new formation minerals (e.g., diopside) can be thus found on the surface, the rest of the clay body shows mineral phases peculiar of a lower firing temperature.

ND [10] was successfully applied to the study of ancient ceramics from different geographical and historical contexts [11].

Time of flight (TOF) neutron diffraction (ND) combined with chemometric tools such as principal component analysis (PCA) and cluster analysis (CA) was carried out on the same set of ancient Sumerian pottery from the archaeological site of Abu Tbeirah (Nasiriyah, Southern Iraq) [12–16], currently under excavation. A total of 36 samples from two excavation campaigns have been analysed. These were also investigated in [17] through neutron resonance capture analysis (NRCA) for isotopic analysis; the NRCA technique [18] uses epithermal neutrons (in the range of 1 eV–1 keV) to investigate the isotopic/elemental composition of the irradiated sample together with a statistical approach: ^{39}K, ^{56}Fe and ^{55}Mn components of the raw materials were identified together with decay products such as sodium chloride ^{23}Na and ^{35}Cl, giving us clues about the homogeneity of the raw materials and manufacturing of the pottery [17].

The site and the archaeological findings such as pottery assemblages date back to the 3rd millennium BC and their studies are still ongoing, evolving and enriched by the new results that each campaign is producing. Although pottery is one of most common materials found in archaeological excavations [1], comprehensive knowledge of Sumerian pottery is still far from completion due to the complex geopolitical context of the Middle East in the last 50 years, which has hampered archaeological research in this area.

Archaeological and Geological Background

The Sumerian culture flourished in Southern Mesopotamia, modern-day Southern Iraq, from 4000 to 2000 BC. Pioneers in agriculture, craft (e.g., metalwork and pottery) and trade, Sumerians are considered as one of the world's first civilizations. Abu Tbeirah was a significant settlement in the constellation of the city-states in which Mesopotamia was divided during the so-called Early Dynastic Period (2900–2350 BC), before the unification of the territory in the first 'World Empire' under the Akkadian King Sargon. The recent resumption of archaeological activities in Southern Iraq [19–23] offers the chance to analyse Sumerian material culture through current methodologies. Previous studies on Mesopotamian pottery [19–24] have demonstrated that the homogeneity of chemical and mineralogical compositions of clay is due to the nature of raw materials found in the alluvial plain, formed by the two great rivers Tigris and Euphrates (see Figure 1) [25].

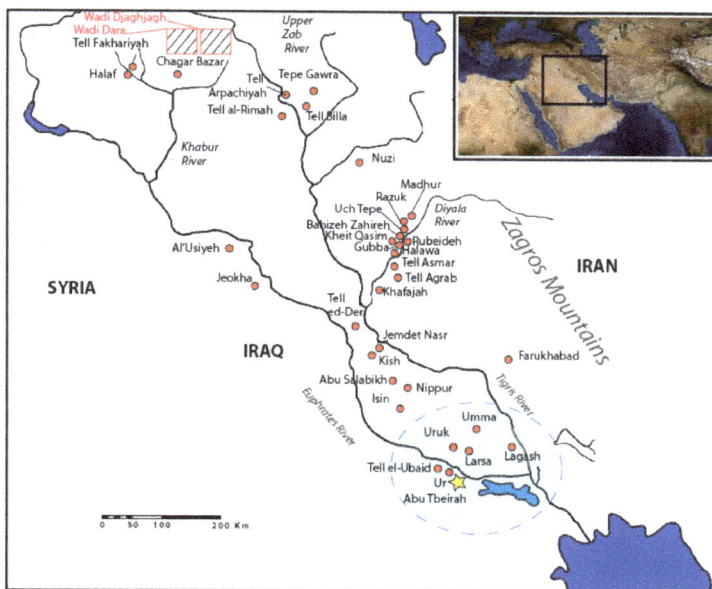

Figure 1. Map of Mesopotamia with the sites, reported in red, whose pottery has been analysed in previous studies. The Abu Tbeirah site is marked with a yellow star.

The Mesopotamian Fluvial Basin, the geological formation context of the clay used by Sumerian potters, belongs to the so-called Zagros Fold-Thrust Belt (ZFTB), a basin enclosed between the Zagros Mountains and the Arabian Platform [26]. Sediments of fluvial, deltaic, and lacustrine origin accumulated in the basin during the Quaternary Period [8,26]. The sediments surrounding Abu Tbeirah are characterised as carbonate and clastic deposits, mainly silty sand and silty clay [27,28]. Local clay in Abu Tbeirah is composed of montmorillonite (main clay mineral $(Na,Ca)_{0.3}(Al,Mg)_2Si_4O_{10}(OH)_2 \cdot n(H_2O)$), muscovite/illite $(KAl_2(Si_3Al)O_{10}(OH)_2$ and kaolinite $(Al_2Si_2O_5(OH)_4)$, while the main non-clay minerals are calcite $(CaCO_3)$, quartz (SiO_2), feldspar and dolomite (dolomite $MgCa(CO_3)_2$) [29]. Al-Mukhtar states [30] that the heavy mineral composition of the sediment in this area consists of 40% of metamorphic origin, 7% of igneous origin and 48% of a different source. These sediments are derived from basic igneous rocks (such as gabbro and basalt), metamorphic rocks (such as schist, gneiss and amphibolites) and are in a minor proportion relative to older sedimentary rocks. Al-Mukhtar also attributes the composition of the sediments to the action of the Euphrates and Tigris Rivers, which transport the sediments of the primary source, the Taurus-Zagros Belt [30]. However, the heavy mineral concentration of the tributaries of the two main rivers differs and some peculiarities in the west area of the Euphrates (where Abu Tbeirah is located) can be related to the provenance of sediments from the Arabian Stable Shelf [31,32]. The local provenance of the clay used for the 3rd millennium BC pottery production was demonstrated by previous studies [19–24] for the Central Iraqi Diyala and Hamrin areas (Khafajah, Tell Agrab, Tell Asmar, Halwa, Gubba, Rubeideh, Bahizeh Zahireh, Uch Tepe, Razuk, Tell Madhur). Data from the Southernmost part of Mesopotamia, in which Abu Tbeirah is located, are still extremely limited: a total of 40 shards coming from several sites (Tell-ed-Der, Jemdet Nasr, Nippur, Kish, Abu Salabikh, Fara, Tell al-Wilayah, Uruk, Larsa, Lagash, Obeid, and Ur) have been analysed [19,20,22]. Mesopotamian 3rd millennium BC pottery has always been regarded as one of the first examples of the wide application of advanced manufacturing techniques [1], such as the use of the wheel for shaping vessels and pottery ovens for firing. The analyses recently carried out in the Ancient Near East, in general, and in Southern Mesopotamia [33,34], in particular, show a more nuanced picture of the evolution of this first pottery technology, with more advanced techniques coexisting with less technological ones [7,35]. Recent excavations [12–16] have documented, especially in the north-east-region of the Abu Tbeirah settlement, the presence of pottery production waste not connected to any visible kiln structures. However, the erosion process that characterizes the site might have hampered the visibility of the original structures. Abu Tbeirah vases often show traces of a non-uniform firing [36], a characteristic that is usually connected to pit-firing [9,37]. The analyses carried out on the 36 samples confirm the local provenance of Abu Tbeirah's pottery and show a range of temperatures that can be attributed to pit-firing. This contributes to the general understanding of Sumerian pottery technology, highlighting the use until the 3rd millennium BC of firing techniques, such as pit-firing, considered out-dated for the period.

2. Materials and Methods

2.1. Description of the Samples

A total of 36 pottery fragments were selected from the early Dynastic-Akkadian contexts of Area 1 [12–16] together with one modern clay sample (sample n. 16), from the canal near the excavated site and used for comparison purposes. The samples were chosen after an autoptic subdivision into fabric-groups. The preliminary classification of ceramic pastes is based on macroscopic compositional features, such as particle size (coarse/fine-granulometry, porosity and colour). These macroscopic analyses showed differences in colour, plausibly linked to the firing temperature and/or intentional adding of temper or other inclusions. Four main macroscopic groups (6 with sub-groups) were identified and used to select samples for the present analyses. The following fabrics are reported in Figure 2.

- *Fabric A:* fine-grained paste with low porosity featured by planar voids. Red-orange paste colour. Firing mainly in an oxidising atmosphere. Self-slipped or sometimes covered with clearer slip (fabric A + slip)—samples n. 2, 3, 6–9, 12, 14, 15, 17–19, 23, 31, 32, 34.
- *Fabric B1:* fine-grained paste with a low porosity featured by planar voids. Yellow paste colour. Firing mainly in an oxidising atmosphere. Sometimes self-slipped—samples n. 5, 10, 20, 27, 35, 36.
- *Fabric B2:* fine-grained paste with a low porosity featured by planar voids. Yellow paste colour with orange inclusions. Firing mainly in an oxidising atmosphere—samples n. 1, 13, 21, 24–26, 28.
- *Fabric C:* coarse-grained paste with abundant sedimentary fragments and angular inclusions. High porosity compared to other groups. Firing mainly in an oxidising atmosphere-sample n. 33.
- *Fabric D1:* fine-grained paste with abundant straw. Red-orange paste colour. Firing mainly in an oxidising atmosphere-sample n. 30.
- *Fabric D2:* fine-grained paste with abundant straw. Yellow paste colour with orange inclusions. Firing mainly in an oxidising atmosphere-samples n. 4, 11, 22, 29.

Figure 2. A selection of the pottery and clay samples analysed, divided according to the fabric classification. Photos of the fragments were acquired through a portable digital microscope. The average thickness of the samples spans between 0.5 cm and 2.3 cm.

The slip is defined as a lighter and fluid suspension of the same clay applied over the vase as a thin superficial layer before firing [7,8] while the self-slip is realised by wiping the surface with a wet hand or cloth, a procedure that brings the finest components of the clay to the surface [9].

2.2. Association between Fabrics and Pottery Shapes at Abu Tbeirah

Abu Tbeirah pottery presents, on the one hand, discontinuous use of the same fabric for a given vessel typology (see, for example, the beakers realized with different fabrics in Figure 3) but, on the other hand, the following general trends can be recognized: (a) drinking vessels, such as quickly wheel-thrown or wheel-coiled beakers and conical bowls, are usually realised in fabric A; (b) medium and large closed vessels are instead mainly realised in fabrics B1–B2 or fabric A, self-slipped/slip but never in fabrics C or D (with the exception of ring bases, added to stabilize jars, that are always realised in fabric D); (c) rare cooking pots are realised in fabric C, suitable for cooking purposes, withstanding the thermal shock caused by the contact with fire, thanks to its coarse-grained and porous paste. Fabric C It is rare at Abu Tbeirah due to the wider use of tannur (typical clay oven) and other peculiar firing installations for food preparation; (d) large containers, such as trays, vats or coffins are always realised in fabric D1–D2.

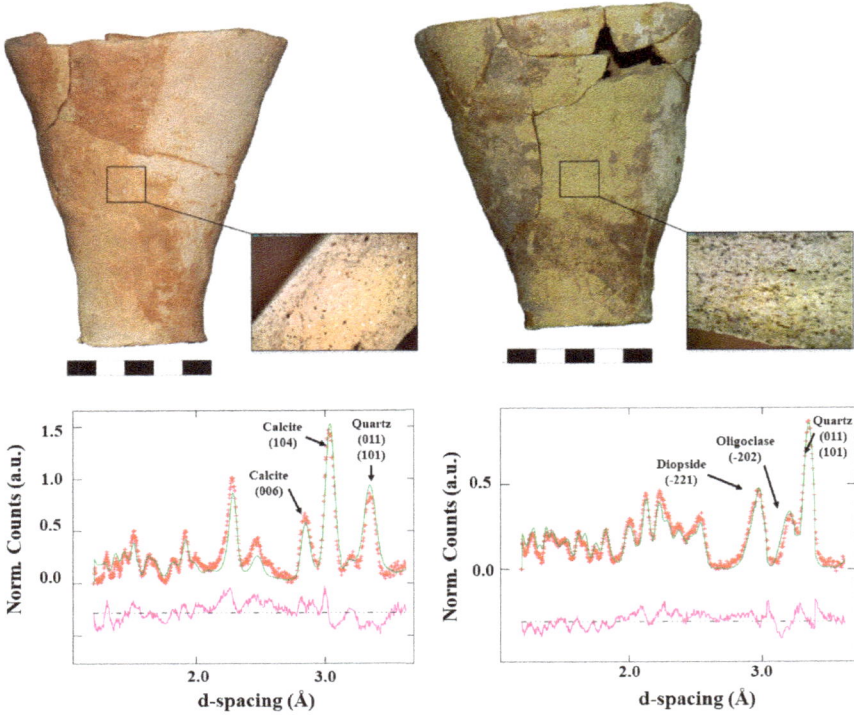

Figure 3. Two Sumerian pottery vessels studied via neutron diffraction; diffraction spectra (bank n. 8) of the two vases are also reported in red (normalised number of counts as a function of d-spacing). Best fit of data (green line) is also shown together with the residue, reported in violet, and the labelling of the main diffraction peaks.

2.3. Experimental Set-Up

Measurements were carried out on the INES beamline, located at the ISIS Pulsed Neutron and Muon Source, Science and Technology Facility Council, UK. The instrument is a general-purpose powder diffractometer [38,39] where time of flight (TOF) neutron diffraction (ND) [39]. Nneutrons are a powerful tool to investigate the microscopic structure of materials in a non-destructive and non-invasive way. INES is equipped with 9 detector banks, each providing a full diffraction spectrum, of which each is in turn composed of 16 squashed ^{3}He tubes with an active volume of $100 \times 12.5 \times 2.5$ mm^3. The wide angular range covered by the detectors (11.60–170.60 degree) together with the thermal neutron beam from the water moderator (incident wavelength range of 0.17–3.24 Å) covers a wide d-spacing range (between d = 0.1 Å and d = 16 Å) with high resolution up to $\Delta d/d = 0.001$.

The neutron diffraction process is described as the elastic scattering of the incident neutron beam by crystal planes (hkl), according to Bragg's law. In the case of Time Of Flight (TOF) measurements, Bragg's law $n\lambda = 2d_{hkl} \sin\theta_0$ can be rewritten as a relation between the TOF of neutrons scattered from a set of planes in the sample $(TOF)_{hkl}$ and the spacing between these planes, d_{hkl}, as follows:

$$(TOF)_{hkl} = (2m_n/h)Ld_{hkl}\sin\theta_0 \tag{1}$$

where m_n is the neutron mass, h is Planck's constant, L is the flight path, d_{hkl} is the d-spacing of the set of crystallographic planes hkl, and θ_0 is a fixed scattering angle of a specific detector. The mean collection time was 3 h per sample and all the data were normalised to the intensity of the incident neutron beam.

3. Results

Recorded neutron diffraction spectra together with photographs of the corresponding samples are reported in Figure 3.

Diffraction data were analysed using the GSAS phase analysis software [40] and the EXPGUI (A Graphical User Interface for GSAS) [41], based on the Rietveld refinement method that provides weight fractions (wt (%)) of the phases present in the sample. Quantitative phase analysis is based on the principle that in a multi-phase sample each phase exhibits a unique set of diffraction peaks, and the peak intensity of a particular phase is proportional to its weight-fraction; the measured diffraction pattern is the weighted sum of all single-phase patterns. Quantitative diffraction results together with the classification and the sample thicknesses are reported in Table 1.

Table 1. Results of neutron diffraction analysis on the Sumerian pottery samples and fabric classification. The weight percentages of the detected phases [wt (%)] are reported. The errors are ±0.1 wt%.

Sample	Fabric	Thickness (cm)	Calcite	Quartz	(Pyroxenes) Diopside	(Feldspar) Oligoclase	(Phyllosilicate/mica) Muscovite
1	B2	0.6	-	15.7	50.5	33.8	-
2	A	0.7	-	29.1	38.9	32.0	-
3	A	0.9	-	29.8	36.5	33.7	-
4	D2	1.9	-	18.9	51.1	30.0	-
5	B1	0.7	-	23.8	43.4	32.8	-
6	A	0.6	17.9	22.4	-	19.5	40.2
7	A	0.8	8.3	12.4	-	22.6	56.7
8B*	A	0.6	11.1	15.2	-	18.7	55.0
8W*	A	0.6	10.5	15.7	-	19.0	54.8
9	A	0.7	-	26.0	34.6	39.4	-
10	B1	0.7	-	22.4	50.7	26.9	-
11	B2	1.2	-	15.7	57.1	27.2	-
12	A	1.0	8.7	13.7	-	20.1	57.5
13	B2	0.5	-	12.4	56.5	31.1	-
14	A	0.8	5.3	18.5	-	31.3	44.9
15	A	0.5	-	24.6	36.8	38.6	-
16	Modern clay	1.0	12.7	8.7	-	16.0	62.6
17	A	0.7	-	24.4	38.3	37.3	-
18L*	A	0.7	12.1	14.2	-	20.1	53.6
18S*	A	0.7	12.5	14.2	-	16.5	56.8
19L*	A	0.6	3.2	9.4	74.7	12.7	-
19S*	A	0.6	2.4	8.0	79.7	9.9	-
20	B1	0.7	2.8	33.0	31.3	32.9	-
21L*	B2	0.7	4.7	19.8	47.1	28.4	-
21S*	B2	0.7	4.7	16.1	51.2	28.0	-
22	D2	1.7	-	17.1	39.5	43.4	-
23	A	1.0	7.2	14.5	-	20.5	57.8
24	B2	0.7	-	-	71.0	29.0	-
25	B2	0.7	-	21.4	51.9	26.7	-
26L*	B2	0.6	-	11.6	58.1	30.3	-
26S*	B2	0.6	-	8.0	59.8	32.2	-
27	B1	0.7	2.3	27.6	40.5	29.6	-
28	B2	0.6	-	10.1	59.3	30.6	-
29	D2	2.3	-	17.3	37.0	42.5	-
30	D1	1.7	11.6	22.5	-	22.5	43.4
31	A	0.9	8.3	17.6	-	24.6	49.5
32L*	A	1.0	5.5	12.4	-	18.0	64.1
32S	A	1.0	10.0	20.0	-	15.0	55.0
33	C	0.8	14.4	13.6	-	18.6	55.0
34	A	0.7	9.0	16.0	-	33.0	42.0
35	B2	0.8	3.5	27.2	36.3	33.0	-
36	B2	0.8	1.6	19.5	45.8	29.3	-

B = black, W = white, L = large, S = small.

In Figure 4, the trend of the amount (wt %) of the detected phases (oligoclase, muscovite, calcite, quartz and diopside) through ND is shown.

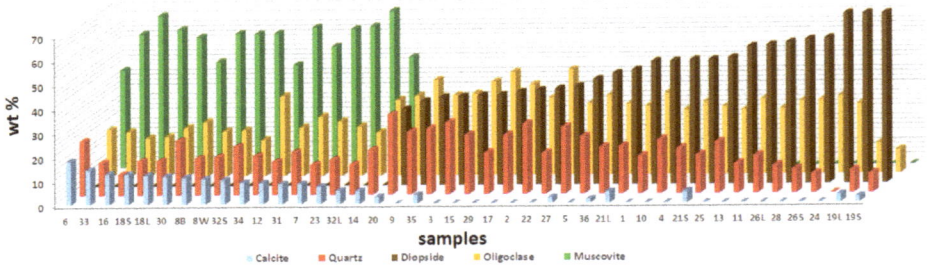

Figure 4. Histogram based on neutron diffraction quantitative results. The x-axis reports the number of the sample as a function of the increasing percentage of diopside; the y-axis reports the weight percentage of the detected crystal phase; z-axis reports the detected phases (yellow—feldspar (oligoclase), green—phyllosilicate/mica (muscovite), blue—calcite, red—quartz, brown—pyroxenes (diopside)).

Among the detected crystalline phases, phyllosilicate/mica (muscovite) and pyroxenes (diopside) represent track markers of firing processes [4]. In Figure 4, two main groups of samples are distinguishable. The first group (samples n. 6, 33, 16, 18S, 18L, 30, 8B, 8W, 32S, 34, 12, 31, 7, 23, 32L, 14) is characterised by the presence of calcite and phyllosilicate/mica (muscovite) while the second group (samples n. 9, 35, 3, 15, 29, 2, 22, 5, 36) is characterised by the presence of pyroxenes (diopside). Quartz presents a constant trend. In the second group, the primary calcite is completely transformed in diopside or calcite of secondary formation above 850 °C. Considering the temperature stability of the above mentioned phases, the range of firing for the set of samples can be ascribed between 800–1000 °C.

Relative weight percentages of the detected phases are analysed through a multivariate statistical approach carrying out correlation analysis namely principal component analysis (PCA) [42,43]. This technique has been successfully applied in archaeological contexts since the 1970s [43,44]. PCA is a projection method, an orthogonal bilinear matrix decomposition, where components or factors are obtained in a sequential way explaining the maximum variance where the distance between data points is the largest. These components are called principal components (PCs) and are orthogonal to each other. They are computed iteratively, in such a way that the first PC is the one that carries most information while the second PC will then carry the maximum share of the residual information (i.e., not taken into account by the previous PC). In order to classify samples and to distinguish among the most important variables to keep in a model (variables that characterise the population), a bi-plot (scores and loadings plotted together) was realised via Unscrambler X v10.3 software [45] taking into account a matrix of 42 × 5 where 42 is the number of the measurements and 5 is the total number of detected phases. In this case, the system is described by three principal components, explaining 99% of the total variance (PC1: 89%, PC2: 8%, and PC3: 2%). Singular value decomposition (SVD), which is a non-iterative algorithm that is generally used when there are few variables, and a cross method were applied to estimate the model stability and prediction ability [43]. Results are reported in Figure 5.

In order to find groups in the data without any predefined class structures, hierarchical cluster analysis (HCA) was carried out employing an average-linkage algorithm [46,47] and applying the Euclidean distance as the relative distance to define the number of the clusters, more appropriate for normalised data [46]. The dendrogram is reported in Figure 6, a visual analysis of the hierarchy classification according to the similarity of the samples: samples with similarities in composition are grouped according to their minimal relative distance (Euclidean relative distance). In the present case, the HCA is used to corroborate the macroscopic fabric classification. Figure 6 shows two main groups of clusters, highlighted in orange and green; the third cluster (coloured in violet and defined singletons), represented by samples number 24, 19S and 19L, is defined as a sub-group of the orange one. The singletons described the samples with a relative weight of diopside higher than 70%. Three clusters are also identified by ellipses in Figure 5. From Figure 6, the most similar samples are 8W and 8B (they are joint for first-minimal distance) as we expected because they are two fragments of the same vase. The following minimal relative distance is represented by samples n. 25 and n. 10, together with samples n. 28 and n. 26L, characterised by a fine-yellow-grained paste with inclusions. The three main groups, identified by the HCA, can be related to different grained pastes identified by macroscopic observations such as: (a) fine-yellow-colour-paste characterised by diopside, (b) fine-red-orange-grained paste characterised by muscovite and primary calcite, and (c) an intermediate group characterised by yellow paste colour with orange inclusions.

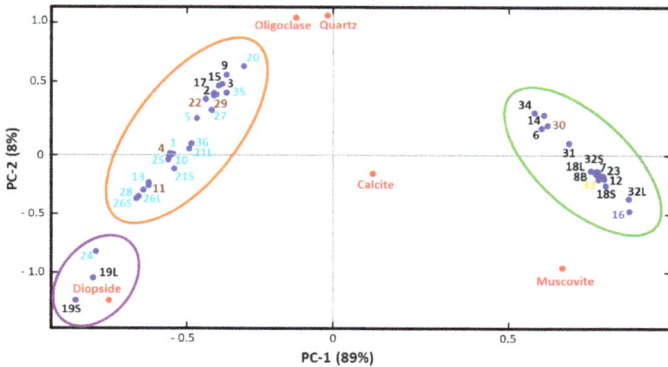

Figure 5. Principal component analysis (PCA). Bi-plot of the investigated samples. The *scores* plot is reported in blue and identifies the measurements while the *loadings* plot is reported in red and identifies the detected phases; the ellipses highlights the three groups identified through the hierarchical cluster analysis. The name of the samples is coloured as a function of the main fabrics (A = black, B1–B2 = light blue, C = yellow, D1–D2 = brown).

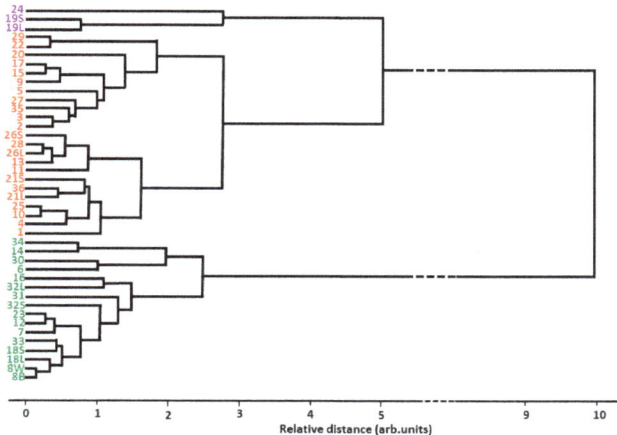

Figure 6. Average linking clustering using Euclidean distance. The dendrogram shows three main groups (numbers underlined as violet, orange and green) among the diffraction dataset (Table 1) as a function of the relative distance, identifying the similarity of the pottery samples.

Therefore, PC1 in Figure 5 is attributed to the firing temperature reached (increasing with the decrease of PC1), and PC2 is attributed to the phase transitions as a function of temperature (i.e., from primary calcite to pyroxenes (diopside)).

4. Conclusions

The paper presents the systematic study, via non-destructive neutron diffraction, of 36 Sumerian pottery fragments from Abu Tbeirah (3rd millennium BC), Southern Iraq. Key results of this study were: (a) the identification of crystalline phases in the samples and (b) their classification by temperature markers through chemometric analyses of phase composition. Integration of such results with archaeological evidence and analyses permit a deeper understanding of Sumerian firing technology of this ancient period. A comparison of pottery samples with modern clay, gathered from the canal near the excavated site, suggests a local origin of the clay used for the Sumerian vases. Temperature marker classification allowed identification of three main clusters. They are characterised by: (a) red-orange paste with crystalline phase composition that indicates a firing temperature below 900°, with the coexistence of primary calcite and phyllosilicate/mica (muscovite); (b) yellow paste with orange inclusions with crystalline composition that indicates a firing temperature over 900°, with the coexistence of pyroxenes (diopside) with primary phases (such as feldspar (oligoclase) and quartz); (c) yellow paste with crystalline composition characterised by more than 70% relative weight fraction of pyroxenes (diopside) that indicates a firing temperature over 900°. The samples analysed are distributed according to an almost continuous temperature gradient.

In general, utilization of kilns has been associated with a firing temperature range between 750 °C and 1150 °C: the results presented here demonstrate that temperatures reached during firing never exceeded 900–1000 °C. Kiln firing is considered a more advanced technology compared to open or pit-fires, which are usually not considered adequate for reaching these temperatures. The presence of the kilns is confirmed in the Mesopotamian archaeological record for earlier periods, but the coeval presence of the other firing methodologies, such as pit-firings, was neglected by archaeologists in the previous century. The indication of the maximum temperatures reached during firing of Abu Tbeirah pottery together with the absence of kiln traces in the archaeological site provides a valid argument to hypothesize the use of pit-firing for pottery production.

Author Contributions: G.F. coordinated the experiments, data analysis and manuscript writing. L.R. wrote the historical account and sample's description (§§ 1.1; 2.2); V.F. made the autoptic subdivision (§2.1) and M.N. G.F. and C.S. performed the chemometrics data analysis and the mineralogical compositional discussion. A.S. contributed to the experiments and the data processing at the INES beamline. C.A. and F.D. contributed to the theoretical discussion and the manuscript. R.S. discussed the results and contributed to the manuscript. The introduction and conclusion are common work of the authors.

Acknowledgments: This research is supported by CNR, within the CNR-STFC Agreement 2014–2020 (N. 3420), concerning collaboration in scientific research at the ISIS Spallation Neutron Source.

Conflicts of Interest: The authors declare no conflict of interest.

References

1. Rice, P.M. *Pottery Analysis: A Sourcebook. The Raw Materials of Pottery Making: Perspectives from Chemistry, Geology, and Engineering*; University of Chicago Press: Chicago, IL, USA, 2015.
2. Bland, C.; Roberts, A.L.; Popelka-Filcoff, R.S.; Santoro, C.M. Early vitrification stages identified in prehistoric earthenware ceramics from northern Chile via SEM. *J. Archaeol. Sci. Rep.* **2017**, *16*, 309–315. [CrossRef]
3. Cardiano, P.; Ioppolo, S.; De Stefano, C.; Pettignano, A.; Sergi, S.; Piraino, P. Study and characterization of the ancient bricks of monastery of "San Filippo di Fragalà" in Frazzanò (Sicily). *Anal. Chim. Acta* **2004**, *519*, 103–111. [CrossRef]
4. De Bonis, A.; Cultrone, G.; Grifa, C.; Langella, A.; Leone, A.P.; Mercurio, M.; Morra, V. Different shades of red: The complexity of mineralogical and physico-chemical factors influencing the colour of ceramics. *Ceram. Int.* **2017**, *43*, 8065–8074. [CrossRef]

5. De Benedetto, G.E.; Laviano, R.; Sabbatini, L.; Zambonin, P.G. Infrared spectroscopy in the mineralogical characterization of ancient pottery. *J. Cult. Herit.* **2002**, *3*, 177–186. [CrossRef]

6. El Ouahabi, M.; Daoudi, L.; Hatert, F.; Fagel, N. Modified Mineral Phases During Clay Ceramic Firing. *Clays Clay Miner.* **2015**, *63*, 404–413. [CrossRef]

7. Moorey, P.R.S. *Ancient Mesopotamian Materials and Industries: The Archaeological Evidence*; Eisenbrauns: Warsaw, IN, USA, 1994; p. 144.

8. Armstrong, J.A.; Gasche, H. *Mesopotamian Pottery. A Guide to the Babylonian Tradition in the Second Millennium B.C*; The University of Ghent: Ghent, Belgium; The Oriental Institute of the University of Chicago: Chicago, IL, USA, 2014; p. 83.

9. Smith, A.L. Bonfire II: The Return of Pottery Firing Temperatures. *J. Archaeol. Sci.* **2008**, *28*, 991–1003. [CrossRef]

10. Squires, G.L. *Introduction to the Theory of Thermal Neutron Scattering*; Cambridge University Press: Cambridge, MA, USA, 1978.

11. Bland, C.; Roberts, A.L.; Popelka-Filcoff, R.; Santoro, C.; Carter, C.; Bennett, J.; Stopic, A. 1500 years of pottery: Neutron activation analysis of northern Chilean domestic ceramics from Caleta Vitor and clay samples from nearby valley and highland contexts. *Archaeometry* **2017**, *59*, 815–833. [CrossRef]

12. D'Agostino, F.; Romano, L. Rediscovering Sumer. Excavations at Abu Tbeirah. Southern Iraq. In *My Life Is Like the Summer Rose*; Cerasetti, B., Lamberg-Karlovsky, C.C., Eds.; Maurizio Tosi e l'Archeologia Come Modo di Vivere. Papers in Honour of Maurizio Tosi for His 70th Birthday; BAR-IS 2690; Archaeopress: Oxford, UK, 2014; pp. 163–167.

13. D'Agostino, F.; Kadhem, A.; Romano, L.; Vidale, M.; Angelozzi, M. Abu Tbeirah. Preliminary Report of the First Campaign (January–March 2012). *Rivista Degli Studi Orientali* **2011**, *84*, 17–34.

14. D'Agostino, F.; Romano, L. Abu Tbeirah. Preliminary Report of the 2012–2013 Campaigns. In Proceedings of the 9th International Congress on the Archaeology of the Ancient Near East, Basel, Switzerland, 9–13 June 2014; Kaelin, R.S.O., Mathys, H.-P., Eds.; Harrassowitz: Wiesbaden, Germany, 2014.

15. D'Agostino, F.; Romano, L.; Khadem, A.; Tafuri, M.A. Abu Tbeirah. Preliminary Report of the Second Campaign (October–December 2012). *Rivista Degli Studi Orientali* **2013**, *86*, 69–92.

16. D'Agostino, F. Rediscovering and Revitalizing the History of Iraq: The Excavations at Abu Tbeirah. In *Strategie e Programmazione della Conservazione e Trasmissibilità del Patrimonio Culturale*; Filipovic, A., Troiano, W., Eds.; Italian Heritage Awards; Edizioni Scientifiche Fidei Signa: Roma, Italy, 2013; pp. 212–221.

17. D'Agostino, F.; Romano, L.; Kadhem Ghanim, A. Abu Tbeirah. Nasiriyah (Southern Iraq). Preliminary Report on the 2013 Excavation Campaign. Available online: https://revistas.uam.es/isimu/article/view/3209 (accessed on 29 January 2019).

18. Nardini, M.; Andreani, C.; Senesi, R.; Scherillo, A.; D'Agostino, F.; Romano, L.; Scatigno, C.; Festa, G. Neutron Resonance Capture Analysis and chemometric tools: An integrated approach. *J. Phys. Conf. Ser.* **2018**, *1055*, 012005. [CrossRef]

19. Postma, H.; Schillebeeckx, P. Non-destructive analysis of objects using neutron resonance capture. *J. Radioanal. Nucl. Chem.* **2005**, *265*, 297–302. [CrossRef]

20. Thuesen, I.; Heydorn, K.; Gwozdz, R. Investigation of 5000-Year-Old Pottery from Mesopotamia by Instrumental Neutron Activation Analysis. In Proceedings of the Second Nordic Conference on the Application of Scientific Methods in Archaeology, Helsingør, Denmark, 17 August–19 August 1981; Mejdahl, V., Ed.; Council of Europe: Strasbourg, France, 1982; Volume 2, pp. 375–381.

21. Mynors, H.S. An Examination of Mesopotamian Ceramics Using Petrographic and Neutron Activation Analysis. In Proceedings of the 22nd Symposium on Archaeometry Held at the University of Bradford, Bradford, UK, 30 March–3 April 1982; Aspinall, A., Warren, S., Eds.; University of Bradford: Bradford, UK, 1983; pp. 377–387.

22. Mynors, H.S.; Al Kaissi, B. Ceramic Analyses of Mesopotamian Wares in the Early Dynastic Period. In *Research on the Antiquities of Saddam Dam Basin Salvage and Other Researches*; Republic of Iraq, Ministry of Culture and Information, State Organization of Antiquites and Heritage: Baghdad, Iraq, 1987; pp. 134–154.

23. Méry, S.; Schneider, G. Mesopotamian Pottery Wares in Eastern Arabia from the 5th to the 2nd Millennium BC: A Contribution of Archaeometry to the Economic History. In Proceedings of the Twenty-Ninth Meeting of the Seminar for Arabian Studies, Cambridge, UK, 20–22 July 1995; Archaeopress Publishing Ltd.: Oxford, UK, 1996; Volume 26, pp. 79–96.

24. McGibson, G. *Uch Tepe II: Technical Reports*; The Oriental Institute of the University of Chicago: Chicago, IL, USA, 1990.

25. Sanjurjo-Sánchez, J.; Montero Fenollós, J.L.; Polymeris, G.S. Technological aspects of Mesopotamian Uruk pottery: Estimating firing temperatures using mineralogical methods, thermal analysis and luminescence techniques. *Archaeol. Anthropol. Sci.* **2018**, *10*, 849–864. [CrossRef]

26. Verhoeven, K. Geomorphological Research in the Mesopotamian Flood Plain. In *Changing Watercourses in Babylonia. Towards a Reconstruction of the Ancient Environment in Lower Mesopotamia*; Gasche, H., Tanret, M., Eds.; Volume I. Mesopotamian History and Environment. Series II. Memoirs V; The University of Ghent: Gent, Belgium, 1998; pp. 159–240.

27. Yacoub, S.Y. Stratigraphy of the Mesopotamia Plain. *Iraqi Bull. Geol. Min.* **2011**, *4*, 47–82.

28. Al-Asadi, M.M.M. The Sedimentary Model of Thi-Qar Governorate During the Holocene. *South West Iraq. J Basrah Res. (Sci.)* **2007**, *33*, 91–101.

29. Sissakian, V.K.; Saeed, Z.B. Lithological Map of Iraq. Compiled Using GIS Techniques. *Iraqi Bull. Geol. Min.* **2012**, *8*, 1–13.

30. Alfatlawi, E.A.M. Swelling Potential and Mineralogy of Thi-Qar University Soil. *J. Thi-Qar Univ.* **2011**, *7*, 1–15.

31. Al-Mukhtar, L.E. Heavy Mineral Analysis of the Quaternary Sediments in the Southern Part of the Mesopotamia Plain, Iraq. *Iraqi Bull. Geol. Min.* **2015**, *11*, 59–73.

32. Ali, M.Y.; Watts, A.B.; Searle, M.P. Seismic stratigraphy and subsidence history of the United Arab Emirates (UAE) rifted margin and overlying foreland basins. In *Lithosphere Dynamics and Sedimentary Basins: The Arabian Plate and Analogues*; Springer: Berlin/Heidelberg, Germany, 2013; pp. 127–143.

33. Erb-Satullo, N.L. Alloys from Anau: The Manipulation of Metallic Properties in 3rd Millennium B.C. Southern Central Asia. *MRS Proc.* **2011**, *1319*. [CrossRef]

34. Courty, M.A.; Roux, V. Identification of Wheel-Throwing on the Basis of Ceramic Surface Features and Microfabrics. *J. Archaeol. Sci.* **1995**, *22*, 17–50. [CrossRef]

35. Courty, M.A.; Roux, V. Identification of Wheel-Fashioning Methods: Technological Analysis of 4th-3rd Millennium BC Oriental Ceramics. *J. Archaeol. Sci.* **1998**, *25*, 747–763.

36. Romano, L. A Fragment of a Potter's Wheel from Abu Tbeirah. *Zeitschrift für Assyriologie* **2015**, *105*, 220–234. [CrossRef]

37. Roux, V.; Rosen, S. An Introduction to Technological Studies in the Archaeology of the Proto-Historic and Early Historic Periods in the Southern Levant. In *Techniques and People: Anthropological Perspectives on Technology in the Archaeology of the Proto-Historic and Early Historic Periods in the Southern Levant*; Rosen, S., Roux, V., Eds.; De Boccard: Paris, France, 2009; pp. 11–22.

38. Imberti, S.; Kockelmann, W.; Celli, M.; Grazzi, F.; Zoppi, M.; Botti, A. Neutron diffractometer INES for quantitative phase analysis of archaeological objects. *Meas. Sci. Technol.* **2008**, *19*, 034003. [CrossRef]

39. Windsor, C.G. *Pulsed Neutron Scattering*; Taylor and Francis: London, UK, 1981.

40. Larson, A.C.; von Dreele, R.B. *GSAS: General Structure Analysis System*; Report LAUR 86-748; Los Alamos Laboratory: Los Alamos, NM, USA, 1986.

41. Toby, B.H. EXPGUI, a graphical user interface for GSAS. *J. Appl. Cryst.* **2001**, *34*, 210–213. [CrossRef]

42. Jolliffe, I.T. *Principal Component Analysis*; Springer: Berlin, Germany, 1986; p. 487, ISBN 978-0-387-95442-4.

43. Husson, F.; Lê, S.; Pagès, J. *Exploratory Multivariate Analysis by Example Using R*; The R Series; Chapman & Hall/CRC: London, UK, 2009; 224p, ISBN 978-2-7535-0938-2.

44. Baxter, M.J. *Exploratory Multivariate Analysis in Archaeology*; Edinburgh University Press: Edinburgh, UK, 1994.

45. Available online: https://www.camo.com/unscrambler/ (accessed on 29 January 2019).

46. Arai, K.; Barakbah, A.R. Hierarchical K-means: An algorithm for centroids initialization for K-means. *Rep. Fac. Sci. Eng.* **2007**, *36*, 25–31.

47. Anderberg, M.R. *Cluster Analysis for Applications*, 1st ed.; No. OAS-TR-73-9; Elsevier: New York, NY, USA, 1973.

geosciences

MDPI

Article

Stratigraphy, Petrography and Grain-Size Distribution of Sedimentary Lithologies at Cahuachi (South Peru): ENSO-Related Deposits or a Common Regional Succession?

Marco Delle Rose [1,*]**, Michele Mattioli** [2]**, Nicola Capuano** [2] **and Alberto Renzulli** [2]

[1] Istituto di Scienze dell'Atmosfera e del Clima, Consiglio Nazionale delle Ricerche, Complesso Ecotekne, Via per Monteroni, 73100 Lecce, Italy

[2] Dipartimento di Scienze Pure e Applicate, Università degli Studi di Urbino Carlo Bo, Campus Scientifico Enrico Mattei, 61029 Urbino, Italy; michele.mattioli@uniurb.it (M.M.); nicola.capuano@uniurb.it (N.C.); alberto.renzulli@uniurb.it (A.R.)

* Correspondence: m.dellerose@isac.cnr.it; Tel.: +39-0832-298816

Received: 13 December 2018; Accepted: 6 February 2019; Published: 8 February 2019

Abstract: Several central Andean pre-Columbian sites struck by hydrogeological disasters due to El Niño-Southern Oscillation (ENSO) events are reported in the literature. The mainstream explanation for the decline and demise of Cahuachi (*pampa* of Nazca, south Peru) implies the damage and burial of such a ceremonial center as a consequence of two catastrophic river floods, which occurred around 600 CE and 1000 CE, respectively. Therefore, geological studies at Cahuachi are mandatory with regard to both the correlations of ENSO-related deposits ("event-strata") among different Peruvian sites and the assessment of the millennium-scale climate variability. In particular, the latter is crucial to evaluate the environmental and economic consequences due to the incoming fluctuations of ENSO. In this paper, stratigraphic, grain-size distribution, and petrographic investigations on a sedimentary section exposed close to one of the main temples of Cahuachi are reported. They represent the first test for the current mainstream explanation. The preliminary finding indicates that the studied stratigraphic interval may belong to the common regional succession of the *pampa* of Nazca rather than the ENSO-related deposits described in the literature. However, further geological research will be necessary to unravel this issue in more detail.

Keywords: Mega El Niño; *pampa* of Nazca; Cuenca Pisco; Rio Grande de Nazca; grain-size; volcaniclastic layer; stratigraphy; petrography

---

1. Introduction

Several central Andean pre-Columbian sites struck by hydrogeological disasters due to ENSO (El Niño-Southern Oscillation) events are reported in both the literature in the Earth sciences [1–3] and archaeology [4–6]. Grodzicki [7,8] claims that the ceremonial center of Cahuachi (Rio Grande de Nazca, South Peru) was first severely damaged, then completely buried by catastrophic river floods as a result of two Mega El Niño events, which occurred around 600 Common Era (CE) and 1000 CE, respectively. According to such a hypothesis, each event would be proved by a conglomerate layer. This is based on an interpretative geological setting of the deposits outcropping around the archaeological site, without, however, either geological characterization or direct dating of the main sedimentary lithologies. The ages assigned by Grodzicki to conglomerates were deduced from uncertain stratigraphic relationships with deposits containing materials dated with the ^{14}C method. Despite this uncertainty, the stratigraphic succession of Cahuachi could be a climate proxy-record of particular interest, especially regarding the millennium-scale climate variability, and thus must be investigated in further detail. Again, several authors have emphasized the correlations of ENSO-related deposits ("event-strata") among different archaeological Peruvian sites [9–11]. These correlations would support the concept of Mega or Super El Niño events [12–14] leading to heavier precipitation phases driven by ENSO, which would hit the central Andean coast with centennial-scale periodicity. It is worth observing that the understanding of the variability of El Niño over geological time is crucial for assessing the environmental and economic consequences due to the incoming fluctuations of ENSO [15–17].

However, before the publication (in 1994) of the *Mapa geologico del cuadrangulo de Palpa* (geological quadrangle map of the Palpa) [18], the geological knowledge of the Rio Grande de Nazca Basin was scarce, and therefore, the thesis of Grodzicki and his group (launched in 1990 and 1992 and then becoming nearly a "paradigm") was difficult to prove. In addition, the lack of large outcrops within the Cahuachi archaeological site hampered the stratigraphic survey of the bedrock. During the 2012 archaeological excavation works at Cahuachi, the geological substratum close to one of the main temple of the site, namely the Piramide Sur, was temporarily exposed. In this way, a stratigraphic study and sampling were carried out, and the grain size features and petrographic analysis were subsequently performed. As a whole, the data collected on the geological bedrock of Piramide (or Temple) Sur first allowed for a comprehensive comparison with nearby Pliocene–Pleistocene deposits, and shed light into the robustness or not of Grodzicki's thesis about the ENSO-related catastrophic river floods, considered for at least three decades as the mainstream and paradigmatic explanation of the demise of Cahuachi as for other pre-Columbian sites for which, however, the hypothesis is better supported by the data [19–22].

2. Geological and Environmental Features

2.1. Regional Setting

Cahuachi is located within the morpho-structural unit named Cuenca Pisco consisting, in turn, of Cordillera de la Costa, Llanuras Costeras, Depresión de Ica-Nasca and Fruente Andino [23]. The Llanuras costeras and the Depresión de Ica-Nasca show flat landscapes, i.e., the *pampas*, due to the sub-horizontal setting of the pre-Quaternary substratum (Figure 1).

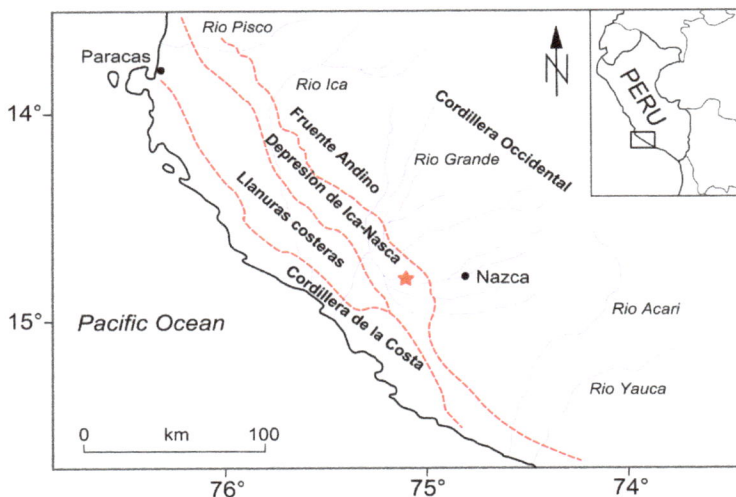

Figure 1. Morpho–structural scheme of the Cuenca Pisco (from [23], modified). The red star indicates the location of Cahuachi.

The Cuenca Pisco gradually emerged from the ocean during the late Pliocene–early Pleistocene, thus evolving from a marginal sea (seaward barred by the ridge of the Cordillera de la Costa) to the present coastal plain. The uplift of the Andes led to abundant supplies of conglomerates that built alluvial fans within the Fruente Andino while planation surfaces and stream terraces formed at the Llanuras Costeras and the Depresión de Ica-Nasca. The lower planation surface has an altitude of about 385 m a.s.l. and a late Middle Pleistocene age (i.e., 200 ka) as determined by cosmogenic ray dating [24]. The regional succession of the *pampa* of Nazca is constituted by the Pisco and Changuillo formations, both of Tertiary age. The plain has an average altitude of about 500 m a.s.l., gently slopes toward the north-west and is bounded by mountain chains of sedimentary and igneous Mesozoic units (Figure 2).

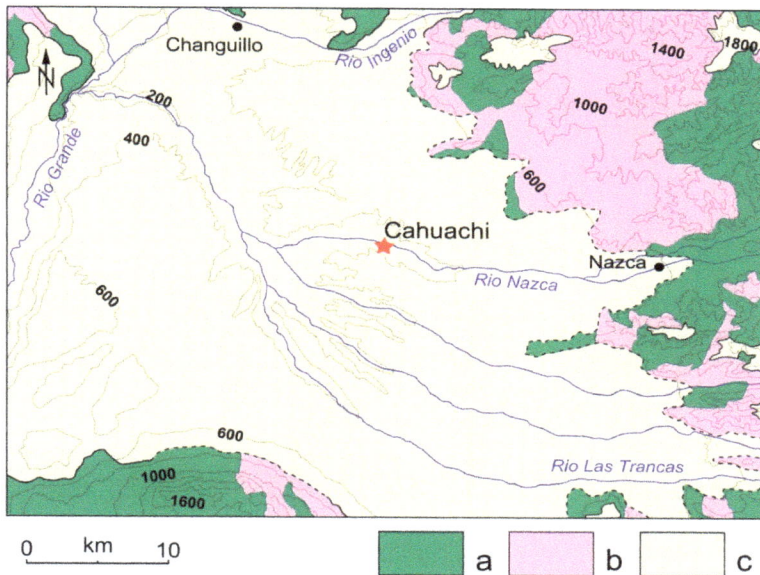

Figure 2. Geological sketch of the *pampa* of Nazca. Legend: (a) marine Jurassic-Cretaceous formations, (b) igneous rocks, (c) marine to continental Tertiary formations (from [18], modified).

The Pisco Formation (Fm) was defined at the beginning of the XIX century during the pioneering geological surveys of the Cuenca Pisco. In contrast, the Changuillo Fm was identified only during the 1980s with the field mapping of the *Mapa geologico del cuadrangulo de Palpa* (geological quadrangle map of the Palpa) [18]. The latter shows a marine to continental transition from the bottom to the top and is constituted by variously-interstratified mudstones, sandstones, breccias and conglomerates. Around Changuillo Village, i.e., the "area-type", some volcaniclastic layers have been found [23]. In several places at the east side of the Cuenca Pisco, conglomerates are prevalent at the top of the Changuillo Fm. Thus, a specific lithostratigraphic unit, named the Canete Fm, has also been defined [23], and is built up by progradational alluvial fans. Along the Cordillera de la Costa, a staircase of terraces represents the corresponding deposits and suggests the episodic uplift of the Dominio Costero. The boundary between the Changuillo and Canete Fms is generally described as transitional while the facies evolve from shallow marine water to shoreface to an alluvial system. Biostratigraphically significant fossils are scarce. The presence of *Carcharodon carcharias*, *Dinocardium nov. sp. Aff.*, and *D. ecuadorialis* suggests a middle–upper Pliocene for the Changuillo Fm [25]. No fossils have been found in the Canete Fm, although an upper Pliocene–lower Pleistocene age can be roughly supposed due to the regional stratigraphic setting [26].

The *pampa* of Nazca belongs to the hyper-arid "chala" life-zone (i.e., the region comprised between the coast and the Andean Precordillera [27]) and constitutes the north-west outlying area of the Atacama Desert. Its environmental conditions are controlled by the SE Pacific anticyclone and the cold Humboldt Current, both inhibiting rainfall. Currently, the annual precipitation is lower than 20 mm, although El Niño-Southern Oscillation may determine significant inter-annual variations. The aridity degree as well as the average impact of El Niño events change along a south–north gradient [28]. Ephemeral or seasonal rivers cross the *pampa* and allow both biological life and human settlements such as the ceremonial center of Cahuachi, which is located on the course of the Nazca River (Figure 2).

Environmental conditions have repeatedly changed over south Peru during the late Quaternary. Starting from the Upper Pleistocene, fluctuations in the moisture transport led to repeated shifts from grassland to desert, and vice versa, on the lower western slopes of the Andes [29]. At the beginning of the Holocene, an increase in easterly precipitation led progressively to persisting vegetation belts in the mountain areas that, in turn, have determined the formation of soils and eolian deposits. About 4.2 ka, the establishment of the modern ENSO atmospheric conditions determined a decrease in precipitation and progressive expansion of deserts. Further environmental changes occurred later as inferred from geomorphological features. Finally, from the end of the last millennium Before the Common Era (BCE) to about 1.3 ka (which is a time interval comprising the rise and decline of the Nasca Culture), alluvial deposits have widely accumulated within the Depresión de Ica-Nasca. Such a process suggests an important hydrological activity [30].

2.2. The Nasca Culture: From the Thriving Period to the Demise

The Nasca Culture arose during the Early Intermediate Period, i.e., around 200 BCE, later spreading its cultural and religious dominance over a large area of southern Peru under the influence of the theocratic capital of Cahuachi [31,32]. The ceremonial center (Figure 3) became a place of periodic pilgrimage from places very far afield, preserving this function until about 450 CE [33].

Figure 3. Cahuachi ceremonial center. The mounds of Piramide Sur (background, on the left) and of the Gran Piramide (centre) are shown.

During this period, intense rains and associated mud-debris flows damaged the structures built with adobe bricks, sometimes involving the reconstruction of buildings among which was the Gran Piramide. Moreover, several authors have stated a strong link between the course of the Nasca civilization and the environmental conditions. As an example, Schittek et al. [34] stressed that the thriving period of the Nasca culture coincided with a relatively humid period while its demise coincided with an abrupt environmental turnover occurring at 1.3 ka. Up to about 0.8 ka, the lack of river activity is interpreted as due to renewed hyper-arid conditions, coinciding with the warm Medieval Climate Anomaly [35]. A further shift to relatively humid conditions is again indicated by the river activity and occurrence of landslides. This moist and hydrologically unstable period would last until the Little Ice Age [30].

3. Material and Method

Cahuachi consists of a number of naturally smoothed mounds forming the core of pyramids and temples made of adobe [36]. According to Grodzicki [7,8,37] the bedrock of Cahuachi is represented by Holocene alluvium, covering the Tertiary substratum, and includes three ENSO-related conglomerates, dated to the first century BCE, around 600 CE, and around 1000 CE, respectively. Each conglomerate would have been deposited by "slimy-crumbly flows" triggered by huge precipitation events. The upper and younger conglomerate was identified by Grodzicki on the Gran Piramide (Figure 4) and correlated with a similar strata visible in the northern area of the site. This latter was dated using organic material placed at its top [7,37]. In this way, this conglomeratic bed should cover all of the structures of Cahuachi up to 408 m a.s.l. [38].

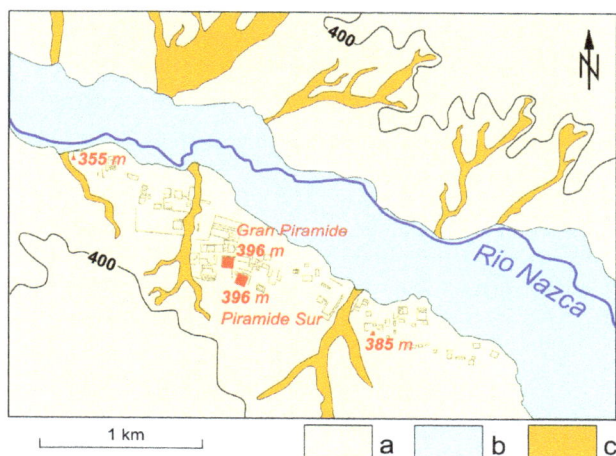

Figure 4. Geological map of Cahuachi. Legend: (a) Mudstones, sandstones and conglomerates; (b) alluvial deposits; (c) colluvium, debris flows; from [39], modified. 400 m a.s.l. contour line is reported. Gran Piramide and Piramide Sur are evidenced with red squares.

As mentioned above, an excavation close to the Piramide Sur (about 70 m south-east of the Gran Piramide, see Figure 4) carried out by the archaeological team of G. Orefici, has temporarily allowed the observation of a stratigraphic column of about 13 m (Figure 5). A close cavity (probably a type of food storage) has enabled us to extend the stratigraphic observations in depth for a total of 17 m [39].

The measured section is capped by a conglomerate corresponding to that described by Grodzicki [8,39] on the Gran Piramide and thus allows us to test his thesis. The stratigraphic observation was performed together with the collection of twelve samples for laboratory analysis.

The grain-size composition of the sediment samples was determined through dry sieving after overnight drying in an oven at 80 °C. Dried sediments were homogenized and classified according to the Wentworth scale grade [40] by passing through a stack of sieves with mesh apertures of 4 mm (granules), 2 mm (very coarse sand), 1 mm (coarse sand), 0.50 mm (medium sand), 0.25 mm (fine sand), and 0.063 mm (silt). These grain-size steps were then transformed into the logarithmic φ scale of Krumbein [41]. Weights (±0.0001 g) of the sediment fractions were used to determine the general grain-size distribution of the sediments. Moreover, the grain size data for all samples were reclassified into gravel, sand, and mud, and presented in a ternary diagram [42].

Mineralogic and petrographic investigations were carried out at the University of Urbino Carlo Bo using a polarized light optical microscope and a scanning electron microscope (SEM) FEI Quanta 200 FEG environmental scanning electron microscope (ESEM) equipped with an energy-dispersive X-ray spectrometer (EDAX) for semi-quantitative chemical analyses.

Figure 5. Panning shot of the excavation used to measure the stratigraphic section.

4. Results (Stratigraphic, Mineralogical and Petrographic Analysis)

4.1. Stratigraphy

The succession measured beneath the Piramide Sur had a sub-horizontal attitude and showed four subvertical joint systems, which were arranged according to angles of about 45°. Three main lithologies were recognized: greyish conglomerates and breccias, light brown sandstones, and light beige mudstones (Figure 6). In addition, a lithoid silty layer, a few centimeters thick, was found and sampled (CH3) at 6.5 m from the base.

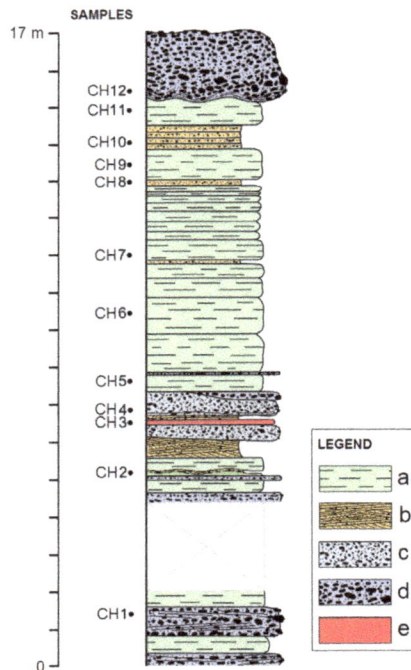

Figure 6. Piramide Sur stratigraphic section (from [39], modified). Legend: a, mudstone; b, sandstone; c, conglomerate with sharp clasts (breccia); d, conglomerate; e, lithoid siltstone.

The section ended just below the foundation of the temple, as indicated by the presence of mortar for adobe (Figure 7b). The stratification was plane-parallel except for some lenticular bodies and a few undulated lithological boundaries. Conglomerates prevailed at the lower half portion of the column while mudstones were more abundant at the upper half. The former presented imbrications of the pebbles, erosive basal contacts, and a sandy-muddy matrix. The pebbles generally showed degrees of sphericity from low to middle (Figure 7a). The thickest conglomerate measured 1.8 m, showed an abrupt undulated lower surface, and was placed at the top of the section (Figure 7b). Mudstones were either laminar or massive, but always poorly consolidated (Figure 7c). Load casts were observed at the upper surface of each stratum of mudstone overlaid by a conglomerate. Sandstones presented cross lamination (Figure 7d).

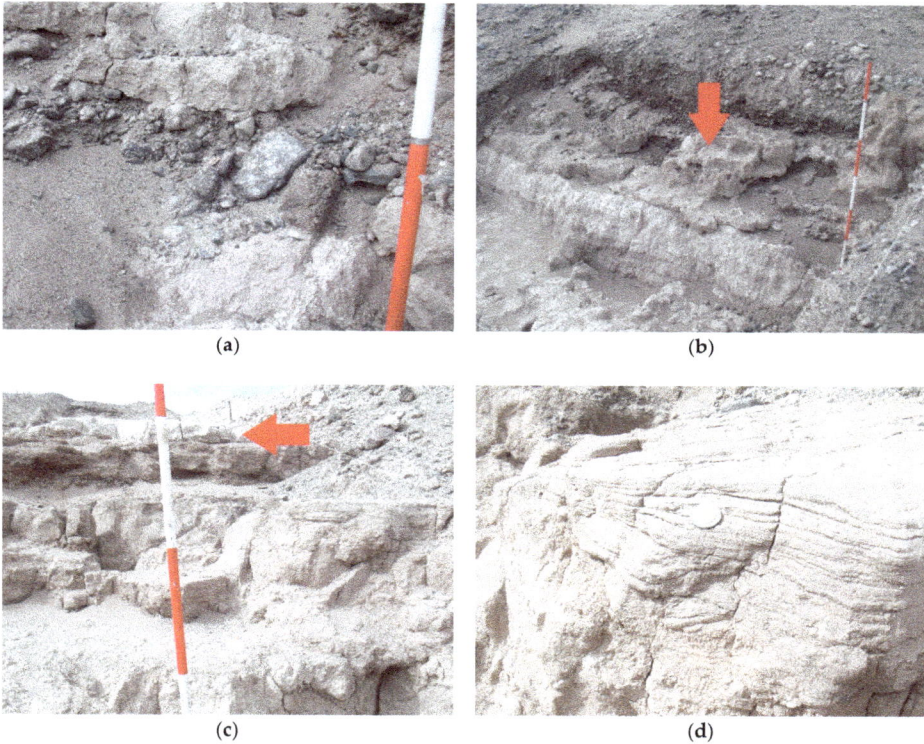

(a)

(b)

(c)

(d)

Figure 7. Stratigraphic features of the Piramide Sur section: (**a**) pebbles concentrations at the base of the upper conglomerate; (**b**) top of the section (1.8 m conglomerate overlays mudstone; mortar for adobes partially covers the contact); (**c**) the portion of the section comprised between 5 and 6.5 m from the section's base (it consists of mudstones and sandstones; note the whitish thin bed in the upper side); (**d**) a detail of cross-lamination in sandstone.

Sandy and silty fractions of the samples were screened on macro- and micro-fossil contents, but all tests gave a negative result.

4.2. Textural (Grain-Size) Characteristics

According to the stratigraphic description reported in the previous section, and applying the textural classification by Folk (1954), our analysis of the sediment grain-size (on the basis of the method of Wentworth, 1922, revisited on the logarithmic φ scale by Krumbein, 1934) shows that the lithologies of Cahuachi are categorized into three groups: "sandy conglomerate", "muddy sandstone", and "sandy mudstone", with one sample (CH2) falling in the slightly conglomeratic muddy sandstone field (Table 1 and Figure 8). The majority of the sediment samples (46 vol.%) belong to the sandy mudstone group, while sandy conglomerate and muddy sandstones are present in a minor and comparable amount (27 vol.%). It is worth noting that one of the collected samples (CH3) was not processed for the grain size analysis because of its lithoid state; for this reason, it will be described as a separate sample (see next section). In addition, the statistic grain-size parameters of the sandy mudstones were not processed as the most abundant grain size fraction >4 φ (i.e., <63 microns) of this textural group was not separated into different grain-size classes for a correct cumulative curve.

Table 1. Grainsize distribution of the statistical parameters of studied sediments. SC = Sandy Conglomerate; SCMS = Slightly Conglomeratic Muddy Sandstone; SM = Sandy Mudstone; MS = Muddy Sandstone; PK_G = Platykurtic; MK_G = Mesokurtic; LK_G = Leptokurtic; VLK_G = Very Leptokurtic; PSo = Poorly Sorted; VPSo = Very Poorly Sorted; MSo = Moderately Sorted. n.d. = not determined.

Sample	Mean (Mz)	Median (Md)	Sorting (σ)	Skewness (Sk)	Kurtosis (K_G)	Remarks		Gravel (%)	Sand (%)	Mud (%)	Texture Group
CH1	0.51	0.55	1.77	0.04	0.74	PK_G	PSo	28.32	68.82	2.864	SC
CH2	1.67	1.73	2.2	0.08	1.25	LK_G	VPSo	8.008	78.77	13.23	SCMS
CH4	0.02	−0.38	1.93	0.46	1.33	LK_G	PSo	33.45	59.58	6.96	SC
CH5	n.d.	n.d.	n.d.	n.d.	n.d.	n.d.	n.d.	-	12.65	87.35	SM
CH6	n.d.	n.d.	n.d.	n.d.	n.d.	n.d.	n.d.	-	43.57	56.43	SM
CH7	n.d.	n.d.	n.d.	n.d.	n.d.	n.d.	n.d.	-	23.78	76.22	SM
CH8	2.88	2.69	1.25	0.4	1.76	VLK_G	MSo	-	86.06	13.94	MS
CH9	n.d.	n.d.	n.d.	n.d.	n.d.	n.d.	n.d.	-	26.41	73.59	SM
CH10	2.71	2.63	1.33	0.26	1.61	VLK_G	MSo	-	87.82	12.18	MS
CH11	n.d.	n.d.	n.d.	n.d.	n.d.	n.d.	n.d.	-	21.5	78.5	SM
CH12	−0.4	−1.10	1.69	0.67	1.04	MK_G	PSo	55.43	40.17	4.4	SC

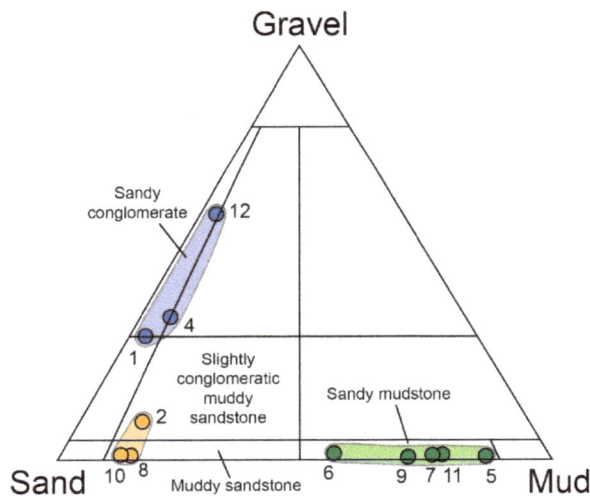

Figure 8. Ternary grain-size classification diagram proposed by Folk (1954), showing the proportion of mud, sand and gravel for the studied sediments collected at the Cahuachi site.

The grain size distribution patterns for the studied samples are graphically illustrated in Figures 9 and 10. According to the histograms in Figure 9, the sandy conglomerates show slightly bimodal distributions, with the main mode coinciding with the coarse modal class (−1 φ) and the second, minor mode moving to the finer classes (2 φ in the CH1 sample, >4 φ in the other samples). Muddy sandstones are typically unimodal (3 φ), except for the CH2 sample, which showed a second, minor mode toward the finer classes (>4 φ). Sandy mudstones are typically fine-grained, always showing a significant mode for classes >4 φ. Mz values range from –0.4 to 0.51 φ in the sandy conglomerates and 1.67 to 2.88 φ in the muddy sandstones.

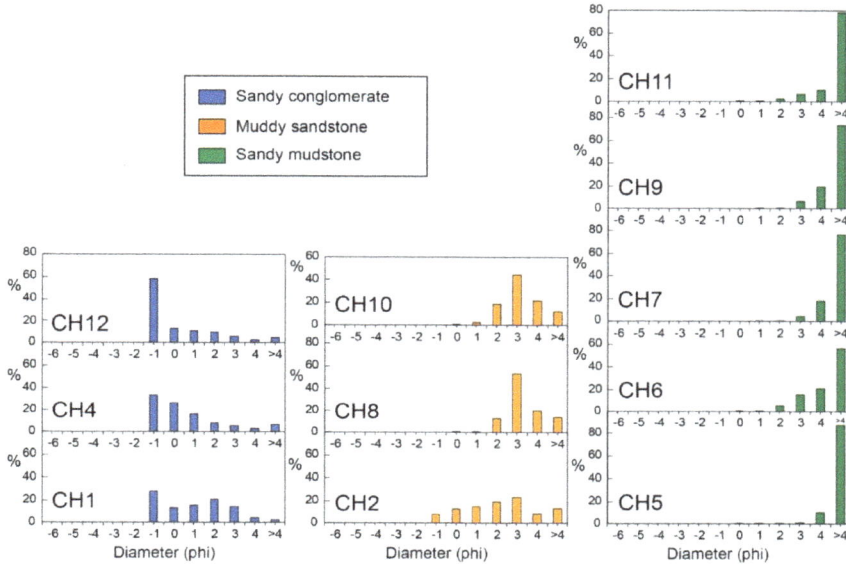

Figure 9. Histograms of the studied samples of Cahuachi.

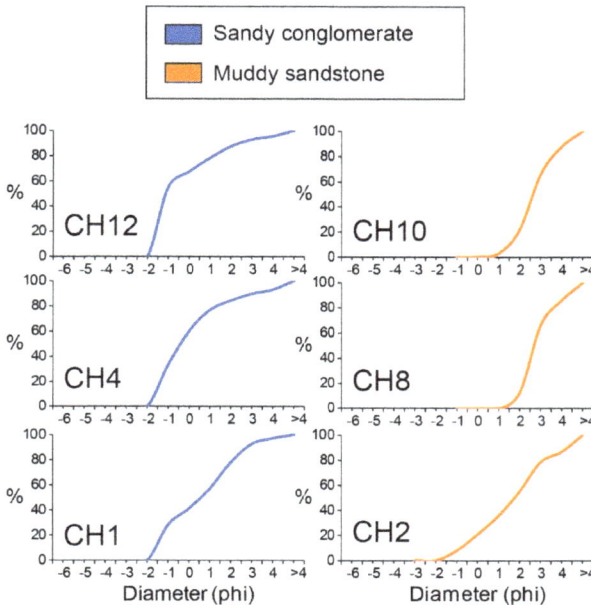

Figure 10. Cumulative curves showing the trends of the coarse-grained samples of Cahuachi.

On the basis of the cumulative curves, the median diameter (Md, i.e., 50th percentile) of the sandy conglomerates varies from 0.55 to –1.10 φ whereas that of the muddy sandstones ranges from 1.73 to 2.69 φ. The sorting value, mostly linked to the supply sources, to the energy of the environment, and the transport processes, varies from poorly sorted (1.69–1.93 φ) sandy conglomerates to moderately sorted (1.25–1.33 φ) muddy sandstones with the slightly conglomeratic muddy sandstone being very poorly sorted (2.2 φ).

The skewness values (Sk, indicating the deviation between Md and Mz) in the Cahuachi lithologies range from 0.04 to 0.67 φ in the sandy conglomerates, and from 0.08 to 0.4 φ in the muddy sandstones (this also includes the slightly conglomeratic muddy sandstone sample). Accordingly, all of these samples vary from near symmetrical to strongly positively skewed.

Regardings the Kurtosis parameter (K_G, showing the measure of the shape of the histogram curve, i.e., plate vs. tip), the studied samples also showed significant variability, with K_G ranging from platykurtic (0.74 φ, CH1), through to mesokurtic (1.04 φ, CH12) and leptokurtic (1.33 φ, CH4) in the sandy conglomerates. The muddy sandstones are very leptokurtic (1.61–1.76 φ), with the slightly conglomeratic muddy sandstone that is leptokurtic (1.25 φ).

4.3. Composition of the Volcaniclastic Layer CH3

Along the stratigraphic succession of Cahuachi, the lithoid silty layer (CH3) has a volcaniclastic origin. It placed above the two sandy conglomerates (CH1 and CH2), at the bottom of the stratigraphic section (Figure 6). This layer is composed of a predominant juvenile volcanogenic glass, associated with minor sialic and mafic volcanic crystals, and a scarce lithic fraction (Figure 11). The juvenile material consists of predominantly white pumices, marked by fluidal, spongy, and minor blocky texture, and fresh glass shards showing typical morphological features of explosive silicic magmatic eruptions (i.e., cuspate, plate, Y-shaped). Frequently, the delicate spines of bubble-wall and pumice shards are undeformed, suggesting that the shards were sufficiently cold at the moment of the depositional process of this volcaniclastic level. Igneous minerals include plagioclase, clinopyroxene, biotite, and Fe–Ti oxides. These generally occur as loose subhedral crystals, but they can also be found as phenocrysts/microphenocrysts in pumices and lithic clasts. The lithic fraction is mainly represented by fine-grained, vitropyric rock fragments. Some lithics contain phenocrysts of plagioclase and pyroxene in a groundmass consisting of the same phases and brown glass.

Semi-quantitative chemical analyses were carried out on selected mineral phases and glass shards in a thin polished section with an SEM (Figure 11). The composition of the analyzed plagioclase crystals roughly ranged from andesine to labradorite. Pyroxene was present as euhedral to subhedral crystals of augitic compositions, while mica was Mg-rich biotite.

This volcaniclastic level could represent a widespread tephrostratigraphic regional marker in the framework of the regional sedimentary sequences.

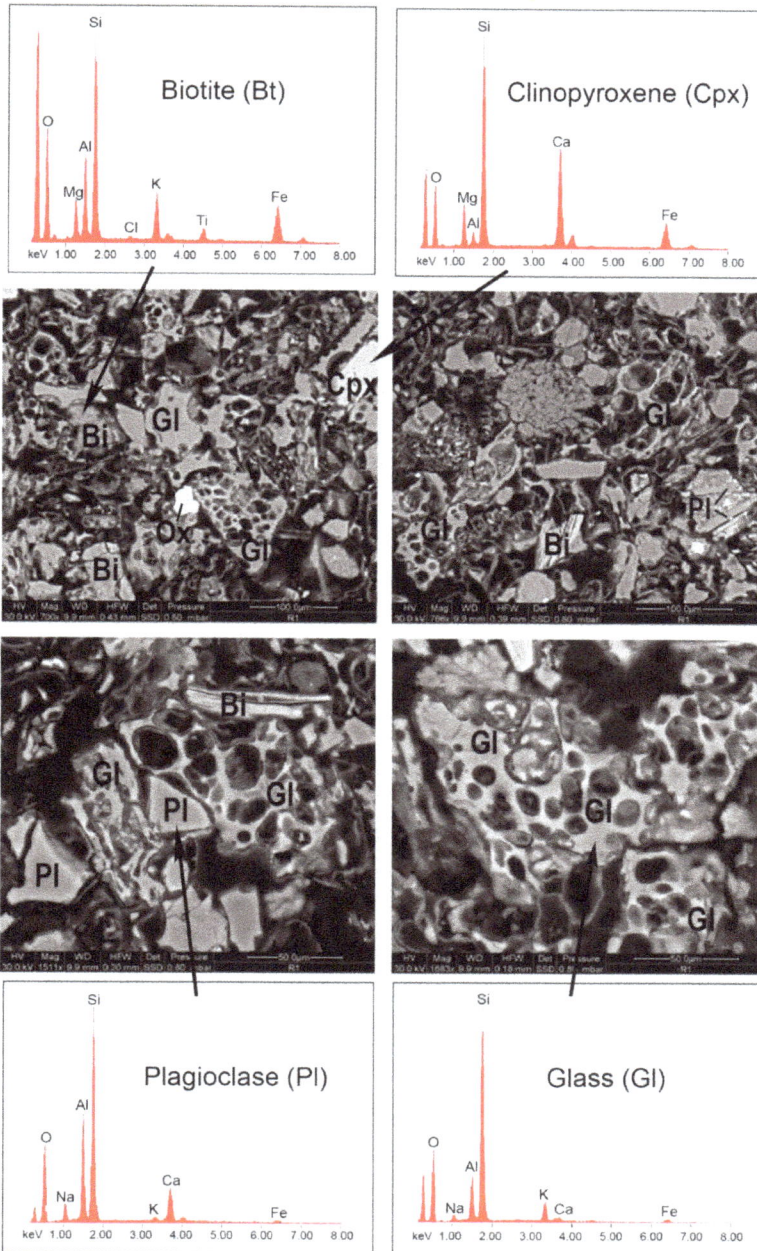

Figure 11. Scanning electron microscope (SEM) microphotographs and energy-dispersive X-ray spectrometer (EDAX) spectra of the studied volcaniclastic layer (CH3) from the Cahuachi succession. Gl = glass; Pl = plagioclase; Bi = biotite; Cpx = clinopyroxene; Ox = Fe-Ti oxides.

4.4. Petrography of Pebbles within the Sandy Conglomerates

Representative pebbles within the sandy conglomerate samples (CH1, CH4 and CH12) of the investigated Cahuachi succession were selected for thin section petrographic study by a polarized light optical microscope to determine their lithology. The sandy conglomerates contain pebbles of similar petrographic composition (Figure 12). The petrotypes observed can be referred to as sedimentary, magmatic, and metamorphic rocks. Sedimentary lithoclasts, mainly represented by siltites, arenites, quartz-arenites, and volcaniclastic arenites; magmatic lithoclasts (plutonic and volcanic) consisting of granites, granodiorites, diorites, andesites, dacites, and rhyolites (also comprising of vitrophyric pyroclasts); and metamorphic pebbles, composed of gneisses and schists. Fragments of vitrophyric pyroclasts are also abundant.

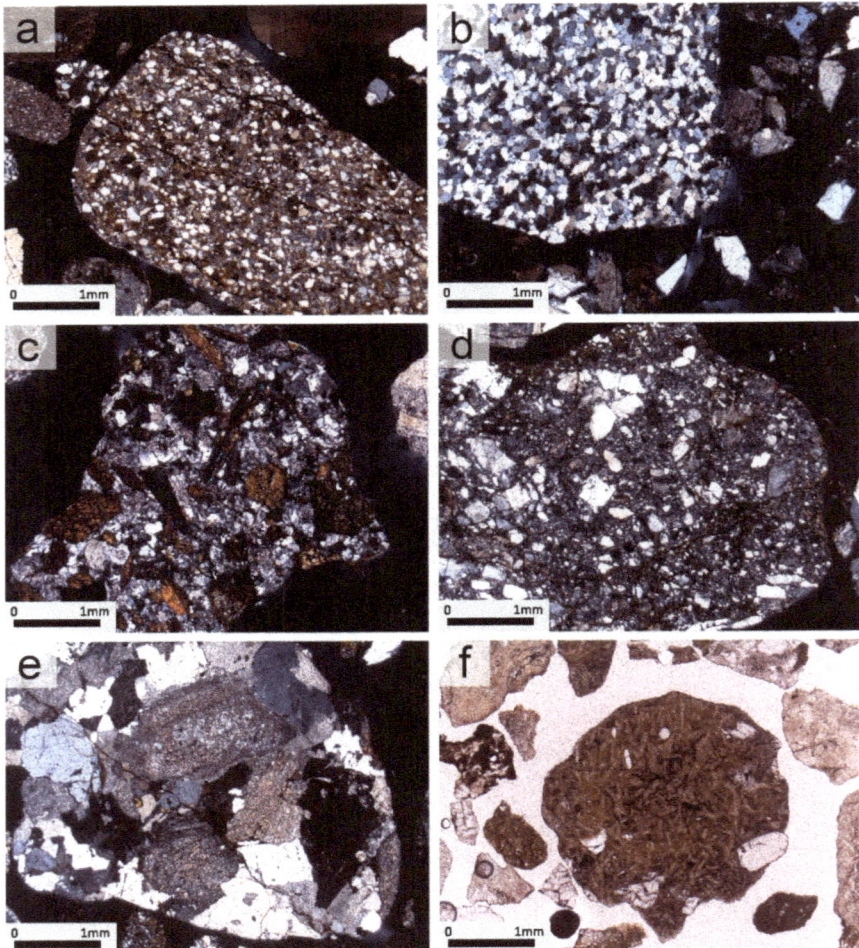

Figure 12. Thin section microphotographs (**a–e**: crossed nicols polarized light; **f**: simple polarized light) in of representative pebbles from the sandy conglomerates. **a** = arenite; **b** = quartz-arenite; **c** = intermediate (andesite-like) volcanic rock; **d** = acid (dacite-rhyolite-like) volcanic rock; **e** = granitoid (with feldspar crystals mostly altered to sericite and clay minerals); **f** = vitrophyric pyroclast.

The sedimentary lithoclasts are dominated by arenites and quartz-arenites, while siltites and volcaniclastic arenites are subordinate. Arenites and quartz-arenites (Figure 12a,b) show a mineralogically homogeneous texture dominated by very fine to coarse-grained quartz grains. The detrital grains are variable, but in general, are of a medium degree of roundness and sphericity. The quartz is predominantly monocrystalline and non-undulatory to weakly undulatory, but some grains exhibit undulose extinction or distinct zones of extinction with sharp boundaries (i.e., polygonized quartz). The amount of variously altered alkali feldspar is scarce and decreases passing from arenites to quartz-arenites. Granitoid rock fragments and carbonate lithics appeared in some thin-sections of the arenites. The siltite clasts show good sorting and appear with a massive and parallel laminated microstructure. The volcaniclastic sandstones, which are predominantly coarse-grained, contain a mixture of framework grain types including variably altered, coarse-grained porphyritic types with large euhedral feldspars, finely crystalline basaltic, and devitrified glassy grains, together with sub-rounded to angular detrital quartz grains.

The magmatic lithoclasts are mainly represented by granite, granodiorite, syenite, and volcanic rocks (andesites, dacites and rhyolites). These clasts occur as pebbles, with various roundness degrees. Pebbles of volcanic origin (Figure 12c,d) are typically sparsely to densely porphyritic, rarely aphyric with a glassy to a fine-grained matrix. Pale brownish clinopyroxene and amphibole are the dominant mafic phenocrysts. Plagioclase phenocrysts are abundant in the phenocrysts population, while either plagioclase or alkali feldspar could be part of the groundmass. The granites and the granodiorites are mostly characterized by the slight to strong alteration of feldspars into secondary minerals such as sericite and clay minerals (Figure 12e). Their microstructures are holocrystalline and hypidiomorphic granular. Diorite lithoclasts are subordinate (with respect to granitoids).

Metamorphic lithoclasts are very subordinate and are mainly represented by gneisses and schists. They show a variable schistose texture characterized by the parallel alignment of fine- to medium-grained mica flakes (mainly biotite, muscovite) intercalated with quartz, plagioclase, K-feldspar, and other minor minerals. Muscovite is generally more abundant than biotite in the schists, where it could be the only phyllosilicate.

Vitrophyric pebbles (Figure 12f) consist of colorless to brownish, vesicle-poor to pumiceous, silicic glass shards, euhedral crystals (feldspars and biotite), and altered volcanic lithic clasts. In some cases, these clasts are almost entirely formed by an assemblage of vitrophyric particles such as Y-shaped glass shards and glass fragments with sinuous or cuspate outlines, with subordinate subhedral to euhedral crystals.

5. Discussion, Summary and Outlook

The stratigraphic study of the Piramide Sur section (Figure 6) did not emphasize any fundamental discontinuity and therefore the section must belong to a single unit. This feature is not in agreement with the interpretation of Grodzicki [8,37] of the sandy conglomerate at the top as an ENSO-related deposit. Moreover, due to the lithological features, the whole succession of Figure 6 can be easily correlated to the regional succession, and in particular, to the upper part of the Changuillo Formation or the transitional stratigraphic level of the Changuillo–Canete Fms [23,43]. The surveyed features did not permit the correlation of the geological section to a well-defined depositional environment, however, one can hypothesize a depositional setting related to the progradation of alluvial fans.

The detection of mortar for adobe (for the constructive characteristics of the Piramide Sur Temple, see Orefici [44]) covering the top conglomerate at the Piramide Sur stratigraphic section (Figure 7b) requires some considerations. Near Gran Piramide, Grodzicki [8] asserted that the conglomerate covered a wall of adobe, thus supporting both its depositional interpretation and dating. Currently, clear contact between the conglomeratic layer and the adobe structure is not visible due to both the restoration works of the archaeological building and the thick level of eluvium constituted by gravel and sand coming from the conglomerate [39]. It is most likely that Grodzicki did not have good exposure of the bedrock and considered the conglomerate was a recent alluvial deposit under the influence of a geomorphological aerial photo interpretation [45]. On the contrary, the top of the Piramide Sur section clearly shows the relation between the sandy conglomerate and the bottom of an adobe wall, thus constraining the pre-existence of the geological layer from which sample CH12 was collected.

An outlook to support the preliminary results of the present study should be to date the sandy conglomerates of the Cahuachi stratigraphic section using the optically stimulated luminescence (OSL) method to rule out the recent age supposed by Grodzicki [8,37]. The optical dating of sediments using OSL signals in mineral grains began in 1985 [46]. This exciting new technique dates deposition back to 200 ka or more and its applications cover many areas of Earth and environmental sciences, comprising archaeological and anthropologic contexts [47,48]. OSL has, in fact, the potential to determine the time elapsed since energy was trapped in the crystal structures as a result of exposure to natural ionizing radiation was last released by exposure to daylight (i.e., the time elapsed since the last transport and deposition of sediment). Minerals most commonly used in OSL are quartz and feldspar, which are virtually ubiquitous in terrestrial surface sedimentary settings. The technique complements other Quaternary dating methods of sedimentary succession including radiocarbon (^{14}C), uranium series methods, and cosmogenic nuclide techniques. Debris flow dating by OSL is currently used (e.g., [49] and references therein), although the opportunity for sufficient light exposure during debris flows is limited, and thus heterogeneous resetting of the latent OSL signal can be expected [50].

The texture and composition of the sandy conglomerates confirm the above lithostratigraphic correlation. On the basis of the petrographic composition of the pebbles, a very similar provenance area of the sandy conglomerates could be inferred for all three of the investigated samples (CH1, CH4 and CH12). Although CH12 is characterized by a relatively higher mean grain-size (Mz −0.4 with respect to 0.02–0.51 of CH1–CH4; Figure 8), all of the other textural parameters (Md, Sorting, Skewness, Kurtosis; Table 1; Figure 9) are comparable for the three sandy conglomerates. In particular, the similarity of the three cumulative curves (Figure 10) emphasize a poorly sorted character (σ 1.69–1.93 φ), indicating the similar energy of the sedimentary environment and transport processes. All of the above features, coupled with similar roundness for all of the pebbles, independent of their sedimentary, magmatic, or metamorphic origin, address the interpretation of the three sandy conglomerates as being derived from similar geological conditions of (i) supply area, (ii) transport, and (iii) kind of siliciclastic rocks, unlike their different position in the investigated stratigraphic sequence at Cahuachi (CH1—CH4, bottom; CH12, top). As a matter of fact, if the thesis of Grodzicki was right, significant textural and compositional differences should have occurred in the sandy conglomerate at the top, with respect to the other similar lithological levels throughout the investigated section. Moreover, the sandy conglomerates of Cahuachi are quite different to the El Niño catastrophic-flood signatures described in the literature [19–22,51].

The lithoid silty layer CH3 has a volcaniclastic origin as shown above. It is worth noting that some volcaniclastic layers have been reported at the upper portion of the Changuillo Fm [23,43], while no volcanic layer was detected in the recent surficial deposits at the *pampa* of Nazca [52]. However, this should be further verified by additional mineralogical and petrographic studies and detailed chronologic and tephrostratigraphic correlations on a regional scale among various sections of the Changuillo Fm. Tephra fingerprinting is widely recognized as a useful method for stratigraphic reconstructions across wide areas, which makes them a straightforward instrument for dating and correlating stratigraphic sections. This method is largely employed in the Quaternary (e.g., [53]) and has been used for correlating deposits of large eruptions of the Central Andes [54,55]. The abundant volcanic ashes interbedded with the basin-filling sedimentary succession provide a snapshot of the long-lived activity of Andean volcanoes of southern Peru, frequently punctuated by large explosive eruptions. In particular, fingerprints to correlate the tephra layer will be addressed with similar geochemical compositions, morphology of glass shards, and their grain-size, together with a detailed chronostratigraphic study using ^{39}Ar–^{40}Ar to date the biotite microlites that are ubiquitous in the erupted pyroclastic products of intermediate to high-silica compositions characterizing subduction-related volcanism. Among the products of the continuous Neogene magmatic activity in southern Peru [56], our tephrostratigraphic investigations will focus on the ash layers derived from the Upper Barroso arc (between 3 and 1 Ma) as defined in the literature [57].

Our preliminary results on the stratigraphic, textural and petrographic study of the geological interval investigated at Cahuachi provide fundamental clues on the matching of the studied section with common regional successions (i.e., Changuillo– or Changuillo–Canete Formations) of the *pampa* of Nazca rather than the ENSO-related deposits described by Grodzicki [7,8]. Further investigation of the fine matrix particles (mostly clay minerals, work in progress) in the lithological units of the section of Cahuachi here reported, together with the study of all the sections reported by Grodzicki, should give additional constraints. The impact of the Mega El Niño events on southern Peru, especially those that occurred around 1000 CE, should be reconsidered in terms of our thesis, which will be reconciled with detailed chronological (e.g., tephrochronology) and biostratigraphic data [58,59].

Author Contributions: M.D.R. conceived the research and conducted the geological survey, the stratigraphic reconstruction and the sample collection work; M.M. and A.R. performed textural, mineralogical and petrographic analysis; N.C. analysed the stratigraphic data; M.D.R., M.M. and A.R. analysed the results, wrote and revised the manuscript.

Funding: This research was funded by the Department of "Scienze Pure e Applicate" (Università di Urbino) through the project "Enhancement of research 2017-2018" entitled "Il contributo delle Scienze della Terra nella definizione della scomparsa della civiltà *Nasca*: il caso di studio geoarcheologico dell'abbandono del sito cerimoniale di Cahuachi (Perù)", responsible A.R. The field survey was performed during the project "Italian Mission of Heritage Conservation and Archaeo-Geophysics (ITACA)", responsible N. Masini (Consiglio Nazionale delle Ricerche).

Acknowledgments: The authors would like to thank M. Bellagamba, S. Galeotti and F. Frontalini (Università di Urbino) for macro- and microfossils-check of the studied samples.

Conflicts of Interest: The authors declare no conflict of interest.

References

1. Magilligan, F.J.; Goldstein, P.S. El Niño floods and culture change: A late Holocene flood history for the Rio Moquegua, southern Peru. *Geology* **2001**, *29*, 431–434. [CrossRef]
2. Sandweiss, D.H. Terminal Pleistocene through mid-Holocene archaeological sites as paleoclimatic archives for the Peruvian coast. *Palaeogeogr. Palaeoclimatol. Palaeoecol.* **2003**, *194*, 23–40. [CrossRef]
3. Brooks, W.E.; Willett, J.C.; Kent, J.D.; Vasquez, V.; Rosales, T. The Muralla Pircada: An ancient Andean debris flow retention dam, Santa Rita B archaeological site, Chao Valley, northern Peru. *Landslides* **2005**, *2*, 117–123. [CrossRef]
4. Craig, A.K.; Shimada, I. El Niño flood deposits at Batan Grande, northern Peru. *Geoarchaeology* **1986**, *1*, 29–38. [CrossRef]
5. Uceda, C.S.; Canziani Amico, J. Evidencias de grandes precipitaciones en diversas etapas constructivas de la Huaca de la Luna, costa norte del Perú. *Bull. Inst. Fr. Études Andines* **1993**, *22*, 313–343. (In Spanish)
6. Contreras, D.A. Landscape and Environment: Insights from the Prehispanic Central Andes. *J. Archaeol. Res.* **2010**, *18*, 241–288. [CrossRef]
7. Grodzicki, J. Las catástrofes ecológicas en la Pampa de Nazca a fines del Holoceno y el fenómeno "El Niño". In Proceedings of the El fenómeno El Niño: A través de las fuentes arqueológicas y geológicas, Warsaw, Poland, 18–19 May 1990; Grodzicki, J., Ed.; Warsaw University; pp. 64–102. (In Spanish)
8. Grodzicki, J. Los geoglifos de Nasca segun algunos datos geologicos. In Proceedings of the Paleo ENSO Records, ORTOM-CONCYTEC Intern. Symp., Lima, Peru, March 1992; pp. 119–130. (In Spanish)
9. Macharé, J.; Ortlieb, L. Registros del fenomeno El Niño en el Perù. *Bull. Inst. Fr. Études Andines* **1993**, *22*, 35–52. (In Spanish)
10. Van Buren, M. The archaeology of El Niño events and others "natural" disasters. *J. Archaeol. Method Theory* **2001**, *8*, 129–149. [CrossRef]
11. Silverman, H. *Ancient Nasca Settlement and Society*; University of Iowa Press: Iowa City, IA, USA, 2002.

12. Moseley, M.; Tapia, J.; Satterlee, D.S.; Richardson, J.B. Flood events, El Niño events, and tectonic events. In Proceedings of the Paleo-ENSO Records Intern. Symp., Lima, Peru, March 1992; Ortlieb, L., Macharé, J., Eds.; Orstom-Concytec: Lima, Peru; pp. 207–212. (In Spanish)

13. Mörner, N.A. Present El Niño-ENSO events and past super-ENSO events. *Bull. Inst. Fr. Études Andines* **1993**, *22*, 3–12.

14. Keefer, D.K.; Moseley, M.E.; de France, S.D. A 38 000-year record of floods and debris flows in the Ilo region of southern Peru and its relation to El Niño events and great earthquakes. *Palaeogeogr. Palaeoclimatol. Palaeoecol.* **2003**, *194*, 41–77. [CrossRef]

15. Laoshuti, T.; Selover, D.D. Does El Niño affect business cycles? *East. Econ. J.* **2007**, *33*, 21–42. [CrossRef]

16. Cai, W.; Borlace, S.; Lengaigne, M.; Van Rensch, P.; Collins, M.; Vecchi, G.; Timmermann, A.; Santoso, A.; McPhaden, M.J.; Wu, L. Increasing frequency of extreme El Niño events due to greenhouse warming. *Nat. Clim. Chang.* **2014**, *4*, 111–116. [CrossRef]

17. Gutierrez, L. Impacts of El Niño-Southern Oscillation on the wheat market: A global dynamic analysis. *PLoS ONE* **2017**, *12*, e0179086. [CrossRef] [PubMed]

18. INGEMMET. *Mapa Geologico del Cuadrangulo de Palpa*; Ministerio de Energia y Minas: Lima, Peru, 1994.

19. Nials, F.L.; Deeds, E.E.; Moseley, M.E.; Pozorski, S.E.; Pozorski, T.G.; Feldman, R. El Niño: The catastrophic flooding of coastal Peru. *Field Mus. Nat. Inst. Bull.* **1979**, *50*, 4–14.

20. Nials, F.L.; Deeds, E.E.; Moseley, M.E.; Pozorski, S.E.; Pozorski, T.G.; Feldman, R. El Niño: The catastrophic flooding of coastal Peru. Part II. *Field Mus. Nat. Inst. Bull.* **1979**, *50*, 4–10.

21. Wells, L.E.; Noller, J.S. Holocene coevolution of the physical landscape and human settlement in northern coastal Peru. *Geoarchaeology* **1999**, *14*, 755–789. [CrossRef]

22. Beresford-Jones, D.G.; Arce Torres, S.; Whaley, O.Q.; Chepstow-Lusty, A.J. The role of *Prosopis* in ecological and landscape change in the Samaca Basin, lower Ica Valley, south coast Peru from the Early Horizon to the Late Intermediate Period. *Latin Am. Antiqu.* **2009**, *20*, 303–332. [CrossRef]

23. Leó, W.; Aleman, A.; Torres, V.; Rosell, W.; De La Cruz, O. *Estratigrafía, sedimentología y Evolución de la Cuenca Pisco Oriental*; Boletín serie D, n. 27; INGEMMET: Lima, Peru, 2008. (In Spanish)

24. Hall, S.R.; Farber, D.L.; Audin, L.; Finkel, R.C.; Mériaux, A.S. Geochronology of pediment surfaces in southern Peru: Implications for Quaternary deformation of the Andean Forearc. *Tectonophysics* **2008**, *459*, 186–205. [CrossRef]

25. Macharé, J. La marge continentale du Pérou: Régimes tectoniques et sédimentaires cénozoïques de l'avant-arc des Andes centrales. Thèse Doct. Sc., Université Paris XI, Orsay, France, 1987. (In French)

26. Sebrier, M.; Macharé, J. Observaciones acerca del Cuaternario de la Costa del Perù. Central. *Bull. Inst. Fr. Études Andines* **1980**, *9*, 5–22. (In Spanish)

27. Pulgar Vidal, J. Geografía del Perú. Las Ocho Regiones Naturales. Editorial Inca SA: Lima, Peru, 1987. (In Spanish)

28. Caviedes, C.N.; Fik, T.J. The Peru–Chile eastern Pacific fisheries and climatic oscillation. In *Climate Variability, Climate Change and Fisheries*; Glantz, M.H., Ed.; Cambridge University Press: Cambridge, UK, 1992; pp. 355–376.

29. Mächtle, B.; Unkel, I.; Eitel, B.; Kromer, B.; Schiegl, S. Molluscs as evidence for a Late Pleistocene and Early Holocene humid period in the northern Atacama desert, southern Peru (14.5° S). *Quat. Res.* **2010**, *73*, 39–47. [CrossRef]

30. Mächtle, B.; Eitel, B. Fragile landscapes, fragile civilizations—How climate determined societies in the pre-Columbian south Peruvian Andes. *Catena* **2013**, *103*, 62–73.

31. Orefici, G. *Nasca—Archeologia per una ricostruzione storica*; Jaka Book: Milan, Italy, 1992. (In Italian)

32. Orefici, G. Nasca historical and cultural analysis. In *Ancient Nasca World. New Insights from Science and Archaeology*; Lasaponara, R., Masini, N., Orefici, G., Eds.; Springer: Basel, Switzerland, 2016; pp. 65–86.

33. Orefici, G. The decline of Cahuachi and the end of the Nasca theocracy. In *Ancient Nasca World. New Insights from Science and Archaeology*; Lasaponara, R., Masini, N., Orefici, G., Eds.; Springer: Basel, Switzerland, 2016; pp. 449–468.

34. Schittek, K.; Forbriger, M.; Mächtle, B.; Schäbitz, F.; Wennrich, V.; Reindel, M.; Eitel, B. Holocene environmental changes in the highlands of the southern Peruvian Andes (14° S) and their impact on pre-Columbian cultures. *Clim. Past* **2015**, *11*, 27–44. [CrossRef]

35. Graham, N.E.; Hughes, M.K.; Ammann, C.M.; Cobb, K.M.; Hoerling, M.P.; Kennett, D.J.; Kennett, J.P.; Rein, B.; Stott, L.; Wigand, P.E.; et al. Tropical Pacific—Mid-latitude teleconnections in medieval times. *Clim. Chang.* **2007**, *83*, 241–285. [CrossRef]

36. Orefici, G. *Cahuachi. Capital Teocrática Nasca*; Fundo Editorial Universidad SMP: Lima, Peru, 2012. (In Spanish)

37. Grodzicki, J. *Nasca: Los Síntomas Geológicos del Fenómeno El Niño y sus Aspectos Arquelógicos*; Cesla series Estudios y Memorias, 12; Warsaw University: Warsaw, Poland, 1994. (In Spanish)

38. Valladares, R.; Denysse, P. El Sector CAH 04, Y8—EXP 108: Las funciones de la Plaza al Este de la Gran Pirámide en los complejos ceremoniales de Cahuachi (Nazca), Período Intermedio Temprano. Graduation Thesis, Pontificia Universidad Católica del Perú, Lima, Peru, 2007. (In Spanish)

39. Delle Rose, M. The geology of Cahuachi. In *Ancient Nasca World. New Insights from Science and Archaeology*; Lasaponara, R., Masini, N., Orefici, G., Eds.; Springer: Basel, Switzerland, 2016; pp. 47–64.

40. Wentworth, C.K. A scale of grade and class terms for clastic sediments. *J. Geol.* **1922**, *30*, 377–392. [CrossRef]

41. Krumbein, W.C. Size frequency distributions of sediments. *J. Sediment. Petrol.* **1934**, *4*, 65–77. [CrossRef]

42. Folk, R.L. The distinction between grain size and mineral composition in sedimentary rock nomenclature. *J. Geol.* **1954**, *62*, 344–356. [CrossRef]

43. Montoya, M.; Carcía, W.; Caldas, J. *Geología de los Cuadrángulos de Lomitas, Palpa, Nasca y Puquio*; Boletín Serie A 53; INGEMMET: Lima, Peru, 1994. (In Spanish)

44. Orefici, G. Recent discoveries in Cahuachi: The Templo Sur. In *Ancient Nasca World. New Insights from Science and Archaeology*; Lasaponara, R., Masini, N., Orefici, G., Eds.; Springer: Basel, Switzerland, 2016; pp. 363–374.

45. Ostaficzuk, S. Development stages of the Nazca morphological features as readible on aerial photos. In Proceedings of the El fenómeno El Niño: A través de las fuentes arqueológicas y geológicas, Warsaw, Poland, 18–19 May 1990; Grodzicki, J., Ed.; Warsaw University; pp. 49–65.

46. Huntley, D.J.; Godfrey-Smith, D.I.; Thewalt, M.L.W. Optical dating of sediments. *Nature* **1985**, *313*, 105–107. [CrossRef]

47. Aitken, M.J. *An Introduction to Optical Dating*; Oxford University Press: Oxford, UK, 1998.

48. Rhodes, E.J. Optically Stimulated Luminescence Dating of Sediments over the Past 200,000 Years. *Annu. Rev. Earth Planet. Sci.* **2011**, *39*, 461–488. [CrossRef]

49. Zhao, Q.; Thomsen, K.J.; Murray, A.S.; Wei, M.; Song, B. Single-grain quartz OSL dating of debris flow deposits from Men Tou Gou, south west Beijing, China. *Quat. Geochronol.* **2017**, *41*, 62–69. [CrossRef]

50. Olley, J.; Caitcheon, G.; Murray, A. The distribution of apparent dose as determined by Optically Stimulated Luminescence in small aliquots of fluvial quartz: Implications for dating young sediments. *Quat. Sci. Rev.* **1998**, *17*, 1033–1040. [CrossRef]

51. Well, L.E. Holocene history of the El Niño phenomenon as recorded in flood sediments of northern coastal Peru. *Geology* **1990**, *18*, 1134–1137. [CrossRef]

52. Zavala, B.; Velarde, T. *Caracterización Geológica de los Depositos Superficiales en las Pampas de Nazca, Lineas y Geoglifos de Nazca, Informe Técnico—Geologia Ambiental y Riesgo Geolócico*; INGEMMET: Lima, Peru, 2008. (In Spanish)

53. Lowe, D.J.; Pearce, N.J.; Jorgensen, M.A.; Kuehn, S.C.; Tryon, C.A.; Hayward, C.L. Correlating tephras and cryptotephras using glass compositional analyses and numerical and statistical methods: Review and evaluation. *Quat. Sci. Rev.* **2017**, *175*, 1–44. [CrossRef]

54. Lebti, P.P.; Thouret, J.C.; Wörner, G.; Fornari, M. Neogene and Quaternary ignimbrites in the area of Arequipa, Southern Peru: Stratigraphical and petrological correlations. *J. Volcanol. Geotherm. Res.* **2006**, *154*, 251–275. [CrossRef]

55. Breitkreuz, C.; de Silva, S.L.; Wilke, H.G.; Pfander, J.A.; Renno, A.D. Neogene to Quaternary ash deposits in the Coastal Cordillera in northern Chile: Distal ashes from supereruptions in the Central Andes. *J. Volcanol. Geotherm. Res.* **2013**, *269*, 68–82. [CrossRef]

56. Thouret, J.C.; Jicha, B.R.; Paquette, J.L.; Cubukcu, E.H. A 25 myr chronostratigraphy of ignimbrites in south Peru: Implications for the volcanic history of the Central Andes. *J. Geol. Soc. Lond.* **2016**, *173*, 734–756. [CrossRef]

57. Mamani, M.; Wörner, G.; Sempere, T. Geochemical variations in igneous rocks of the Central Andean orocline (13°S to 18°S): Tracing crustal thickening and magma generation through time and space. *Geol. Soc. Am. Bull.* **2010**, *122*, 162–182. [CrossRef]

58. Delle Rose, M.; Renzulli, A. Overcoming the paradigm of the destruction of Nasca culture due to a Mega-El Niño event: A clue from the stratigraphic survey at Cahuachi (Peru). *Rendiconti online della Società Geologica Italiana* **2014**, *31*, 96.

59. Delle Rose, M.; Orefici, G.; Capuano, N.; Galassi, G.; Mattioli, M.; Santini, S.; Spada, G.; Renzulli, A. The decline of the Nasca culture (Peru) as the result of an increasing environmental stress: Overcoming the paradigm formulated at Cahuachi of catastrophic mega-floods due to El Niño-Southern Oscillation. In Proceedings of the 2018 SGI-SIMP Congress, Catania, Italy, 12–14 September 2018.

geosciences

MDPI

Article

Geospatial Sciences and Space Law: Legal Aspects of Earth Observation, Remote Sensing and Geoscientific Ground Investigations in Africa

Gbenga Oduntan

Kent Law School, University of Kent, Canterbury CT2 7NS, UK; O.T.Oduntan@kent.ac.uk

Received: 11 November 2018; Accepted: 5 January 2019; Published: 29 March 2019

Abstract: Geospatial sciences play crucial roles in and have effects on the socioeconomic, political and security fortunes of states. Earth observation, remote sensing and geoscientific ground investigation increasingly occupy vantage positions in the legal order of states, particularly in evidential terms and in the verification of facts under international law. How then do these aspects of space law and space sciences affect contemporary Africa and the commercial fortunes, as well as international relations among some African states? What impact do they have in relation to: (a) international boundaries disputes and demarcation activities; (b) management and the preservation of the African heritage; (c) disaster and conservation management? The paper will test the hypothesis that it is crucial for the development of the continent especially in the areas mentioned above that states should sustain and increase investment in the following areas: archaeological prospection, condition assessment of heritage assets; Geographic Information System (GIS) analysis of spatial settlement patterns in modern landscapes and assessment of natural or human-induced threats to conservation.

Keywords: space law; disaster and conservation management; Geographic Information System (GIS); international boundaries; Africa; Cameroon-Nigeria Mixed Commission; satellite imagery; Boundary Demarcation; international law; relict boundaries

---

1. Introduction

It is a truism that geospatial sciences have crucial roles and effects on the socioeconomic, political and security fortunes of states. Earth observation, remote sensing and geoscientific ground investigation increasingly occupy vantage positions in the legal order of states, particularly in evidential terms and in the verification of facts under international law. How then do these aspects of space law and space sciences affect contemporary Africa and the commercial fortunes as well as international relations among some African states? What impact do they have in relation to: (a) international boundaries disputes and demarcation activities; (b) management and the preservation of the African heritage; (c) disaster and conservation management?

The paper will test the hypothesis that it is crucial for the development of the continent especially in the areas mentioned above that states should sustain and increase investment in the following areas: archaeological prospection, condition assessment of heritage assets; Geographic Information System (GIS) analysis of spatial settlement patterns in modern landscapes and assessment of natural or human-induced threats to conservation.

The paper will thus, establish how the geospatial sciences, earth observation, remote sensing and geoscientific ground investigation in their interactions with space law can aid the significant thirst for development in Africa. The obvious facts are that the space sciences are the progenitors of space activities which are all governed by space law. Thus, space law governs earth observation and remote sensing activities among others. Which specific principles of space law govern these sciences and towards what developmental ends is an enquiry worth having? It is suggested that certain legal problems may arise in the gathering of geodetic information particularly in the use of remote sensing technology. The paper examines some of these in relation to Africa as a developing region. Our methodology is critical, comparative and socio-legal but includes a focus on the space active African states. In the area of demarcation of boundaries and geodetic ground investigation, we will examine specifically the adoption of space based solutions to the resolution of the Cameroon-Nigeria boundary demarcation process.

2. The Scope of Space Law in Africa

It is necessary to lay out the province of space law especially as it affects our discussion herein this paper. Space law has been defined by the United Nations Office for Outer Space Affairs as the body of law applicable to and governing space-related activities. The term "space law" is most often associated with the rules, principles and standards of international law appearing in over a dozen international treaties, standards and rules governing outer space which have been elaborated since the first space flight in 1957 [1]. It is also important to emphasize the broader conceptualization of space law so as to capture the multidimensional discipline it has become, extending among others to satellite telecommunications, intellectual property, environmental as well as military and security applications [2]. Space law has preponderantly been drawn up under the leadership of the United Nations Organization especially through the development of the five UN-originating space treaties. It however, includes and is enriched by other treaties, conventions, international agreements, rules and regulations of international/intergovernmental organizations (e.g., the International Telecommunications Union (ITU), ITU is the United Nations specialized agency for information and communication technologies–ICTs. The organisation allocates global radio spectrum and satellite orbits, develop the technical standards that ensure networks and technologies seamlessly interconnect and strive to improve access to ICTs in underserved communities worldwide [3] and the Committee for Peaceful Uses of Outer Space (COPOUS: was set up by the General Assembly in 1959 to govern the exploration and use of space for the benefit of all humanity: for peace, security and development. The Committee was instrumental in the creation of the five treaties and five principles of outer space.) [4] as well as national laws, rules and regulations, executive and administrative orders and judicial decisions. The increasing formulation of national space laws, space policies and the establishment of national space agencies create another level of regulatory framework which must be reckoned with in understanding legality of space related activities [5].

Overall it is fair to state that contemporary rules of public international space law are based on time tested principles of law recognised by civilised nations [6].

Particular mention must be made of the following ten instruments for which it may be said constitute the *corpus juris spatialis*. They are best understood in two groups as follows:

- The "five United Nations 'originating' treaties on outer space" (a) Treaty on Principles Governing the Activities of States in the Exploration and Use of Outer Space, including the Moon and Other Celestial Borders (1967) [7];
- (b) Agreement on the Rescue of Astronauts, the Return of Astronauts and the Return of Objects Launched into Outer Space (1968), U.K.T.S. 56 (1969), Cmnd. 3997; (1969) 63 A.J.I.L. 382. In force 1968. 86 parties, including the five permanent members of the Security Council [8];
- (c) Convention on International Liability for Damage caused by space Objects (1972), U.K.T.S. 16 (1974), Cmnd. 5551; 961 U.N.T.S. 187; 10 I.L.M. 965. In force 1973. 76 parties, including the five permanent members of the Security Council [9];
- (d) Convention on Registration of Objects Launched into Outer Space (1975), UNTS 187; 14 ILM 43; UKTS 70 (1978); In force 1976 39 parties including the five permanent members of the Security Council [10];
- (e) Agreement Governing the Activities of States on the Moon and other Celestial Bodies (1979), G.A. Res. 34/68, U.N. GAOR, 34th Sess. Supp. No. 46 at 77, U.N. Doc. A/34/664 (1979) [11].

The five declarations and legal principles

- (a) The Treaty Banning Nuclear Weapon Tests in the Atmosphere, in Outer Space and Under Water 1963 [12];
- (b) The Principles Governing the Use by States of Artificial Earth Satellites for International Direct Television Broadcasting (1982) [13];
- (c) The Principles Relating to Remote Sensing of the Earth from Outer Space [14];
- (d) The Principles Relevant to the Use of Nuclear Power Sources in Outer Space [15];
- (e) The Declaration on International Cooperation in the Exploration and Use of Outer Space for the Benefit and in the Interest of All States, Taking into Particular Account the Needs of Developing Countries [16].

In addition to the above there are other national and bilateral agreements relating to activities in outer space that all together provide legal cover for all of man's endeavours in outer space. It is a remarkable feature of space law and a silent success of international relations that most of the regime governing outer space has been drawn up several decades ago. Another underreported success of this law is that African states have a very strong showing in terms of acceptance of these laws and they have signed and/or ratified them in very large numbers. Indeed our table below reveals that 53 African States have signed and/or ratified space related treaties. This primarily indicates that African governments have embraced the idea of international scientific exploration of outer space and that they do have national aspirations thereto. It also reveals their readiness to partake in space policy, cooperation and exploration activities both directly and indirectly as equal partners and as sovereign states.

The following; Table 1 is revealing of the impressive level of African States integration into the general field of space law and legal ordering.

An increasing number of states worldwide are responding to the opportunities of active outer space exploration by implementing dedicated legislation. States which have national legislation governing space-related activities in a holistic manner include inter alia Argentina, Australia, Canada, Finland, France, Germany, Hungary, Indonesia, Japan, New Zealand, Philippines, Republic of Korea, Russian Federation, Slovakia, Sweden, South Africa, Ukraine, the United Kingdom of Great Britain and Northern Ireland and the United States of America.

Table 1. African States and the Status of International Agreements on Activities in Outer Space (as at 1 January 2017).

State	OST 1967	ARRA 1968	LIAB 1972	REG 1975	MOON 1979	NTB 1963	BRS 1974	ITSO 1971	INTR 1971	ESA 1975	ARB 1976	INTC 1976	IMSO 1976	EUTL 1982	EUM 1983	ITU 1992
Algeria	R		R	R		S		R	R		R		R			R
Angola								R								R
Benin	R	R	R			R	R	R								R
Botswana	S		R			R		R								R
Burkina Faso	R					S		S*								R
Burundi	S					S										R
Cameroon	S	R				S		R					R			R
Cape Verde	S					R		R								R
Cent. Afr. Rep.	S		S			R		R								R
Chad						R		R								R
Comoros								R					R			R
Congo								R								R
Cote d'Ivoire						R	S	R								R
Dem. Rep. Of Congo	S	S	S			R		R								R
Djibouti											R					R
Egypt	R	R	S			R		R			R		R			R
Eq. Guinea	R					R		R								R
Eritrea																R
Ethiopia	S					S		R								R
Gabon		R	R			R		R					R			R
Gambia	S	R	S			R		R								R
Ghana	S	S	S			R		R					R			R
Guinea								R								R
Guinea-Bissau	R	R				R		R								R
Kenya	R		R			R	R	R					R			R
Lesotho	S	S				R										R
Liberia						R							R			R
Libya																R
Madagascar	R	R				R		R								R
Malawi						R		R								R
Mali	R		R			S		R								R
Mauritania						R		R			R		R			R
Mauritius								R					R			R
Mozambique								R					R			R
Namibia								R								R

Table 1. *Cont.*

State	OST 1967	ARRA 1968	LIAB 1972	REG 1975	MOON 1979	NTB 1963	BRS 1974	ITSO 1971	INTR 1971	ESA 1975	ARB 1976	INTC 1976	IMSO 1976	EUTL 1982	EUM 1983	ITU 1992
Niger	R	R	R	R		R		R					R			R
Nigeria	R	R	R	R		R		R					R			R
Rwanda	S	S	S			R	R	R								R
Sao Tome and Principe																R
Senegal	R	S	R			R	S	R					R			R
Seychelles	R	R	R	R		R										R
Sierra Leone	S	S	S			R										R
Somalia	S	S				S		R	R							R
South Africa	R	R	R	R		S		R			R		R			R
South Sudan																R
Sudan						R		R			R					R
Swaziland		R	S			R		R								R
Tanzania			R			R		R					R			R
Togo	R		R			R	R	R								R
Tunisia	R	R	R			R		R			R		R			R
Uganda	R					R		R								R
Zambia	R	R	R			R		R								
Zimbabwe								R			R					R

R means treaty has been ratified. S means treaty has been signed.

Emergent Patterns in African Domestic Space related Policies and Space Dedicated Legislation

Our research here shows that there are four categories of states in Africa in relation to Space Activities and space regulation.

- First, there are the majority of states that participate in space treaties but do not have active space investments and/or domestic space specific regulations.
- Second, there are those countries with active space interest, aims and aspirations but which however, do not have any comprehensive legal text regulating outer space activities. These group consist of Libya (A membership of a regional agency Centre Régional de Télédétection des Etats de l'Afrique du Nord (CRTEAN) (North African Centre for Remote Sensing) and the international organization—ITU), Sudan (A member -CRTEAN and the international organizations–Inter Islamic Network on Space Sciences and Technology—(ISNET) and the ITU.), Senegal (A member of ISNET and the ITU.), Kenya (Kenya's progress in space sector has taken major strides over the last decade with the formation of the Kenya National Space Agency. In addition Kenya has established a National Space Secretariat (see further http://www.mod.go.ke/?p=1932). Kenya interestingly is not in any African cooperation organisation but is in the Asia-Pacific Regional Space Agency Forum (APRSAF) which was established in 1993 to enhance space activities in the Asia-Pacific region. In addition Kenya is a member of UNCOPUOS and ITU. Kenya is also currently considering its own space law), Niger (A member of ISNET and the ITU), Mauritania (A membership of CRTEAN and the ITU.) and Morocco (It has its own AGENCY–the CRTS and it has a pertinent Decree creating the Royal Remote Sensing Centre (CRTS), of 17 January 1990. Morocco has membership of UNCOPUOS, CRTEAN, ISNET, ITU.).
- Third group of states consist of Algeria and Tunisia that have no comprehensive legal text at the moment but have an impressive array of space connected or related legislation. Algeria has established an Algerian Space Agency and several space related decrees. These Include Presidential Decree No. 02-49 "Creation, Organization and Functioning of the Algerian Space Agency (ASAL)" of 16 January 2002; Presidential Decree No. 06-225 "Ratifying the Convention for Damage Caused by Space Objects" of 24 June 2006 and Presidential Decree No. 06-468 "Ratifying the Convention on Registration of Objects Launched into Outer Space" of 11 December 2006 [17], Tunisia has a National Outer Space Commission (CNEEA), created by Decree n° 84-1125 (24 September 1984) and also established a National Mapping and Remote Sensing Centre of Tunisia (CNCT), (Established under Law 88-83 of June 11 1988. Conformity with international and domestic telecommunications law is maintained through laws such as the Tunisian Order of the Minister of Communication Technologies dated 11 February 2002, approving the National Radio Frequency Plan. See the Tunisian -Order of the Minister of Communication Technologies dated 11 February 2002, approving the National Radio Frequency Plan. The Minister of Communication Technologies, passed this Order having regard inter alia to Law No. 2001-110 of 9 November, 2001, ratifying the Final Acts of the Congress of Plenipotentiaries of the International Telecommunication Union) [18]. The Country retains membership of UNCOPUOS, CRTEAN, ISNET, ITU and UNIDROIT and the ITU, for more information on this organisation see www.cnt.nat.tn visited 15 December 2018. It has domesticated several international space treaties into national law [8,19,20], also has a Law on Telecommunications of 2008 and has created its own national laws of a limited purpose for space activities, Such as the Order on Radio Frequencies of 2002, Minister of National Defense Order on the National Remote Sensing Center Fees. Tunisian law importantly identifies the state's strategic interest and provides mechanisms for registration processes for space objects and provides the functions of state of registry.
- Fourth category consist of those states that have shown a high degree of political and infrastructural commitment to space exploration backed up with elaborate dedicated domestic space law regime. The two countries that currently satisfy these characteristic the most are Nigeria and South Africa.

African countries, both individually and as a regional bloc have at least in the last decade began seriously expressing with increasing interest the development of space programmes. These programmes aim at harnessing and optimising the general societal benefits derived from space applications. In ideal circumstances, space programmes would be carefully designed through carefully crafted space policies, strategies, technology roadmaps and dedicated space laws. However, the development of dedicated national space policies and legislation in Africa has been rather slow and insufficient given the appetite for space technological products in all parts of the continent. It has been correctly noted that much of African space policies, to the extent, they exist at all, appear to have been drawn up without the benefit of specialist knowledge [21].

There are however, a couple of models of good example principally led by Nigeria and South Africa, which both have National Space Policies, instituted in 2001 [22] and 2008 respectively [23]. It is significant to note that at the continental level, African Heads of States and Government in January 2016 adopted the African Space Policy and Strategy [24]. Overall Etim and Munsami aptly described the tenuous state of elaboration of space policies and legislation:

> " . . . an underlying challenge is the requisite skills and experience needed for drafting such Space Policies. Much of the space policy developments on the continent have happened through on-the-job learning and a look elsewhere-and-adopt approach. In some cases, a foreign company is contracted for sector analysis. Although this informal learning and experience is valuable for the ongoing developments of national and regional space programmes, it is not the most efficient and effective means of achieving the end product. Reliance on foreign parties may not be the nation's interest" [21].

The existence of a Space Policy provides an overarching interdisciplinary framework that informs the development of a space sector, based on its connections to the national ecosystem of corporate, industrial, educational and other sectors as well as human needs. It draws from and feeds into national economic policies, social development policies, national security policies and foreign policies, among other policy frameworks [21,25].

In terms of legislation Africa is already very well subscribed to the main space treaties and other relevant international instruments. From the Table 1 above we can see that 53 out of the existing 57 African states are subscribed to the multilateral space treaties but the question is whether these states have backed up their readiness to accept space law on the international level with adequate domestic regulations.

South Africa has perhaps the most comprehensive national space legislation/regulatory environment in Africa. Applicable legislation in force include: Statutes of the Republic of South Africa—Trade and Industry, Space Affairs Act, No. 84 (1993); Space Affairs Amendment Act, No. 64 (1995); South African National Space Act 36 of 2008; No. 21 of 2007: Astronomy Geographic Advantage Act, 2007; Spatial Data Infrastructure Act No. 54 of 2003; Electronic Communications Act No.36 of 2005 (ECA). The South African National Space Agency (SANSA) came into being in December 2010 and is guided by the National Space Policy, 2008 and National Space Policy, 2009 [26], SANSA operations fall into four programme areas: Earth Observation, Space Engineering, Space Operations, and Space Science. South Africa is also member of UNCOPUOS, APRSFAF and ITU and UNIDROIT.

It is notable that South Africa alone has made elaborate provisions jointly for laws over persons in space (jurisdiction *ratione personae*: refers to the personal jurisdiction over natural and corporate persons bearing the nationality of a state.) [27] and over space objects (jurisdiction *ratione materiae* which is the jurisdiction of states over vessels and objects like ships, aircraft, spacecraft that bear the flag of a state and are therefore, subject to the jurisdiction of the flag state) [27]. In South Africa's jurisdictional powers as a state are expressed in its space law with respect to personal application, territorial application and extraterritorial application [28].

Nigeria's comprehensive national space legislation/regulation is the National Space Research and Development Agency (NASDRA) Act 2010 [29] which among other things created its own space agency.

Nigeria is a member of UNCOPUOS, ITU and UNIDROIT. The scope of the NASRDA Act covers regulation of Space objects launch and return; Space objects control; Ground segment operation and Space technology. The country has legal mechanisms in place to monitor and control; Authorization and licensing; Continuous supervision of non-governmental activity; Liability and insurance; Transfer of ownership in space or space objects control; Safety or space debris removal; State strategic interest; Registration process or registry.

Nigeria for its part similarly provides for jurisdiction over materials placed in space and space objects. Indeed only Nigerian Law specifically provides for space objects control. However, by leaving out provisions *rationae personae*, the country's space regime arguably falls short of its ambitious space programme which includes running its own indigenous astronauts programme. Furthermore, specific rules regarding space object launch and return, space technology and ground segment operation provisions are to be found in the dedicated, domestic legislation of Nigeria and South Africa. Nigeria, alone has fairly suitable provisions on the transfer of ownership in space for space objects control as well as for the safety of operations and space debris control. South Africa uniquely has provisions providing for an insurance ceiling [28].

Both Nigeria and South Africa commendably provide specific rules for authorisation and licensing, as well as mandatory rules prescribing continuous exercise of supervision. Both countries laws control non-governmental entities and the duty of operators to maintain liability and insurance. With respect to provisions indicating the State's strategic interests, the policies and laws of Algeria, Nigeria and South Africa are presently the clearest and most ambitious. This includes transparent provisions covering the registration process for all materials in space and establishment of registry for space objects launched on the account or with the permission of the state, such as Licensing and Duties and Liabilities of Licensee of the Space Affairs Act (respectively) (South Africa, 1993) Statutes of the Republic of South Africa–Trade and Industry No. 84 of 1993 (Assented to 23 June 1993) (Date of Commencement: 6 September 1993). For Nigeria's licensing and registration regimes see S. 9 and 10 of the NASRDA Act (2010)

3. Technical Features and Socio-Legal Characteristics of Africa's Satellite Constellation

Geospatial products and the air- and space-based platforms from which they are derived are useful in a myriad of ways for all modern states. Information and communications, derived from them are vital for imaging, telecommunications, telebroadcasting and information technology. In recent years microwave remote sensing devices and other meteorological applications have been very useful in monitoring cyclones, rainfall, floods and cold waves, which are regularly occurring phenomena in many developing countries, for example the Bangladesh communication to COPUOUS of 27 October 2009 as quoted in United Nations, General Assembly, Committee on the Peaceful Uses of Outer Space, Questions on the definition and delimitation of outer space: replies from Member States January 2010, A/AC.105/889/Add.5 11 [30]. Satellite technology has applications in medical aid, international healthcare and telemedicine. With satellite link up technology, such as the European Healthware project, people living in remote areas where health care professionals are unavoidable or unable to leave their homes may receive advanced health care. Medical professionals also discuss cases and tap into specialisations without expensive transfers. Success stories in the area of teleconsultation healthcare have been demonstrated in Cyprus, France, Greece, Czech Republic, Poland and United Kingdom, The Healthware system uses a communications format known as Digital video broadcasting return Chanel via Satellite (DVB-RCS) which do not depend on a physical cable. See GMES; European Commission, DG enterprise and Industry, EU: Connecting People through satellite-based Telemedicine Connecting People Through satellite-based Telemedicine Solutions: Scenarios and Practical Experiences from the Healthware project available at (http://www.epractice.eu/files/media/media1916.pdf).

Apart from the relatively towering achievements of South Africa and Nigeria, a few other African States have taken a keen interest in space technology in furtherance of their national goals and other commercial and security interests. Earth observation, remote sensing and telecommunications are the centrepieces of African satellite technological requirements and therefore investments. The technical specifications of Nigeria's stable of satellite acquisition, for instance, betray the country's needs at its stage of development, in 1976, Nigeria declared its space ambition to members of the Economic Commission for Africa and Organization of African Unity during an inter-governmental meeting in Addis-Ababa. In 1988 the National Council of Ministers' approved the establishment of a National Centre for Remote Sensing, to be located at Jos; with a Ground Receiving Station, at Kerang, in Mangu LGA of Plateau State (site of the defunct Aerostat Balloon Project). NASRDA's focus today is in Rocketary and Propulsions, Earth Observations and Space Communications. For more information on NASRDA see its website (http://nasrda.gov.ng/en/). Nigeria has taken impressive lead in Sub-Saharan Africa by launching a total of four satellite missions—NIGERIASAT-1, NIGERIASAT-2, NIGCOMSAT-1 and NIGERIASAT-X [31].

Reconstruction of African Territories from Space

The conclusion that "Border is Fate" [32] is true to the extent that the fate of African peoples are continuously changing with court and tribunal judgments and awards since the end of the colonial era. African territories are very often subject to disputed ownership and are continuously before international tribunals for adjudication, the docket of the International Court of Justice is even as of present still grappling with fresh cases and crises arising out of colonial actions over African territories. A legal challenge to British sovereignty of the Chagos Islands has been instituted by Mauritius based on allegations of coercion leading to the giving away of a large swathe of Mauritius territory before independence. See *"Legal consequences of the separation of the Chagos Archipelago from Mauritius"* in 1965 (Request for Advisory Opinion) available at: https://www.icj-cij.org/en/case/169; Owen Bowcott, "Chagos Islands: UK used secret threats to keep Chagos Islands, court hears" *The Guardian* Monday 3 September 2018 available at https://www.theguardian.com/world/2018/sep/03/mauritius-takes-uk-to-court-over-chagos-islands-sovereignty. Even in peace time boundary management of vast territories require joint and sometimes separate teams of technical personnel who have to faithfully follow and trace the boundary line through difficult terrain or on maritime zones. Indeed it is widely recognised that European colonialism continues to underlie most territorial disputes in Africa. Recent examples include the Nigeria–Cameroon dispute over the Bakassi Peninsula; the Gabon–Equatorial Guinea dispute over the islands of Mbanié, Cocotiers, and Conga in the Corisco Bay; the Mauritius–United Kingdom dispute over the Chagos Archipelago; and the Comoros–France [33].

With geospatial sciences and space satellite technology, legally significant geographic facts can be determined with exactitudes and relatively cheaply. Legally significant facts and truths increasingly turn on evidence derived from geodesy, remote sensing and other relevant geoscientific ground data. The geospatial sciences have several levels of manifest legal usefulness. Geodesy, the science concerned with the shape of the earth, for instance, has international, regional and of course, national benefits [34]. Thus, the need for geospatial products in Africa is of an urgent and increasing nature. Verification of locations by traditional surveying means is fraught with many limitations. Without space applications traditional surveying is comparatively costlier and fraught with quantitative difficulties. Multiple teams of bilateral surveying and boundary commissions are at work simultaneously across the continent. Surveying vast territories has more than its fair share of dangers. Surveyors are often caught between hostile villagers and communities, security agencies of neighbouring states and wild animals. The prospect of getting irretrievably lost in wild terrain are real and has occurred quite too often, for a particularly harrowing tale of starvation while on a surveying mission see Alistair Macdonald *Mapping the World* [34].

Geospatial products also now afford African states the crucial luxury to reconstruct boundary lines in a way more faithful to impartial geographic and scientific realities. This is particularly important for Africa because much of African territories were drawn up by people who never had any intimate knowledge of the terrain they were delimiting, Thus, African historians like Mohammed Ahmed have rightly concluded that "At both the continental and regional levels, as represented by the African

Union and ECOWAS Commission, African Boundaries inherited at independence have been accepted as fait accompli" [35]. Colonial delimitation and demarcation of Africa was deeply flawed in very many ways. As far back as 1890, Lord Salisbury admitted:

> "We have been engaged ... in drawing lines upon maps where no white man's feet ever trod; we have been giving away mountains and rivers and lakes to each other but we have only been hindered by the small impediment that we never knew exactly where those mountains and rivers and lakes were" [36].

Thus, the providence of older maps much of which are relied on till date is notably shakeable and weak. Traditional demarcation without the use of geospatial sciences and precisely space imagery and GPS technology was understandably much less reliable. Experts warn that using an old map as a means of relocating disputed boundary markers or lost boundaries would likely give a leeway of up to 25 km on the ground. In other words mapping technology has its inherent difficulties; worse still if the map has no geodetic datum this leeway can be greater by a factor of ten or more, which could be 25 km more or less on the ground. This assessment was given by Dr. Gary Jeffress, Registered Professional Land Surveyor (Texas) and Professor of Geographic Information Science in reaction to queries raised by the author on the International Boundaries Research Unit website "Re Legal Effects of Illustrative maps" available at (http://osdir.com/ml/culture.discuss.boundary-point/2004-08/msg00225.html). Not only were previous mapping exercises far from a precise art, but as Botswana successfully advanced in relation to the maps in *Kasikili Sedudu case*, early maps show too little detail or may be too small in scale, to be of value. The World Court has also significantly admitted of colonial maps as follows: "maps merely constitute information which varies in accuracy from case to case; of themselves and by virtue solely of their existence, they cannot constitute a territorial title" [37]. Nigeria also argued in its written submissions to the ICJ in the *Land and Maritime Dispute case*, that it was not unknown for colonial surveyors "to round things up" in order to save themselves from more detailed but difficult work. Brownlie, in his seminal work on African boundaries, noted of the Benin-Niger border as follows: "the precise division of the rivers and thus the allocation of islands, remains the subject of doubt since the relevant French instruments are not sufficiently precise" [38].

Inherited land boundaries notoriously have ambiguous features, whereas space based solutions bring much transparency to the tasks before decision makers. Thus, on mapping and cartographic grounds alone earth observation technology could not have come soon enough [39].

The determination of the African Union to address comprehensively the problems of boundary and frontier determination and demarcation could also not have come too soon as the tensions, skirmishes, Sudan Liberation Army fighters and nomads frequently have deadly skirmishes involving humans and even camels over boundary positions [40] and outright war over boundaries have plagued African states and its peoples, Saïd Djinnit, Commissioner for Peace and Security at the African Union in a speech had reason to note that since African States gained independence, borders inherited from colonisation have been a factor of recurrent conflicts, adding that most of these borders were ill-defined and un-demarcated. AU, "Report of The Meeting: Preventing conflicts, Promoting integration" Preparatory Meeting of Experts on the African Union Border Programme Addis Ababa, Ethiopia Conference of African Ministers (Available at: http://www.africa-union.org/root/au/Confe rences/2007/june/PSC/7/Report_final.doc) [41]. The deployment of modern day tools in the form of satellite imagery and other geodetic data is therefore, an imperative in the quest for ascertainment of legal positions during boundary negotiations and dispute settlement procedures [42–44].

The work to be done is immense. Africa has approximately 28,000 miles of international boundaries. The national boundaries are recognisably of high level of porosity with up to 109 of its international boundaries characterised by permeability. It is significant that experts agree that up to 25% of African international boundaries are completely undemarcated, African international boundaries are 'protected' by about 350 official road crossing points, or one for every 80 miles of boundary [45]. Less than 50% of the world's maritime boundaries have been agreed upon, in Africa, the number is even less than the average in other continents. Africa has 27 mainland coastal states and maritime delimitation is fraught with ample grounds for dispute. In light of these, it is quite commendable that African States are collectively determined to address their cartographic, earth observation and Remote Sensing (AOCRS) needs and to bring these to bear on the important tasks of delimitation and demarcation of African territories [46,47].

4. Management of Boundary Lines and Conservation of State Boundaries: Case Study of the Cameroon-Nigeria Process

It may be acknowledged that there are few international legally significant events that manage to capture the attention of both the government and the people of independent states like the rendering of a verdict of the International Court of Justice (Hereinafter referred to as the ICJ, the Court or the World Court.). The judgment of 10 October 2002 in the *Land and Maritime dispute Case* (Cameroon v. Nigeria) was the first by the Court in the 21st Century to decide upon a territorial dispute on the African continent. It served as a poignant reminder of the lingering effects of a colonial era fast receding in popular memory [48–50], see particularly para.30. Three notable decisions of the ICJ preceded the Land and Maritime dispute in the new century and each had generated significant interests in their own rights [51–53], many of the Cameroon-Nigeria Mixed Commission and several sub commission documents referred to in this section are held on file by the author. The ongoing attempt to implement this judgment by both states in collaboration with the United Nations very quickly opened up new vistas for understanding the full benefits of earth observation technology to maintenance of state boundaries in modern Africa. The coverage of the implementation process involves international boundaries of about 2000 km. The process has also opened up some criticisms of the severe short comings and limitations of legal and judicial work of the ICJ. This is in relation to the realisation that some of the delimitation may have been without adequate reliance on the aid of geospatial science.

The parties soon discovered that the lack of modern geospatial data can complicate the job of demarcation of international boundaries. The dearth of Geoscientific Ground data apparently complicated matters both at the judicial hearing and implementation stages of the Land and Maritime Boundary case. At the implementation stage of the Court's judgment it was discovered by the parties that there were fundamental discrepancies in the calculations by the Court's cartographers particularly in relation to the maritime sector and that these were apparently part of the Court's evaluation. When Nigeria superimposed the Court's basepoints used in calculating the coastline against the backdrop of actual satellite imagery map, this resulted in those basepoints falling deep into water much against the technicalities of the establishment of national baselines ("Baseline" refers to the line from which the seaward limits of a State's territorial sea and certain other maritime zones of jurisdiction are measured.) The Nigerian team during the negotiations concluded that it was essential in accordance with the principles of maritime delimitation that the median line as prescribed by ICJ, must lie on the low-low waterline (on land), For this assertions Nigeria specifically relied on United Nations Division for Ocean Affairs and the Law of the Sea, *UN Handbook on the Delimitation of Maritime Boundaries* [54], Note may be taken that both Article 3 of the territorial Sea Convention and Article 5 of the LOSC provide in identical words that 'the normal baseline for measuring the breadth of the territorial sea is the normal baseline for measuring the breadth of the territorial sea is the low-water line along the coast as marked on large-scale, charts officially recognised by the coastal State as marked on large-scale charts officially recognised by the coastal State' Churchill and Lowe explain that "[t]he effect of choosing the low-water line, rather than the high-tide line is to push the outer limit of the territorial sea and other zones farther seawards, particularly on coasts where there is an extensive tidal range" [55]. Issues were also raised regarding the effect of substantial erosion on the Nigerian coastline and the effects that should have on the demarcation. Only consistent satellite imagery maintained over the decades could resolve the effects of erosion issue but those were obviously not available and could not be acquired in retrospect. Consequently, the delegation of Nigeria proposed that an independent consultant be appointed to repeat the calculations based on the application of the rule of equidistance, in order to determine the accurate coordinates of the basepoints as well as those of point X and to conduct a ground trothing.

Ground-truthing is common parlance in cartography with significance in meteorology, analysis of aerial photographs, satellite imagery and a range of other remote sensing techniques in which data gathered at a distance is verified through further processes. For instance, a pixel on a satellite image is compared to what is there in reality (at the present time) in order to verify the contents of the pixel on the image. Ground-truthing is basically a validation process to confirm correspondence between places and their abstractions, whether digitally coded or paper based.

4.1. The Role of Geospatial Products in Boundary Demarcation Contracts

Much of the strides made by Nigeria in the Space sector described above happened after the judgment of the World Court. Its satellite infrastructure and has been taking place slowly and piecemeal manner. Thus, there has been very little opportunity to incorporate its growing competence during the implementation stages of the judgment. Other issues that have prevented such desirable justification for Nigeria's investments in space satellites include administrative 'bottle necks,' the disappearance of Nig-CommSAT from space in November, 2008 and the delays in launching of NigSAT-2 initially proposed for 2007 [56]. Perhaps more significantly from an economic point of view, the inability to utilise developing homegrown satellite technological opportunities created the conditions for the prohibitive costs which have become a valid criticism of the Cameroon-Nigeria implementation process. Two surveyors from both countries closely involved in that process persuasively hint at the loss of huge savings by contracting out what could have been done with domestic capabilities. They wrote that, " . . . the UN cartographic consultant's Specifications and Technical Guide based on 'international standards' made the cost of the demarcation exercise prohibitive" [57].

Typically after states resolve to complete the demarcation of an international boundary a series of commercial geoscientific data based contracts will be awarded. The demarcators would always place reliance on the data acquired and/or retrieved. The Mixed Commission approved the specifications for all the contracts for the demarcation of the boundary and requested that certain key contracts should be announced on the U.N website [58]. The contracts that were announced to international bidders included.

- (a) Contract for Spot 5 Satellite Imagery mapping;
- (b) Contract for GPS Ground survey Control for Spot 5 imagery;
- (c) Contract for Geodetic Datum Stations Emplacement;
- (d) Contract for Quality Assurance of Geodetic Datum Stations;
- (e) Contract for Boundary pillar emplacement;
- (f) Contract for: As-built Survey of Boundary Pillars [59].

The Mixed Commission adopted other assignments for other joint technical teams, including the field verification with the contractor for the imagery of the pillar emplacements shown on the ortho-imagery. Pillar designs were then approved for both the geodetic datum stations and for the final demarcation of the boundary. It was upon these precise mix of scientific products and legal contractual mechanisms that the seventeen yearlong processes for the demarcation of approximately 2000 km of land boundaries between the two African states has proceeded.

4.2. Demarcation of Inaccessible Zones

It is a principle of law that the law must not mandate the doing of the impossible-*Lex non cogit ad impossibilia*. Sometimes a Court's judgment may be easy to interpret yet implementing it may be impossible ab initio or become impossible as things develop. This can occur as it did in the Cameroon–Nigeria process where the attempt to give effect to the delimitation instruments as prescribed by the World Court meant that surveyors will have to traverse inaccessible territories. A clear advantage of space applications in demarcation exercises is their singular advantage in making it possible for demarcators to deal with inaccessible terrain. The boundary between states tend to be delimited by features like water bodies or treacherous natural features such as dense forests, range of mountain peaks, volcanoes or marshland. As a result verification by traversing the territory for surveys may be impracticable or even impossible.

A recurring feature in the lengthy judgment of the ICJ and therefore, in the Cameroon-Nigeria boundary demarcation project is the feature of watersheds as boundary. Note the occurrence of watersheds in the Judgment Paragraphs 79, 86, 103–112, 116–122, 124, 135–140, 147, 161, 162, 164–167, 168, 170, 174, 179, 183. Indeed it was decided that the demarcation exercise must deal with watershed line for approximately 400 km of the boundary. The *Encyclopaedia of International Boundaries* defines a watershed as "the elevated area separating headstreams that are tributaries to different river systems or basins. Watershed therefore, is the area of land where all of the water that is under it drains off it and goes to the same place [60]. The Mixed Commission soon found out that in many cases the watershed feature referred to in the applicable treaties and the World Court's judgment are inaccessible or can be accessed only at great cost or danger to personnel. The parties thus, adopted a very modern space science solution in order to find the watershed line.

Three options were identified: (i) manual delineation from existing topographic maps with contour lines; (ii) automatic extraction from a Digital Elevation Model (DEM) data using a software application (ESRI, 2002 (a); and (iii) direct identification in the field. The UN Cartographic section successfully proposed the use of DEM data produced by the Shuttle Radar Topography Mission (SRTM) as the source data for the extraction of watershed line as well as for the ortho-rectification (See Cartographic section, department of Peacekeeping Operations, United Nations, Data and methodology Used in watershed line Extraction for Cameroon-Nigeria Preliminary mapping, 17 January 2005. The SRTM uses imaging radar technology to view the Earth's surface. The radar signals can see through clouds, but it does not see through thick vegetation canopies and human-made features. It uses a wave length of 5.6 cm which does not penetrate vegetation competently. The shuttle spacecraft emits radar signals that are reflected on the earth surface. Two antennas receive the returning signals. The radar waves are reflected by the ground features, including bare soil, dense vegetation dense and man-made features. The topography of the terrain causes phase difference between the signals received by the two antennas. This phase difference is translated into topographic (elevation) data -Jet Propulsion Laboratory, Jet Propulsion Laboratory, "Shuttle Radar Topographic Mission" California Institute of Technology.) [61]. The Sub Commission on Demarcation selected the option of automatic extraction and specified that the boundary line for watersheds be extracted from "a fully rectified DEM" using "stereo imagery and ground control points" [62]. In the event the UN Cartographic Section was able to produce only limited coverage of stereo imagery over the boundary area that produced the DEM data required according to the parties specifications [63]. In this way GIS and DEM data purchased at considerable cost was eventually deployed to assist other field data gathered during the exercise. As a result the system proved invaluable for places like the extensive summit of the treacherous Alantika Mountains which form a watershed in the north east of the boundary line, The Atlantika Mountains are an extension of the Cameroon line of volcanic mountains, spanning the border between Nigeria and Cameroon. Certain points were however, identified as suitable for helicopter landing for the purpose of verification and construction of boundary pillars, the impressive computer generated imagery wraps the few white clouds around the actual contours of the mountainous watershed so that it does not distort the imagery or the plotting of the boundary line by the software. For those areas that cannot be reached and for which the parties have had to settle for (Digital Elevation Model) DEM watershed line extraction, it may be cynically concluded that modern surveyors continue to give new manifestation to Oppenheim's depiction of boundaries as "imaginary lines" [64,65].

4.3. Earth Observation and Remote Sensing in the Maritime Sectors

The work of the parties in the maritime sector exposed a current legal problem which will perhaps continue to emerge in these sort of processes. The question is how to deal with maritime maps used for delimitation of territory that were made before geospatial data was available. The positions derived from satellite navigation systems such as Global Positioning Systems (GPS) are based on World Geodetic System (WGS) 1984 Datum, The WGS 84 may be described as an Earth-centred, Earth-fixed terrestrial reference system and geodetic datum. This system is based on a consistent set of constants and model parameters that describe the Earth's size, shape, and gravity and geomagnetic fields. It is also is the standard U.S. Department of Defense definition of a global reference system for geospatial information as well as the reference system for the Global Positioning System (GPS). WGS 84 is indeed compatible with the International Terrestrial Reference System (ITRS) [66]. The differences between satellite derived positions and positions on older charts and maritime maps that states may have inherited as boundary positions from their ex colonisers may be significant to navigation let alone being relied upon as accurate measurement of modern boundaries. The maritime boundaries of Africa were also largely drawn up before the availability of modern space based mapping techniques and before the ubiquitous use of the GIS. In this way the ready availability of geospatial products may open up a new set of controversies in international relations.

A controversy erupted early in the Cameroon-Nigeria negotiations as to how the World Court arrived at the delimitation it decided upon in the maritime sector. This is because the treaties it utilised in the judgment which mentioned features and positions were drawn up prior to the introduction of the GIS System. Thus, how did the Court approved delimitation interface with the GIS based system? Specifically which system did the Court rely on behind doors during its deliberations and while considering its ruling. The judgment itself was silent on its reliance on non-reliance upon any available geospatial data. Indeed it became clear that the maritime coordinates on the available charts were not in terms of WGS 84. The task of conversion was also compounded because a certain crucial Map-Chart 3433 which was relied upon by the Court had no horizontal reference datum. As a result there was no computational way of deriving a correction factor converting the coordinates from the chart to WGS 84. An Expert Determination procedure was incorporated into the negotiations which witnessed the appointment of a former Surveyor General of New Zealand to look into the facts. He concluded that: "If the current coordinates are assumed to be WGS84 coordinates, without any conversion, when they are reproduced on the sea or inlet they will not be at the exact position intended by the Parties or the ICJ … " (Report of Bill Robertson annexed to Report of the Third meeting (2005) held on file by author.) Difficult compromises had to be made to resolve the scientific ambiguities which involved establishing new points using GPS receivers.

First the conversion of coordinates to WGS84 datum was achieved through surveying the position of a minimum of 20 identifiable reference points above the surface of the sea on either side of the maritime boundary, which have been previously identified by mutual agreement on the Admiralty Chart 3433 mentioned in the ICJ judgment. Second, where certain features cannot be identified at the location, other physical features above the surface of the sea which appear on the Admiralty Chart 3433 were surveyed to ensure that an adequate number of points had been surveyed for geo-referencing purposes, See Annex III Terms of Reference and Modalities for the Delineation of the Maritime Boundary in Report of the Fifth Meeting (2006). On the basis of these new compromises the parties developed a method for converting the available coordinates to WGS84 [67].

The field methodology for surveying and processing points adopted by the Working Group during its missions and operations are testaments to the utility of contemporary application of geoscientific ground investigation in African boundary demarcation. The precise methodology adopted is presented below and it may indeed be instructive to other teams working on similar delicate assignments in the African continent and elsewhere, See Fieldwork Methodology Annex III –I in Report of the Fifth Meeting (2006):

1. Reference points above the surface of the sea on the 1994 edition of the British Admiralty chart 3433 were to be identified and approved by the parties. The coordinates for these points were to be extracted from chart 3433.

2. The equipment to be used to carry out this field work was to be the equipment already being used by the Joint Technical Team (JTT) for the land boundary field assessment, namely the Trimble Pro XRS GPS equipment.

3. The agreed chart coordinates for the locations to be surveyed (provided from the points selected from Admiralty Chart 3433) were to be uploaded onto the Trimble Pro XRS GPS equipment already configured for WGS84 datum.

4. Using the agreed points as a guideline, the Mission was tasked with using the Trimble Pro XRS GPS equipment to navigate on the sea to the physical features identified from Admiralty Chart 3433.

5. Where the physical feature existed and is accessible, the JTT's task included recording the position using the Trimble Pro XRS GPS equipment (working with sub-meter accuracy) and with the observed position recorded onto the template developed for that purpose.

6. If a physical feature above the surface of the sea exists but the central point is not accessible then four points were observed and recorded on the feature using the Trimble Pro XRS GPS equipment from which the central point corresponding to the feature are then calculated. Each of these observed points and positions were recorded onto a prescribed template.

7. In all cases, twenty (20) epochs were to be used to record the data onto the Trimble Pro XRS GPS equipment and all measurements were in terms of the WGS84 datum in geographical coordinate format.

8. If a physical feature does not exist above the surface of the sea, the Mission's task was to move to the next position provided that a minimum of twenty (20) points were surveyed.

9. If there is a need to supplement the features identified for measurements, that is, if a minimum of twenty (20) physical features/locations are not available, the JTT could agree additional physical features above the surface of the sea and identifiable on the chart which may be surveyed in addition to those physical features listed in the Fieldwork Methodology (An agreed copy of the Admiralty Chart 3433 with physical features was attached as Annex III.).

10. Data collected by the mission were processed using GIS software adopted by the parties for purposes of geo-referencing the British Admiralty Chart 3433 into the WGS84 datum. The maximum geo-referencing errors were by agreement not to be allowed to exceed seven (7) meters. Coordinates of the twenty-one points on the maritime boundary between Cameroon and Nigeria defined in the judgment of the International Court of Justice of 10 October 2002 were extracted digitally from the British Admiralty Chart 3433 and geo-referenced in WGS84 datum.

5. Legal and Evidential Implications of Emerging Satellite Imagery of Ancient African Relict Treasures and Features

Africa is the world's oldest continent and her nations, institutions and peoples are humanity's first. Ancient African civilisations are responsible for founding the original logic, structure and method of statecraft for which modern human civilisation is structured. Africa's contributions to human civilisation are indisputable and vast, spanning, for example, the areas of agriculture, arts, government, law, medicine, monotheistic religion and science [68].

Remote sensing activities are one of the core areas of the utilization of outer space science. It is difficult to overestimate the usefulness of this application to commercial and exploitative activities on earth. Such activities have led to the discovery of sites of archaeological value including an ancient Peruvian pyramid [69], ancient canals in the heart of an Arizona City [69], new Egyptian pyramids [70,71] and about 14,000 Mesopotamian settlement sites, which span some 8000 years [72]. One of the most sensational aspects of satellite imagery and remote sensing applications in relation to Africa has been the astonishing discovery of long lost wall features which indeed constitute evidence of pre-existing relict boundaries [73]. Evidence has been emanating from recent satellite imagery and orthorectified imagery as well as archaeological studies that provide overwhelming evidence of other culturally significant heritage sites and features all over Africa. These include magnificent city walls, mounds and man built ditches as well as other very precise boundary features and markers separating pre-colonial African cities, states and political groups. City walls as archaeological sites add significant tourism value to cities and the discovery of new sites can be of serious economic value to modern states [74–76]. Africa's current boom in the discovery of walls through earth observation is highlighted by the fact that records now reveal between Lake Chad and the Atlantic Ocean, in West Africa, about 10,000 town and city walls at least 25% of which are on presently deserted sites. This makes it possible to easily tap into the commercial and touristic value of these sites if the governments of Africa and the private sector choose to invest in these discoveries.

It is particularly important to consider the socio-legal aspects of the practice of remote sensing on heritage conservation with respect to the birthplace of mankind. Apart from the conservation and tourism value, relict wall discoveries can be of immense evidential value in law. These include their usefulness in dealing with on-going and future self-determination struggles as well as resolution of international and national boundary disputes. The existence of impressive and significant relics all over Africa arguably indicate pre-existing African indigenous sovereignties which flies in the face of popular assertions that Africa was largely unmanned and undeveloped before relatively recent European colonization, Sovereignty is both a legal and political concept of universal significance across human cultures and with manifestation in time and space. The concept is political in conception and is popularly symbolised by the Leviathan of Hobbes. It implies the supreme authority of a state, which recognises no higher authority in the region. Sovereignty of an independent state, therefore, sums up the essence of statehood, and the power it expresses within a territory is at least equal to but often even greater than the sum of the power symbolised in the person of the sovereign [77,78].

5.1. Discovery of Relict Boundaries through Satellite Imagery

The connection between delimitation and demarcation tasks on one hand and archaeological and heritage sites on the other can and often come into sharp focus in Africa as much as elsewhere. This is because heritage sites along with other significant natural features quite often become adapted as marks of boundary alignment. One of the classic rendition of this problem in international relations is to be found however, in the notorious dispute between Cambodia and Thailand over the religious and heritage sites of the Temple of Preah Vihear. This dispute has been adjudicated upon by the ICJ [79]. Despite the existence of a definitive judgment unfortunately, severe problems periodically flare up as a result of religious and cultural implications on the affected population. The *Temple of Preah Vihear case* concerned a boundary conflict between Cambodia and Thailand (formerly known as Siam). The disputed area contained an old temple of great cultural significance. It had been built by the Khmer Peoples, the ancestors of the present Cambodian population, at the high point of their power; although since then the Khmer People have been forced back into smaller areas. The considerations the parties wished the Court to pronounce upon included: to which of the two countries' history is the temple more related. Despite the Court's decision in 1962 conflict persist between the parties in relation to the temple and border skirmishes occurred as recent as 2008 [80,81].

Relict boundaries refer to antecedent boundaries which have been abandoned for political purposes but are still evident in the cultural landscape. Relict boundaries manifest themselves in space by, among other features, direct border remains such as border stones, mounds, ancient walls, border roads, clearings, customs houses and watch-towers [82–84]. It is fortunate that evidence emanating from satellite imagery and orthorectified imagery as well as archaeological studies is providing overwhelming evidence of very precise boundary features and markers separating pre-colonial African political groups. Only a handful of the 10,000 walls and relict features discovered all over Africa have been surveyed so far. This portrays immense opportunities for geoscientific ground investigation well into the future. Fulfilling those tasks will be particularly fulfilling both from a scientific and socioeconomic point of view. Both old aerial photographs and other more modern remote-sensing methodologies like Lidar technology, LIDAR—Light Detection and Ranging—is a remote sensing method used to examine the surface of the Earth" [85] continue to offer an opportunity to record much of this evidence all over Africa. The Kano City walls (a 24 km long, 20 m high perimeter) were considered the most impressive monument in West Africa as of 1903 but that achievement now pales into absolute insignificance in comparison with other recent discoveries of older demarcated boundaries [6].

Of particular note are the following:

a. The 160 km long Sungbo's Eredo wall;
b. The fieldwork surveys and inspections that have revealed 1600 km of the 16,000 km long Benin earthwork complex;
c. The 45 km long Orile Owu walls;
d. Walls of Old Oyo;
e. Old Egbe wall and walls completely surrounding pre-European-influence cities of Kwiambana, Old Ningi, Gogoram, Pauwa, Old Rano, Old Sumaila [86–90];
f. The emerging picture is that since at least the 8th Century AD enormous systems of walls and ditches have been used to demarcate state territorial control in the areas of contemporary Benin and Western Nigeria.

These walls are heritage sites of immense cultural and touristic value and the ignorance of their existence is being removed by rediscoveries through space activities. Eredo for instance represents a system of walls and ditches dug in laterite, a typical African soil consisting of clay and iron oxides. The total length of these fortifications is approximately 160 km. The height difference between the bottom of the ditch and the upper rim of the bank on the inner side can reach 20 m. The diameter of this enormous fortification in a north-south direction is approximately 40 km and in an east-west direction, 35 km. The walls of the ditch are recorded as unusually smooth. The system of walls are thought to encircle the ancient Ijebu state. The total length of the discovered fortifications in this area alone is said to exceed 6000 km [87].

It was not just that Kings of Benin such as Ewuare, Oba (king) Ewuare reigned between 1440–1473. Enjoying full sovereignty, he traded with foreigners including the Portuguese [91] built and maintained secure walls; it is more importantly significant that they maintained their empire very much in the tradition of progressive and sophisticated societies found elsewhere in Asia, Europe and the Americas. It is suggested that further evidence of this can emerge through careful remote sensing investigations of African cities. Quite unbecoming inscriptions of 'otherness' may be found in the writings of older authorities like Hegel, who wrote: "The Negro, exhibits the natural man in his completely wild and untamed state. We must lay aside all thought of reverence and morality—all that we call feeling—if we would rightly comprehend him; there is nothing harmonious with humanity to be found in this type of character [...] They have no knowledge of the immortality of the soul [...] the devouring of human flesh is altogether consonant with the general principles of the African race" [92].

Nothing defeats the idea that Africa was a black hole of underachievement in terms of spatial awareness and technical matters or that it has little cultural heritage sites to offer, more than the discoveries of actual sites through aerial imagery recorded by scientific methods. The promise of the discovery of even more relics and archaeological sites is certainly deserving of further investigation and more investment into earth observation and remote sensing by African states. Through these aerial imagery scholars can retrace and find corroboration and validation for the accounts of some of Europe's first explorers. The Dutch who visited the city of Benin in present day South-Western Nigeria described a highly civilised town with sophisticated spatial lay out and city planning:

The town seems to be very great. When you enter into it, you go into a great broad street, not paved, which seems to be seven or eight times broader than the Warmoes street in Amsterdam .... The king's palace is a collection of buildings which occupy as much space as the town of Harlem and which is enclosed with walls. There are numerous apartments for the Prince's ministers and fine galleries, most of which are as big as those on the Exchange at Amsterdam. They are supported by wooden pillars encased with copper, where their victories are depicted and which are carefully kept very clean.

The town is composed of thirty main streets, very straight and 120 feet wide, apart from an infinity of small intersecting streets. The houses are close to one another, arranged in good order. These people are in no way inferior to the Dutch as regards cleanliness; they wash and scrub their houses so well that they are polished and shining like a looking-glass [93,94].

Similarly the accounts given by British colonial officers of their encounter with Old Ningi debunks the myth of an architecturally undistinguished pre-colonial Africa. The conquest of Old Ningi was described thus:

Old Ningi was a nineteenth century cult settlement opposing Kano, Zaria and Bauchi from its hill fortress base using up to 4000 cavalry. Its mud walls were built on stone-based parapets and presented a complex defence strategy, which the larger kingdoms were unable to breach. It was captured by the British using a local traitor to show a secret way in near the beginning of the 20th Century [87].

A deliberate effort to continue reconstructing and rediscovering the evidence of such African indigenous engineering wonders can and should be systematically embarked upon through the use of the space technology. A policy of investment in this direction is bound to be richly rewarding for African states. Such investment is indeed consistent with the indigenous intellectual curiosity of traditional African societies as detailed studies show us that Africans have long being interested in astronomy, cosmology and cartography, Landscape features have from time immemorial been depicted in Khoisan rock art from the Brandberg, Namibia. They are believed to represent the natural resources of the group's "exploitation space" [95], The *Encyclopaedia Britannica* records that King Njoya of the Barmum Kingdom in Cameroon who reigned c. 1895–1923 CE not only invented a system of writing in 1895 but had a map of the kingdom of Barnum in ink and crayon. Oriented to the west, this map had place-names written in the traditional Mfemfe alphabet. This map was later acquired by a museum in 1937 [96].

5.2. Satellite Evidence of Relics and Self Determination

Apart from the possible economic and intellectual benefits that geospatial discoveries may bring there are other significant utilitarian purposes it may serve. For instance, geospatial imagery may provide backing for secessionist or irredentist movements. The ability to show the link between seceding people and their land is often crucial to their struggle. The relationship of people to their monuments, shrines, burial grounds and other items of cultural value may be of crucial value in proving distinctiveness as a people and affinity with the land. Evidence of Satellite imagery may assist in many ways to throw light upon the situation. As a writer correctly explains:

High-resolution satellite imagery can be used for human rights-related documentation, monitoring and advocacy efforts. Imagery is particularly useful for assessing the extent of violent conflict, forced displacement and other human rights concerns in remote, inaccessible or otherwise tightly controlled areas of the world [97].

Theoretically therefore, the discovery of hundreds of kilometres of walled settlement (the Sungbo wall) in Ijebu land may prove significant to the Ijebu people of South-Western Nigeria in many interesting ways. It may be used on the one hand to disprove claims that the erstwhile colonial power Great Britain met a disorganised Ijebuland or that the people ineffectively occupied the land making it *terra nullius*. There may be implications for modern day national politics as well. The same relic feature may also set up a claim by the Ijebus of a right to distinctive existence within present day Nigeria. It is indeed the case that the Ijebus have been clamouring for a separate state within the federation of Nigeria and this relic may be of assistance to their claims. In this scenario the relict boundary of the Ijebu people may be used today to reassert their pre-existing political independence in order to obtain a favourable treatment under inter-temporal laws, Inter temporal law can be broadly described as the branch of law which governs the usage of treaties, codifications and legal acts to the cases and situations which occurred before their creation or entry into force. As eloquently stated in the *Islands of Palmas Arbitration*, (Netherlands v US, 1928) it explains the legal situation that "a juridical fact must be appreciated in the light of the law contemporary with it."

A country that wants to contest a boundary may bring to the attention of the court or arbitrators evidence of the existence of the relict walls in order to show pre-colonial territorial suzerainty of kingdoms which may now be straddling international boundaries. At any rate there is much scope for rigorous multidisciplinary enquiries to incorporate this emerging phenomenon of discovery of relict walls and other archaeological features in imagining and reimagining African territorialisation.

The newly discovered boundary walls may produce the following effects:

- Evidence of international boundary marking between various pre-colonial and sometimes pre-modern African nations and kingdoms.
- Evidence to shore up claims for self-determination by existing separatist groups and peoples.
- Evidence to refute the claim that an area of Africa was *terra nullius*, thus creating the possibility of nullifying the original claim of title to the territory by the pertinent erstwhile colonial power. This is because territories inhabited by tribes or peoples having a social and political organization were not according to inter-temporal law ever treated as *terra nullius* [98].
- Within independent African states the evidence of these relict boundaries may be used to abolish, create and redraw internal boundaries in the course of the nation-building process.
- The discovery of these cultural treasures and features contribute to the much needed evidence of a glorious African past which defeats the pessimism that pervades much of coverage of the continent and its peoples.

National Space agencies such as the NASRDA and the Algeria Space Agency [99] are particularly suited for the task of developing African states competence in this area. In Nigeria for instance, the power to act proactively is located in the function of its space agency to "develop national strategies for the exploitation of the outer space and make these part of the overall national development strategies and implement strategies for promoting private sector participation in the space industry."

The Department of Strategic Space Applications and Department of International Cooperation as identified empowered in S. 8 (2) of the NASRDA Act could play a more prominent role in studying and articulating Nigeria's capabilities in the area of identification of relict boundaries and in adopting a more systematic approach to their detection and interpretation.

For those African states with no legislation or national bodies dedicated to space technology their capabilities to act independently is obviously limited. However, given the history of loss of Africa's archaeological heritage into the hands of the erstwhile colonial powers, south-south cooperation in respect of heritage research ought to be one of the central planks of cooperation within platforms such as CRTEAN and APRSAF.

6. Earth Observation, Archaeological Research and Spatial Privacy

Archaeological research via space satellites and remote sensing is very usefully lifting the lid over the mysteries of Africa's past. The use of remote sensing generally is likely to have increasing value in the management and exploitation of archaeological research. It has a clear role in discovery of heritage sites across a continent that played host to the very beginnings of mankind. Certain questions suggest themselves at this stage of our analysis: Will African states grasp the opportunities of this area of archaeological research through indigenous efforts? Will they do so in collaboration with the advanced space faring states and their private corporations? Indeed in the area of earth observation and remote sensing will they needlessly remain under the situation of absolute reliance upon foreign investment in this area; such as the American Landsat system or the European Spot system? These questions are germane in light of the issues of privacy, security and national sovereignty that are raised by geospatial technology even in Archaeological/Heritage Research?

The right to privacy is one that is commonly recognised in all legal systems. Yet there appears to be a lacunae in relation to the privacy principle with respect to remote sensing over archaeological and heritage sites. We have settled above that invaluable archaeological artefacts and sites may be identified from space. Something as basic as a simple Google Earth searches has produced accurate identification of important Roman era ceramic artefacts. An Italian computer programmer recently discovered the remains of an ancient Roman Villa in his town after browsing maps and photographs downloaded from the internet site Google Earth. He correctly interpreted curious rectangular shadows he spotted nearby an ancient river as a buried archaeological structure. In this case he alerted the local archaeological museum but it cannot be concluded with certainty that similar course of action will always be adopted by other astute experts from developed countries where there is a higher appreciation of the value of archaeological material much less to expect similar results in poorer regions of the world [98]. This question of archaeological privacy is pertinent given that the history of archaeology has never been free of cloak and dagger practices and that developing states have fallen so often to their artefacts being constructively extracted permanently from their territories. Note the various accounts of treachery in relation to the discovery of Cambodia trail Pichau [100–103]. The developing states of Africa are particularly vulnerable to inadvertent or deliberate surveillance of their archaeological and heritage sites without permission. This has both commercial and security implications to the sovereign states of Africa. Privacy over archaeological treasures and heritage sites may have been lost in a fundamental way to remote sensing capabilities of a few technologically capable states whereas access to the technology is slow in coming to even those developing states that are actually showing interest in space activities. Although, many archaeological sites and monuments are located in inaccessible areas significant numbers also are to be found in cities and urban centres [104]. The use of medium- and high-resolution remotely sensed data of features, water bodies, habitats and vegetation of state territory is necessarily an issue of privacy and sovereignty. Whereas the interests of foreign investigators in Africa is quite high particularly " . . . in hyper-arid Africa, where collected field data are scarce, the areas to be covered are large and the visibility of certain classes of archaeological evidence is high, since it contrasts with the background barren environment" [105].

Cheng in 2004 gave a description of the impact of remote sensing activities upon modern states which may be considered impressive in its analogical and figurative value. He developed the idea that hitherto a state's territory is its castle. According to this analogy:

A state's territory is its castle. No one is allowed to enter it without its permission . . . although aviation has added an extra dimension to the problem of states in controlling what goes on in their territory, their grip in law and in fact remains unaltered, so much so that from the legal point of view the world resembles a series of immense airtight petroleum storage tanks representing the various national States with their three-dimensional sovereignty . . . the arrival of the space age was as if the lid on the tank was suddenly ripped off. And, if we can change the image, it was like opening up an ant-hill with all the ants inside scurrying round wondering how to cover themselves and their secrets and stores" [106].

Where remote sensing is conducted by the mutual agreement between states or between private corporations and states there would be little controversy. The legality of such activities would be manifest as long as the activities fall within the express or implied agreement between the pertinent parties. Thus, for instance, the exact geographic position to be remote sensed would usually be identified. The particular features or resources to be searched for may also have been provided for. In this way the body in charge of conducting the remote sensing operations is bound by this agreement and ought not to go beyond it or retain information that is derived in the course of the remote. It is thus suggested that remote sensing contracts over national territory constitutes a kind of contract of agency. In which case the principal is the state acquiring the required data and/or imagery and the agent is the state or corporate body conducting the scientific aspects of the task and owning or operating the requisite satellites. In this manner the normal duties of a commercial agent to his principal ought to apply and in performing his activities a commercial agent must look after the interests of his principal and act dutifully and in good faith [107–113].

In relation to areas within state territory or jurisdictional competence, the prohibition of clandestine gathering of data is irrefutable and the privacy of states is supported by international law and state practice. Clandestine as used here is not exactly the same as illegal but is wider in meaning to include bad practice and policy of hostile intentions. The general slant of international law is towards protection of the sovereignty and privacy of states. This is further buttressed by the important restrictions in the international regulation of various easements granted to foreign states in areas and zones that expose the territorial state to security concerns. Article 38 of the LOSC (1982) which regulates transit passage in straits used for international navigation includes an important caveat namely—that in relation to research and survey activities during transit passage, foreign ships, including marine scientific research and hydrographic survey ships, may not carry out any research or survey activities without the prior authorization of the States bordering straits. Article 40 which deals with research and survey activities further entrenches an anti-clandestine approach to collection of data by stating that during transit passage, foreign ships, including marine scientific research and hydrographic survey ships, may not carry out any research or survey activities without the prior authorization of the States bordering straits. Relevant state practice may be seen in the reaction of the US to the U2 incident witnessing the flight of US aircraft over Soviet territory and the subsequent downing of the erring aircraft, the US not only apologized for the event but made diplomatic assurances that such reconnaissance aerial flights will no longer be made [106,114].

Remote sensing operations conducted through clandestine means, however, is the new bug bear of international relations. Archaeological spying may just be the new flavour of this age-old problem of International Relations. The problem of clandestine remote sensing may emanate in two basic ways: First and perhaps most significant from the security of the territorial state's point of view is espionage by remote sensing. This problem was first dramatically exposed during the twilight years of the cold war era when a picture emerged in the *Jane's Defence Weekly* in August 1984 and in an edition of the widely circulated UK *Times newspaper* in the same month showing with extraordinary clarity the Soviet shipyard at Nikolaiev on the Black Sea [115]. Inset in the picture was the emerging jigsaw of the construction of the premier Soviet naval 75,000-ton nuclear-powered aircraft. A second way in which clandestine remote sensing is employed is in the form of the gathering of data and images of the natural resources and environment of a territorial state. This includes collation of economically sensitive information about energy resources, hydrocarbon reserves, agriculture, fisheries, forestry, mining, shipping, cartography, wildlife migration patterns, heritage sites and human traffic flows [106,116,117]. Note also the controversies in international space law as to the spatial demarcation boundary plane issue whereby there is no certainty as to where in spatial terms the sovereignty of each state ends. Thus, it is difficult (among other issues) to show where exactly illegality of remote sensing begins.

It is necessary to grapple with the existence of contrary views and interpretations according to which remote sensing is illegal only if violating some specific prohibitions (such as the prohibition to enter foreign airspace without the consent of the relevant state). In short: remote sensing or spying by satellites is not 'illegal' since to sense other countries without their consent and/or even without providing them with derived data is legitimate as long as such states have been given access to data 'on a non-discriminatory basis and on reasonable cost terms' -within the provisions of UN Resolution 41/65 (Principle XII). This particular provision is indeed considered as customary international law. The U.S. as a space power obviously championed the so called "Open Skies" policy according to which there is international availability of remotely sensed data on a non-discriminatory basis. Indeed the UN Remote Sensing Principles (UN Principles) largely follows this policy. It is also important to note the complete absence of protest against the overflight of Sputnik-I and following Soviet space objects [118]. It is however, not incontrovertible that clandestine remote sensing is good policy in international relations. The fact that sensed states have the right to access the primary data, processed data and analysed information on a non-discriminatory basis and on reasonable cost terms is an indication of the instinct of international policy on this sensitive issue.

It is must however, be conceded that further arguments have to be to fine-tune the law in this area of space practice. This will be necessary in order to forestall the issue becoming a flashpoint in the occurrence of military tensions and disputes. The situation is indeed, 'an equal opportunity challenge.'

As technological prowess and commercial successes of private entities becomes more democratized, even powerful states are vulnerable. A writer admits this conclusion when he wrote:

"There is a dark side, however. Just as the military will have access to high-resolution commercial imagery, so too will the general public and foreign entities, allies and adversaries alike. Without proper protections, military movement and build-up, the lay-out of military facilities and even the locations of individual pieces of military equipment could be made available to the public eye within a matter of hours. Obviously, this circumstance could have grave consequences for military operations and U.S. national security" [118].

One of the options open to states that are apprehensive of spying over their archaeological and heritage sites from above is that they should adopt domestic laws with extraterritorial effects that make the remote sensing of certain places illegal. This may include the prohibition of dissemination of any unauthorised information derived therefrom. Presumably such laws will apply to offenders as Cheng said "wherever in the world the offences may have been committed" although the ability to enforce jurisdiction will be severely limited if perpetrators do not come within territorial jurisdiction of the offended states [106]. Furthermore the ability of a state to enforce it such cases will inevitably depend on its diplomatic, political and economic clout.

Cheng one of the fathers of modern day international space law himself had noted the limitations provided by the *Lotus case* (1927) to the extent that the case while seemingly allowing states to criminalize acts committed outside territory does so upon the proviso that enforcement may be attempted only when such persons come within their territorial boundaries, Cheng introduced two helpful concepts of enduring importance. *Jurisfaction* he says denotes the normative element of jurisdiction and it represents the powers a state has to adopt valid and binding legal norms and to concretise them with binding effect through its appropriate organs, whether judicial or otherwise. The spheres of validity or operative force of these norms may be realised *ratione loci* (territorial), *ratione instrumenti* (quasi territorial) or *ratione personae* (personal). *Jurisaction* on the other hand, is the formal element of state jurisdiction and it encompasses the powers a state possesses to, at any place or time, physically perform the acts of making, concretising or enforcing laws. That is it can hold legislative assembly, set up courts or tribunals or even arrest wanted persons. From this point of view, "the validity of jurisaction presupposes jurisfaction, but it is possible to have jurisfaction without jurisaction" [119]. Cheng has persuasively argued that; "Military reconnaissance satellites have not only become a fact of international life that states just have to learn to live with but also a vital instrument in the process of arms control and the preservation of international peace" [106].

In a similar manner a view may be taken that aerial/spatial archaeological research without the underlying state's permission may have become just another manifestation of the shrinking of state sovereignties and a 'nuisance' which states must bear. That view however, is certainly not universal. There is therefore, a call to be made to introduce legal certainty in this area of activity. The final position on the matter will, of course, be reliant upon emergent state practice and the opinion of eminent scholars among other sources of law.

African practice and view in this area is arguably consistent with those of most of the developing states. The sheer fact of placement of cadastral authority under ministries of defence such as in Tunisia is arguably an indication of the importance some African states attach to remote sensing activities. The Minister of National Defence has the power of fixing the administrative and financial organization as well the operating modalities of the centre for national cartography and remote sensing. Tunisia Decrees No 75–671 of 25 December 1975 and Decree No 2006–1902 of 10 July 2006, Space Legal Tech, op.cit. Accordingly, even the, legal powers over the fixing of rates of the services to be rendered in Tunisia is granted to the Minister of Defence the National Centre of Cartography and remote sensing Order of the Minister of National Defence of 22 July 2016. See also the following Laws: Law No. 74–100 of 25 December 1974, creation of the Office of Topography and Cadastre, as amended by Law n° 2009–26 of May 11, 2009; Law No. 88–83 of 11 July 1988, creation of the National Centre for Cartography and remote sensing, as amended by Law No. 2009–24 of May 11, 2009.

There is ample evidence in treaties to indicate that unilateral lifting of the aerial veil of a state without its consent may be seen as hostile, offensive and/or even aggressive conduct in certain instances. A good place to begin the analysis will be the UN Charter. The Charter in Article 74 provides that member states agree that their policy in respect of all territories under their jurisdiction "must be based on the general principle of good neighbourliness, due to account being taken of the interests and well-being of the rest of the world, in social economic and commercial matters." In this way one would struggle to see how clandestine study of another state's commercial and economic resources from outer space will fit within the concept of good neighbourliness. Indeed Bin Cheng's conclusion on the matter is apposite stating that;

"There is no reason why this principle does not apply to data gathering from outer space....Although the import of the principle may not be altogether clear, its relevance to remote sensing is patent, especially in relation to the problems of dissemination and misuse of remote sensing data and information" [106].

In this manner there is nothing about the 'hi tech' phenomena of remote sensing activities that ought to disturb the decisions reached on the binding nature of the obligation of good neighbourliness' and respect for equitable conduct in international relations as may be found in International case law such as the *Trail Smelter Arbitration* (1935) [120], *Corfu Channel case* (Merits) (1949) [121] and the *Fisheries jurisdiction case* [122]. Admittedly it is a difficult proposition to state that specific rules exist under customary international law which completely prohibits archaeological espionage from outer space. States may however be encouraged to institute rules mandating that the result of archaeological research by foreign states and corporations must be freely shared with the underlying state in all circumstances. This may be subject of course to reasonable contributions towards the expenses incurred although such charges ought to be heavily subsidised particularly for developing states.

Disaster Regulation and Assessment of Natural or Human-Induced Threats to Conservation from Space

Natural or human induced disasters include floods, tsunamis, volcanic eruptions, forest fires, landslides and war damage. It is trite observation as stated by the COPUOS that:

"Space applications related to Earth observation, telecommunications and global navigation can play a vital role in supporting disaster risk reduction, response and recovery efforts, by providing accurate and timely information for decision-makers" [123].

Global navigation satellite systems perform truly global tasks such as contributing to an improved understanding of the relative motion of tectonic plates and to the delivery of humanitarian assistance to areas affected by disasters [124].

The potentials of the geospatial sciences and international space regulation as tools of assessment of natural or human-induced threats to conservation is just beginning to be understood in clearer terms [125–128]. Most of the grave environmental problems facing the world are of international significance and demand coordinated scientific responses often requiring space applications.

The following table culled and collated from information available from a Space Application Matrix developed by the United Nations Office for Space Affairs, shows how earth observation and remote sensing applications have been brought to bear upon many of Africa's pressing environmental, conservation, health and safety concerns within the last decade [129–133].

Air transportation is central to global economic activities and even the conduct of modern life generally. The reliance of airspace transportation on space based solutions in monitoring natural disasters was dramatically underlined by the events of the natural disaster of volcanic eruption in early 2010. Satellites are included in regional and international monitoring capabilities of volcanic activities and regulatory arrangements [134–139]. The international legal regime of regulation of air transportation was successfully deployed in reducing the severity of this unique natural disaster. Volcanic ash was monitored through space based platforms and the data gathered informed coordination of international legal responses. There was close coordination and communication between various information providers in fashioning appropriate responses against this disaster and the legal responses were robust and coordinated [140].

Coordination and cooperation between states in an atmosphere of scientific brotherhood will always be key to goal attainment in the area of disaster response and management. From coping with the spread of epidemics and diseases (Vibrio Cholerae, malaria, West Nile Virus, tsetse flies and so forth, as shown in Table 2 above) to the monitoring of air quality and global warming space based solutions blended together with international regulation and cooperation is the future of international relations. This was precisely what the drafters of the Outer Space Treaty envisaged when they referred to:

Table 2. Geospatial Products, Space Applications and Disaster/Conservation Management over Africa.

Space Activity	Used Sensor	Case Study	Location and Date
Earth Observation and Remote Sensing	IKONOS Landscape	Tracking determinants of the Anopheline Mosquito Larval Habitats in the Western and schistosomiasis control in Kenya	Kenya Highlands 2002
EO & RS	LANDSAT 7 (ETM+)	Landscape Determinants of Anopheline Mosquito Larval Habitats	Western Kenya Highlands
EO, GIS & RS	LANDSAT 7 (ETM+)	Mosquito Larval Habitat Mapping West Nile Virus.	
EO & RS	Meteosat-8 (SEVIRI)	Predicting distribution of tsetse flies using temporal Fourier processed meteorological satellite data.	West Africa
EO & RS	NOAA 15-17 (AVHRR 3)	Detection of Vibrio Cholerae by Indirect Measurement/ Predicting the distribution of tsetse flies in using temporal Fourier processed meteorological satellite data.	West Africa
EO & RS	QuickBird-2	Dynamics and Risk Mapping Rift Valley Fever in a Zone Potentially Occupied by Aedes Vexans:	Senegal 2003–2007
EO & RS	RADARSAT-1	The Use of Radar Remote Sensing for Identifying Environmental Factors Associated with Malaria Risk.	Coastal Kenya
EO & RS	SPOT 5 (HRG)	Dynamics and Risk Rift Valley Fever in a Zone Potentially Occupied by Aedes Vexans.	Senegal 2003
EO & RS	Terra (MODIS)	Vector Borne Disease: Risk Mapping West Nile Virus WNV.	
EO & RS	TRMM (PR)	Dynamics and Risk Mapping: Rift Valley Fever in a Zone Potentially Occupied by Aedes Vexans.	Senegal
EO & RS	TRMM (TMI)	Dynamics and Risk Mapping: Rift Valley Fever in a Zone Potentially Occupied by Aedes Vexans.	Senegal
EO, RS & Damage Assessment Map	QuickBird-2	Application of High-Resolution Optical Satellite Imagery for Post-Earthquake Damage Assessment: The 2003 Boumerdes (Algeria).	Boumerdes (Algeria). 2003
EO, RS & Open Water Pollution Map	SPOT 1,2,3 (HRV)	Spectral Enhancement of the SPOT Imagery Data to Assess Marine Pollution near Port Said, Egypt.	Egypt 2006
GPS Infectious diseases risk map	SATNAV	Schistosomiasis in coastal Kenya. Water sources that the residents use (ponds, spring fed rivers and a stream and manmade open wells and boreholes) entered into a GIS.	Kenya 2000–2004
EO/RS	Aqua (MODIS) EO-1 (ALI) EO-1 (Hyperion) Terra (MODIS)	Volcanic Activity Monitoring The autonomous Model-based Volcano Sensor Web (MSW), based at JPL, proved its worth during a volcanic crisis at Nyamulagira, Congo, in 2006.	Congo 2006
EO-1	MODVOLC thermal detection	Detection and monitoring a carbonatite eruption at Oldoinyo Lengai, Tanzania.	Tanzania 2007

"the great prospects opening up before mankind as a result of man's entry into outer space . . . Believing that the exploration and use of outer space should be carried on for the benefit of all peoples irrespective of the degree of their economic or scientific development" See the Preamble to the treaty. Article 1 of the Moon Agreement also provides: "The exploration and use of outer space, including the

Moon and other celestial bodies, shall be carried out for the benefit and in the interests of all countries, irrespective of their degree of economic or scientific development and shall be the province of all mankind [7,11].

It is indeed good international policy that all nations big and small be encouraged towards engagement in geospatial research, earth observation and remote sensing activities. An account shows that: "The fastest growing city in China in economic terms is Shanghai. Since 1996, it has shown an increase in tropospheric nitrogen dioxide of 29%, which is not good news for anyone in China, Europe or the rest of the world." Part of the solution to this universal concern appears quite interestingly to have come from the unlikely effort of space scientists in Netherlands where the Royal Netherlands Meteorological Institute (KMNI) has developed an integrated information system for monitoring and forecasting tropospheric pollutants over China [141].

Space applications are of crucial value particularly to developing states in conservation and disaster management. This is because operational production, distribution and the effective use of environmental, remote sensing and earth observation data is of enormous benefit to sustainable development in the developing world. The VGT4 Africa project for instance, is but one of the developmental projects that has allowed African states to successfully access satellite technology for vegetation monitoring, food security, early warning crop estimation, livestock modelling and better management of biodiversity [142,143].

The VGT4Africa partners are VITO, JRC-IES and MEDIAS-France. The system has been collecting and communicating data derived from the VEGETATION instruments on board SPOT Satellites to states including countries within the Southern Africa Development Community (SADC), Sudan, Rwanda, Ivory Coast, Kenya, Ethiopia, Senegal, Congo-Brazzaville, Mauritania, The Gambia, Tunisia and the Democratic Republic of Congo.

Africa is of course, home to some of the world's fastest growing cities and severe challenges to conservation and pollution concerns will emerge in the course of this century; hence the need for more up to date geospatial information. Africa will of course, have to shore up its competences through regional and cooperative arrangements in space but international scientific aid will also be crucial to development. Going it alone for any country will be expensive, wasteful and perhaps even counter-productive. Large scale separate engagement in space activities by potentially 57 independent states of Africa involving hundreds of private enterprises will perhaps be wasteful and may even present unprecedented dangers. The dangers include contributions to environmental pollution, climate change and international conflict. Scientific projections of the effects of space tourism industry alone shows that with a conservative estimate of 1000 suborbital flights per year rubber-burning engine crafts would produce soot that could disrupt the stratosphere and send temperatures soaring at the poles [144]. Joint cooperation in the acquisition and running of geospatial platforms for earth observation and remote sensing ought therefore, to be encouraged among African states.

As earlier on discussed a couple of African Domestic Space Acts lend themselves readily useful for Geospatial, Remote Sensing Disaster Management. In terms of legislative and institutional readiness the Nigerian situation is relatively enviable. The NASRDA Act under Section 11 (1) created certain institutions pertinent to the development of indigenous expertise. Such competences have the potentials of sub regional and regional usefulness if investments are sustained. The particular 'Development Centres' created and provided for by law include- the (a) National Centre for Remote Sensing, Jos; (b) Centre for Space Science and Technology Education, Ile Ife; (c) Centre for Satellite Technology Development, Abuja; (e) Centre for Geodesy and Geodynamics, Toro; and the (j) Centre for Basic Space Science and Astronomy, Nsukka. The existence of NASRDA is clearly of value in developing expertise in all the areas of geospatial sciences discussed in this work. Particular relevance lies in the obligations of NASRDA to (a) develop satellite technology for various applications and operationalize indigenous space systems for providing space services and shall be the government agency charged with the responsibility for building and launching satellites; (b) enhance the development and entrenchment of research, development and production tradition in the Agency, so as to achieve a high output and make the desired impact on national economic and social development; (c) promote the co-ordination of space application programmes, for the purpose of optimizing resources and develop space technologies of direct relevance to national objectives; The work of the Centre for Geodesy and Geodynamics, Toro, in Bauchi State is particularly of value to surveying tasks that would rely on space applications given its mandate to (d) facilitate and sustain the growth in capacity for geodesy surveying and mapping; (e) monitor crustal detonation and subsidence, due to excessive oil and gas exploitation, global mean sea level rise and other related seismic and geodynamic phenomena; (f) implement international agreements with regard to- (i) Satellite Laser Ranging (SLR), (ii) very Long Baseline Interferometry (VLB), and (iii) Co-operative International Geo-Physical Survey Network.

7. Conclusions

The uses and benefits of space technology are myriad and too many to itemise. Earth observation, remote sensing and geoscientific ground investigations specifically are revolutionising the application of scientific method to problem solving and conduct of international relations in Africa. Three of such important areas are (a) boundaries delimitation, demarcation, management and conservation; (b) discovery, protection and monitoring of relict and other archaeological features and heritage sites; (c) Disaster regulation and management of natural and human-induced threats to health and conservation. In all these areas international laws and specifically Space Law is yet to develop an impressive body of jurisprudence and provisions. Yet there is a good number of international legal instruments to which most countries of the world and impressively nearly all African countries are well subscribed to. The benefits to African countries are manifest in many areas but particularly in the areas discussed and highlighted above.

Space applications now provide means of verification of truth and practical solutions to legal problems beyond the wildest imagination and expectations of even 20th Century lawyers, judges, diplomats, statesmen and administrators. Space science thus, enriches lives in fulfilment of much of the dreams of space exploration as envisaged by the drafters of the space treaties. The manifestation of progress in the areas highlighted above within Africa is exactly what UN member states had in mind when they agreed that there are " ... great prospects opening up before mankind as a result of man's entry into outer space" and expressed the belief that "the exploration and use of outer space should be carried on for the benefit of all peoples irrespective of the degree of their economic or scientific development."

Space science and technology will furthermore unleash tremendous growth in many African states over the next few years and decades. A bottom up approach to the teaching and practice of science and technology in general is required. Education relating to the geospatial sciences will be crucial to solving the future challenges in all the major areas of activities explored in this paper. Africa must therefore, rapidly improve upon the numbers of graduates of space sciences and indeed in space law. Whether a state is complying with international obligations or not will be decided upon through earth observation, remote sensing and geoscientific ground investigations. In May 2018 North Korea's leader, Kim Jong-un as a show of good faith of his commitment to building "a nuclear-free peaceful world" allowed a select group of journalists from Britain, China, Russia, South Korea and the United States to witness the destruction of a nuclear test site. It was really the emergence of before and after satellite imagery pictures of the test site that granted popular verification of the event. Further Satellite imagery showing the apparent dismantling of facilities at the Sohae satellite launching station, North Korea, emerged on 22 July 2018. These indicate North Korea has begun dismantling key facilities at a site used to develop engines for ballistic missiles in further proof of its denuclearisation policies. Choe Sang–Hun "Destruction of Test Site is viewed by Journalists" *The New York Times International* Friday 25 May 2018, p. A8.; Matthew Pennington, "Satellite images indicate North Korea has begun dismantling key facilities at a site used to develop engines for ballistic missiles" [145,146]. The spread of and extent of ownership of resources around and across state boundaries will be routinely determined via remote sensing just as the evaluation of threats to these reserves will be done on real time basis.

Judges are not geospatial experts and they are not expected to be; however, courts will under their pertinent legal rules of procedure take evidence and/or judicial notice of the basic scientific facts that are adduced before them by experts. There is however, the need for more knowledge and expertise in space science and technology among lawyers and the judicial staff of international organisations. This is because the number of cases that will require knowledge about the geospatial sciences will inexorably increase in the course of this century.

Africa has a rich and noble history of achievements in the arts, science, technology and culture. In order for Africa to properly harness its full potentials in archaeological prospection, manage its boundaries and effectively assess its natural or human-induced threats to conservation, it must invest heavily in procurement of geospatial products. This is best done by African States individually and collectively investing in outer space activities. With the benefit of the emerging satellite imagery the discovery of more of Africa's relict walls, heritage and archaeological sites will continue. Their touristic value can then be systematically secured and exploited.

African states would be wise to act counter intuitively by focusing on outer space research and applications in these areas in order to leap frog aeons of development gap that currently separates it from the more developed states in the world. Africa must resist the temptations of ignoring or abandoning investment in space aspirations and space applications for reasons of scarcity of funds. In a sense in this century all states must realise that truly salvation comes from above. This is why the UN was correct in advancing the principle: "All States should be encouraged to contribute to the United Nations Programme on Space Applications and to other initiatives in the field of international cooperation in accordance with their space capabilities and their participation in the exploration and use of outer space", Declaration on International Cooperation in the Exploration and Use of Outer Space for the Benefit and in the Interest of All States, Taking into Particular Account the Needs of Developing Countries [16].

The increasing severity of natural disasters demands the utilisation of space applications in ways that were previously thought impossible. In 2008 for instance, there were historic levels of natural catastrophes. Chinese victims of the earthquake in the province of Sichuan, the cyclone in Myanmar, and several other natural catastrophes in the first part of the year alone claimed the lives of more than 150,000 people. The previous year had recorded 960 disasters and events which was the largest number in recorded history [147] Note should also be made of the planned framework of the United Nations Platform for Space-based information for Disaster Management and Emergency Response (A/AC.105/929). The COPOUS endorsed the work plan of the United Nations Platform for Space-based Information for Disaster Management and Emergency Response (UN-SPIDER) for the biennium 2010–2011 (A/AC.105/937) [148]. Assistance to African states in the areas under review can also be a valid instrument of economic diplomacy. Finally it may be suggested that Africa's development partners ought to explore greater use of economic aid in the form of cooperation in space activities with African states.

Apart from the sheer value of the service renderable to the country through investment in space technology, which will directly and indirectly affect the growth of GDP in African states there are also tariff and other beneficial fees to be made by national governments. The few African states that have invested in this area already establish this point. In Tunisia, the National Center for Cartography and remote sensing approves expenses relating to the realization of the geographical information systems for the benefit of the State, local authorities and institutions and public enterprises for amount fixed at dinars rates for each operation excluding taxes. Under Art. 7 of the Minister of National Defense Order on the National Remote Sensing Center Fees—The National Center for Cartography and remote sensing approves the technical compliance of equipment and materials and charges in part 10% of the amount of equipment excluding taxes presented for approval, for each approval operation. See the Journal Official of the Republic of Tunisia. Tunis, 22 July 2016. Although Nigeria also has modern and elaborate commercial arrangements for use of its satellite facilities by private companies, it is the opinion of this paper that the clear transparency in relation to charges displayed in the pertinent

Tunisian domestic legislation for remote sensing services is missing in the Nigerian NASRDA Act and allied legislation. Although Nigeria also has modern and elaborate commercial legal arrangements for use of its satellite facilities by commercial ventures, the transparency in cost of services found in the Tunisian domestic legislations missing in the Nigerian NASRDA Act. Such as Decrees No 75–671 of 25 December 1975; Decree No 2006–1902 of 10 July 2006 and particularly the Minister of National Defense Order on the National Remote Sensing Center Fees is missing in other African states including those with comparable capacities. See Order of the Minister of National Defense of 22 July 2016, fixing the rates of the services of the National Centre of Cartography and remote sensing. Despite the manifest usefulness of space technology to Africa and its peoples as well as the apparent willingness to exploit these technologies, domestic legal regulation of space related activities in Africa is at this stage rudimentary. Even those African states with regulation in this area need to update their laws to make them fit for purpose in spurring technological and entrepreneurial innovation. What can or must not be done with data legitimately collected from space platforms? What confidentiality rules bind external operators in relation to remote sensing and geoscientific round investigations data? These are just a few of the many questions left unanswered. The trend currently in the few available African domestic legislation is to presume that astronomical observations and satellite operators will always be buying data from governmental sources (e.g., Tunisia and Nigeria -NIGCOMSAT) [149]. Available regulations appear unprepared for privately launched satellites from within the state. With at least one African state –Nigeria planning manned missions into space, regulations will soon be needed on use of planetary resources during space missions. Yet it is South Africa only that makes provisions on personal jurisdiction in space. This is a reflection of the haphazard nature of legislation and the fact that national legislation are not yet expertly aligned with national policies and perhaps even national interests.

This work is in agreement with the groundswell of opinion that Africa needs an African Space Policy Institute, which will among other things champion the cause for capacity building in space policy, promote an African Space Policy and Strategy and conduct appropriate policy research whilst educating decision-makers and various stakeholders in the promotion of Africa's interest [21]. There is a strong requirement for regulation and acquisition of institutional capacity for conducting geospatial operations and investigations for development in Africa. The imperatives are clear -it is important to develop and maintain agencies that have oversight and regulate aspects of outer space exploration and its new applications. The option of not having any domestic regulations at all whilst having signed up liberally to the space treaties is not only insufficient but can be dangerous to the state's legal position as there are important duties to the international community to perform.

Funding: This research received no external funding.

Conflicts of Interest: The author declares no conflict of interest.

References

1. United Nations Office for Outer Space Affairs, "Space Law". Available online: http://www.unoosa.org/oosa/en/ourwork/spacelaw/index.html (accessed on 25 February 2019).
2. Von der Dunk, F.; Tronchetti, F. *The Handbook of Space Law*; Edward Elgar: Cheltenham, UK, 2015; p. xxvi.
3. Materials and Information about the ITU. Available online: https://www.itu.int/en/Pages/default.aspx (accessed on 25 February 2019).
4. The Committee on the Peaceful Uses of Outer Space (COPUOS). Available online: http://www.unoosa.org/oosa/en/ourwork/copuos/index.html (accessed on 25 February 2019).
5. The Treaty Banning Nuclear Weapon Tests in the Atmosphere, in Outer Space and Under Water. The Nuclear Test Ban Treaty, UNTS, 1963; Volume 480, p. 177. Available online: https://www.britannica.com/event/Nuclear-Test-Ban-Treaty (accessed on 25 February 2019).
6. Oduntan, G. *Sovereignty and Jurisdiction in Airspace and Outer Space: Legal Criteria for Spatial Delimitation*; Routledge: London, UK, 2011; p. 180.

7. The Space Treaty. 18 UST 2410, 610 UNTS 205. Available online: http://www.eisil.org/index.php?sid=6597 70401&id=838&t=link_details&cat=637 (accessed on 25 February 2019).

8. The Astronaut Agreement or Rescue Agreement. Available online: http://www.unoosa.org/oosa/en/our work/spacelaw/treaties/introrescueagreement.html (accessed on 25 February 2019).

9. The Liability Convention. Available online: http://www.unoosa.org/oosa/en/ourwork/spacelaw/treaties /introliability-convention.html (accessed on 25 February 2019).

10. The Registration Convention. Available online: http://www.unoosa.org/oosa/en/ourwork/spacelaw/trea ties/introregistration-convention.html (accessed on 25 February 2019).

11. The Moon Treaty or Moon Agreement. Available online: http://www.unoosa.org/oosa/en/ourwork/space law/treaties/intromoon-agreement.html (accessed on 25 February 2019).

12. The Declaration of Legal Principles. *The Nuclear Test Ban Treaty*. 1963, Volume 480, p. 45. Available online: https://www.britannica.com/event/Nuclear-Test-Ban-Treaty (accessed on 25 February 2019).

13. The Broadcasting Principles, General Assembly Resolution 37/92 of 10 December 1982. Available online: http://www.un.org/documents/ga/res/37/a37r092.htm (accessed on 25 February 2019).

14. The Remote Sensing Principles, General Assembly Resolution 41/65 of 3 December 1986. Available online: http://www.un.org/documents/ga/res/41/a41r065.htm (accessed on 25 February 2019).

15. The Nuclear Power Sources. Principles General Assembly resolution 47/68 of 14 December 1992. A/RES/47/68 85th Plenary Meeting. Available online: http://www.un.org/documents/ga/res/47/a 47r068.htm (accessed on 14 December 1992).

16. The Benefits Declaration. General Assembly Resolution 51/122 of 13 December 1996. Available online: http://www.un.org/documents/ga/res/51/a51r122.htm (accessed on 13 December 1996.).

17. Dempsey, P. National Legislation Governing Commercial Space Activities. *J. Space Eng.* **2014**, *1*, 45, 55. [CrossRef]

18. Minéapolis. 1988. Available online: http://search.itu.int/history/HistoryDigitalCollectionDocLibrary/4. 16.43.en.100.pdf (accessed on 25 February 2019).

19. The Ratification of the Treaty on the Principles Governing the Activities of States in the Exploration and Use of Outer Space, Including the Moon and Other Celestial Bodies. Available online: https://www.state.gov/t/ isn/5181.htm (accessed on 25 February 2019).

20. The Convention on International Liability for Damage Caused by Space Objects. Available online: https: //treaties.un.org/pages/showdetails.aspx?objid=08000002801098c7 (accessed on 25 February 2019).

21. Offiong, E.; Munsami, V. Towards a Space Policy Institute for Africa. *Space Pol.* **2018**, *46*, 23. [CrossRef]

22. James, G.K.; Joseph, A.; Shaba, A.H. The Nigerian Space Program and Its Economic Development Model. *New Space* **2014**, *2*, 23–29. [CrossRef]

23. Munsami, V. South Africa's national space policy: The dawn of a new space era. *Space Pol.* **2014**, *30*, 115–120. [CrossRef]

24. Assembly of the African Union, Decisions, Declarations and Resolution Assembly/AU/Dec.589(XXVI), Twenty-Sixth Ordinary Session, 30–31 January 2016, Addis Ababa, Ethiopia. Available online: https://au.int /sites/default/files/decisions/29514-assembly_au_dec_588_-_604_xxvi_e.pdf (accessed on 11 April 2018).

25. Reibaldi, G.G. The importance of Space Policy Teaching in Communicating Space Activities to Society. *Acta Astronaut* **2003**, *53*, 998. [CrossRef]

26. Information and Materials on SANSA. Available online: http://www.iafastro.org/societes/centre-national- de-la-cartographie-et-de-la-teledetection-cnct/ (accessed on 15 December 2018).

27. Cheng, B. The Extra-Territorial Application of International Law. *Curr. Leg. Probl.* **1965**, *18*, 135.

28. Space Legal Tech, Explore National Space Legislation. Available online: http://spacelegaltech.chaire-sirius .eu/use-the-map/# (accessed on 18 December 2018).

29. NASRDA Act. Available online: https://nass.gov.ng/document/download/5892/ (accessed on 11 September 2018).

30. Committee on the Peaceful Uses of Outer Space, Questions on the definition and delimitation of outer space: Replies from Member States Note by the Secretariat. Available online: http://www.unoosa.org/pdf/reports /ac105/AC105_889Add5E.pdf (accessed on 18 June 2010).

31. NASRDA, Space Missions. Available online: http://nasrda.gov.ng/en/missions/ (accessed on 11 September 2018).

32. Julius, V. *Border Is Fate: A Study of Mid-European Diffused Ethnic Minorities*; Australian Carpathian Federation Inc.: Sydney, Australia, 1982.

33. Yoon, M. European Colonialism and Territorial Disputes in Africa: The Gulf of Guinea and the Indian Ocean. *Mediterr. Q.* **2009**, *20*, 77–94. [CrossRef]

34. Macdonald, A. *Mapping the World: A History of the Directorate of Overseas Surveys 1946–1985*; HMSO: London, UK, 1996; p. 8.

35. Ahmed, M.B. Cross Border Governance and Security Coordination: Solution to the North East Crisis. *West Afr. Insight* **2014**, *4*, 10.

36. Memorial of Libya in the Territorial Dispute (Libyan Arab Jamahiriya/Chad), Vol.1, ICJ Rep 25, para.3.01, quoted from The Times, 7 August 1860, Quoted in 1994 *ICJ Reports* 6, 53. Available online: https://www.icj-cij.org/files/case-related/83/6691.pdf (accessed on 25 February 2019).

37. Frontier Dispute, Burkina Faso/Republic of Mali. I.C.J. Reports. 1986, p. 582. Available online: https://www.icj-cij.org/en/case/69 (accessed on 25 February 2019).

38. Brownlie, I. *African Boundaries: A Legal and Diplomatic Encyclopedia*; University of California Press: Berkeley, CA, USA, 1979; p. 161.

39. Strassoldo, R. The State of the Arts in Europe. In *Borderlands in Africa: A Multidisciplinary and Comparative Focus on Nigeria and West Africa*; Asiwaju, P.O., Adeniyi, A.I., Eds.; University of Lagos Press: Lagos, Nigeria, 1989; p. 359.

40. The AU Mission in Sudan Ceasefire Commission, "CFC Ceasefire Violation Report No. 37/05: Alleged SLA Attack on Nertiti on 16 Feb 05" AMIS/CFC/G/VIO/1 March 2005. pp. 1–2. Available online: http://www.africa-union.org/DARFUR/Reports%20of%20the%20cfc/37-05.pdf (accessed on 9 December 2008).

41. Samuel, M.; Makinda, F.; Wafula, O. *The African Union: Challenges of Globalization, Security and Governance*; Routledge: Oxon, UK, 2008; pp. 75–76.

42. King, R.; Vujakovic, P. Peters Atlas: A new era of cartography or publisher's con-trick? *Geography* **1989**, *74*, 247–254.

43. Vujakovic, P. Mapping for world development. *Geography* **1989**, *74*, 97–105.

44. Vujakovic, P. Arno Peters' cult of the 'New Cartography': From concept to world atlas. *Bull. Soc. Univ. Cartogr.* **1989**, *22*, 1–6.

45. Okumu, W. Border Security in Africa. In Proceedings of the African Union Border Programme Regional Workshop, Windhoek, Namibia, 22–23 October 2009; p. 3.

46. Pierre Engel, "Assessment of African policies for the use of satellite remote sensing for development purposes", African Monitoring of the Environment for Sustainable Development Report. 2012, pp. 3, 37–39. Available online: https://www.academia.edu/12006735/Assessment_of_African_policies_for_the_use_o f_satellite_remote_sensing_for_development_purposes. (accessed on 25 February 2019).

47. The Meeting "Preventing Conflicts, Promoting Integration" Preparatory Meeting of Experts on the African Union Border Programme Addis Ababa. Available online: http://www.peaceau.org/uploads/report-bam ako-en.pdf (accessed on 26 February 2019).

48. Land and Maritime Boundary between Cameroon and Nigeria. (Cameroon v. Nigeria: Equatorial Guinea Intervening), Judgment, Preliminary Objections. Available online: www.worllii.org/int/cases/ICJ/1998/2 .html (accessed on 26 February 2019).

49. Request for Interpretation of the Judgment of 11 June 1998 in the *Case Concerning the Land and Maritime Boundary between Cameroon and Nigeria* (Cameroon v. Nigeria), Preliminary Objections (Nigeria v. Cameroon). Available online: www.icj-cij.org/icjwww/idocket/icn/icnjudgment/icn_ijudgment_19990325_frame.htm (accessed on 26 February 2019).

50. Land and Maritime Boundary between Cameroon and Nigeria (Cameroon v. Nigeria: Equatorial Guinea Intervening), Judgment, Merits. 10 October 2002. Available online: www.icj-cij.org/icjwww/idocket/icnju dgment/ (accessed on 29 February 2019).

51. Maritime Delimitation and Territorial Questions between Qatar and Bahrain (Qatar v. Bahrain). Judgment of 16 March 2001. Available online: https://www.icj-cij.org/en/case/87 (accessed on 26 February 2019).

52. LaGrand (Germany v. United States of America). Judgment of 27 June 2001. Available online: https://www.icj-cij.org/en/case/104/judgments (accessed on 26 February 2019).

53. Arrest Warrant of 11 April 2000 (Democratic Republic of the Congo v. Belgium). Judgment of 14 February 2002. Available online: https://www.icj-cij.org/en/case/121/judgments (accessed on 26 February 2019).

54. *UN Handbook on the Delimitation of Maritime Boundaries*; United Nations Publications: New York, NY, USA, 2000; Available online: https://www.un-ilibrary.org/international-law-and-justice/handbook-on-the-de limitation-of-maritime-boundaries_cc72cd88-en (accessed on 26 February 2019).

55. Churchill; Lowe, A.V. *The Law of the Sea*; University Press: Manchester, UK, 1988; p. 27.

56. Chijioke, E. An Appraisal of the Challenges of Resource Exploration and Exploitation for Socio-economic Developments in Nigeria. p. 9. Available online: https://www.fig.net/resources/proceedings/fig_proceed ings/fig2011/papers/ts01d/ts01d_eze_5219.pdf (accessed on 19 December 2018).

57. Alit, T.; Isa, M. Post–Conflict Demarcation of African Boundaries: The Cameroon Nigeria Experience' in African Union Commission. In *Delimitation and Demarcation of Boundaries in Africa: General Issues and Case Studies*; African Union: Addis Ababa, Ethiopia, 2013; p. 54.

58. The Mixed Commission approved the specifications for all the contracts for the demarcation of the boundary and requested that certain key contracts should be announced on the U.N website. In Proceedings of the Sixth Meeting of the Sub-Commission on Demarcation, Abuja, Nigeria, 24–27 October 2003.

59. Cameroon Nigeria Mixed Commission. In Proceedings of the Sixth Meeting of the Cameroon-Nigeria Mixed Commission, Abuja, Nigeria, 29–30 October 2003.

60. Biger, G. *The Encyclopedia of International Boundaries*; Fact on Files: New York, NY, USA, 1995.

61. Jet Propulsion Laboratory, Instrument: The Shuttle Radar Topography Mission. Available online: http://www2.jpl.nasa.gov/srtm/instr.htm (accessed on 21 July 2018).

62. Annex V of the Report of the Tenth meeting of the Sub-Commission on Demarcation. Available online: http://www.unesco.org/new/fileadmin/MULTIMEDIA/HQ/CI/CI/pdf/mow/10th%20meetin g%20subcommittee%20Tech%20Alexandrina%20Egypt.pdf (accessed on 25 February 2019).

63. UN Cartographic Section. *Summary of Preliminary Accuracy Validation of Satellite Image Maps for the Cameroon-Nigeria Sub-Commission on Demarcation, Annex II*; UN Cartographic Section: New York, NY, USA, 2004.

64. Oppenheim, L. *Oppenheim's International Law*; Watts, R.J., Ed.; Longman: London, UK, 1992; p. 661.

65. Shawt, M.N. The Heritage of States: The Principle of Uti Possidetis Juris Today. *Br. Yearb. Int. Law* **1996**, *67*, 77–78, 81–85. [CrossRef]

66. The Website of UNOOSA, "World Geodetic System 1984". Available online: www.unoosa.org/pdf/icg/201 2/template/WGS_84.pdf (accessed on 26 February 2019).

67. Cameroon-Nigeria Mixed Commission. In Proceedings of the Fifth Meeting of the Working Group on the Maritime Boundary Report Yaoundé, Yaoundé, Cameroon, 9–10 October 2006.

68. Levitt, J.I. *Africa: Mapping New Boundaries in International Law*; Hart Publishing: London, UK, 2010; pp. 532–538.

69. Website of Satellite Discoveries Eye in the Sky. Ancient Peru Pyramid Spotted by Satellite. 23 June 2012. Available online: http://satellitediscoveries.typepad.com/ (accessed on 11 September 2018).

70. Frances Cronin. Egyptian Pyramids Found by Infra-Red Satellite Images. *BBC News*. Available online: www.bbc.co.uk/news/world-13522957 (accessed on 25 May 2011).

71. Rossella Lorenzi, Long-Lost Pyramids Found? Available online: http://news.discovery.com/history/archa eology/long-lost-pyramids-found-130715.htm (accessed on 11 September 2018).

72. Menze, B.H.; Ur, J.A. Mapping Patterns of Long-Term Settlement in Northern Mesopotamia at a Large Scale: Proceedings of the National Academy of Sciences. 2012, Volume 109. Available online: https://www.pnas.org/content/109/14/E778 (accessed on 11 September 2018).

73. Oduntan, G.T. Legal and Evidential Implications of Emerging Satellite Imagery of Ancient African Relict Boundaries. *Chin. J. Int. Law* **2017**, *16*, 77–108. [CrossRef]

74. Fouseki, K.; Sandes, C. Private Preservation versus Public Presentation: The Conservation for Display of in Situ Fragmentary Archaeological Remains in London and Athens. *Pap. Inst. Archaeol.* **2009**, *19*, 37–54. [CrossRef]

75. Riegl, A. The Modern Cult of Monuments: Its essence and its development. In *Historical and Philosophical Issues in the Conservation of Cultural Heritage*; Stanley Price, N., Kirby Talley, M., Mellucco Vaccaro, A., Jr., Eds.; The Getty Conservation Institute: Los Angeles, CA, USA, 1996; pp. 69–83.

76. Sandes, C.A. The Conservation and Presentation of 'Fragmentary' Archaeological Sites in Modern Urban Contexts: Post-War Redevelopment in London, Berlin and Beirut. Unpublished Ph.D. Thesis, University of London, London, UK, 2007. Smith, L., The Uses of Heritage, London, Routledge 2006.

77. Bodin, J. *On Sovereignty: Four Chapters from the Six Books of the Commonwealth*; Franklin, J.H., Ed.; Cambridge University Press: New York, NY, USA, 1992.

78. Oduntan, G. *International Law and Boundary Disputes in Africa*; Routledge: New York, NY, USA, 2015; pp. 23–24.

79. International Court of Justice (ICJ) Report. 1962, p. 14. Available online: https://www.icj-cij.org/files/case-related/49/049-19620720-ADV-01-00-EN.pdf (accessed on 25 February 2019).

80. Bell, T. Thailand Steps Back from Cambodia Conflict. *Telegraph*. 6 January 2010. Available online: http://www.telegraph.co.uk/news/worldnews/asia/cambodia/3195213/Thailand-steps-back-from-Cambodia-conflict.html (accessed on 6 January 2010).

81. Parry, R.L. Thailand and Cambodia Teeter on Edge of Conflict at Cliff-top Temple. *Times*. 19 July 2008. Available online: http://www.timesonline.co.uk/tol/news/world/asia/article4360257.ece (accessed on 6 January 2010).

82. Hartshorne, R. Suggestions on the Terminology of Political Boundaries. In *26 Annals of the Association of American Geographers*; Taylor & Francis: Abingdon, UK, 1936; pp. 56–57.

83. Newman, D. Boundaries, borders and barriers: Changing geographic perspectives on territorial lines. In *Identities, Borders, Orders: New Directions in International Relations Theory*; Albert, M., Jacobson, D., Lapid, Y., Eds.; University of Minnesota Press: Minneapolis, MN, USA, 2001; p. 140.

84. Sobczynski, M. Studies on Relict Boundaries and Border Landscape in Poland, Università degli Studi di Trento. 2006, Volume 1, p. 3. Available online: http://web.unitn.it/archive/events/borderscapes/download/abstract/SOBCZYNSKI_paper.pdf (accessed on 26 February 2019).

85. National Oceanic and Atmospheric Administration. What Is LIDAR? Available online: https://oceanservice.noaa.gov/facts/lidar.html (accessed on 28 September 2018).

86. Stone, P.G. *Cultural Heritage, Ethics and the Military*; The Boydell Press: Woodridge, IL, USA, 2011.

87. African Legacy—School of Conservation Sciences, Bournemouth University. Hausaland Walled Cities & Towns—Remote Sensing Studies. 2013, p. 158. Available online: http://apollo5.bournemouth.ac.uk/africanlegacy/kano_walls.htm (accessed on 17 March 2019).

88. Ayandele, E.A. Ijebuland 1800–1891: Era of splendid isolation. In *Studies in Yoruba History and Cultures: Essays in Honour of Professor. S.O. Biobaku*; Olusanya, G.O., Ed.; University Press: Ibadan, Nigeria, 1983; pp. 88–105.

89. Darling, P. Benin Earthworks: Some Cross-Profiles. *Nigerian Field* **1975**, *44*, 164–165.

90. Darling, P.J. Sungbo's Eredo, Southern Nigeria. *Nyame Akuma* **1998**, *48*, 55–61.

91. Sandomirsky, L.N. Benin, Empire: Oba Ewuare, Trade with the Portuguese. In *Encyclopedia of African History*; Woodward, D., Lewis, G.M., Eds.; Routledge: Abingdon, UK, 2004; p. 133.

92. Poliakov, V.L. *The Aryan Myth: A History of Racist and Nationalist Ideas in Europe*; Barnes & Noble Books: New York, NY, USA, 1974; p. 241.

93. Maggs, T. Cartographic Content of Rock Art in Southern Africa. *Cartography in the Traditional African, American, Arctic, Australian, and Pacific Societies*. Woodward, D., Lewis, G.M., Eds.; 1998, Volume Two. Available online: www.press.uchicago.edu/books/HOC/HOC_V2_B3/Volume2_Book3.html (accessed on 25 February 2019).

94. Adler, F. "Karty pervobytnykh narodov" (Maps of primitive peoples), Izvestiya Imperatorskago Obshchestva Lyubiteley Yestestvoznanya, Antropologii i Etnografii: Trudy Geograficheskago Otdeliniya (Proceedings of the Imperial Society of the Devotees of National Sciences, Anthropology and Ethnology: Transactions of the Division of Geography). 1910. no. 2. pp. 119, 177. Available online: www.worldcat.org/title/karty-pervobytnykh-narodov/oclc/37279291 (accessed on 25 February 2019).

95. Smith, A.B. Metaphors of Space: Rock Art and Territoriality in Southern Africa. In *Contested Images: Diversity in Southern African Rock Art Research*; Dowson, T.A., Lewis-Williams, D., Eds.; Wits University Press: Johannesburg, South Africa, 1994; pp. 373–384.

96. Photograph Courtesy of the Museum of Ethnography, Geneva (Gift of Jean Rusillon, 1966; no. 33553) in: David Woodward and G. Malcolm Lewis (eds.). Available online: https://www.press.uchicago.edu/.../HOC_VOLUME2_Book3_gallery1.pdf (accessed on 25 February 2019).

97. The American Association for the Advancement of Science. High-Resolution Satellite Imagery Ordering and Analysis Handbook. 2016. Available online: www.aaas.org/page/high-resolution-satellite-imagery-ordering-and-analysis-handbook (accessed on 26 February 2019).

98. Evans, A.E. Judicial Decisions. *70 American JIL*. 1976, p. 367. Available online: https://heinonline.org/HOL/LandingPage?handle=hein.journals/ajil72&div=12&id=&page= (accessed on 25 February 2019).

99. Presidential Decree No. 02-49 "Creation, organization and functioning of the Algerian Space Agency (ASAL)". Available online: http://www.asal.dz/mission.php (accessed on 16 January 2002).

100. Adam, D. Ancient Villa Discovered Thanks to Internet Maps. *The Guardian* **2005**, *21*, 19.

101. Eisner, P. Who Discovered Machu Picchu? Controversy Swirls as to Whether an Archaeologist's Claim to Fame as the Discoverer of Machu Picchu Has Any Merit. Available online: https://www.smithsonianmag.com/history/who-discovered-machu-picchu-52654657/#MrpVMg8bDHr7GGij.99 (accessed on 30 September 2018).

102. McIntosh, M.M. Exploring Machu Picchu: An Analysis of the Legal and Ethical Issues Surrounding the Repatriation of Cultural Property; 17 Duke J. Comp. & Int'l L. 199 2006–2007. pp. 199–201. Available online: https://scholarship.law.duke.edu/djcil/vol17/iss1/6 (accessed on 26 February 2019).

103. Adebiyi, B. *Legal and Other Issues in Repatriating Nigeria's Looted Artefacts*; Lulu: Lagos, Nigeria, 2009; pp. 15–24.

104. Hadjimitsis, D.; Agapiou, A.; Alexakis, D.; Sarris, A. Exploring natural and anthropogenic risk for cultural heritage in Cyprus using remote sensing and GIS. *Int. J. Digit. Earth* **2013**, *6*, 115–142. [CrossRef]

105. Biagetti, S.; Merlo, S.; Adam, E.; Lobo, A.; Conesa, F.C.; Knight, J.; Bekrani, H.; Crema, E.R.; Alcaina-Mateos, J.; Madella, M. High and Medium Resolution Satellite Imagery to Evaluate Late Holocene Human–Environment Interactions in Arid Lands: A Case Study from the Central Sahara. *Remote Sens.* **2017**, *9*, 351. [CrossRef]

106. Cheng, B. *Studies in International Space Law*; Claredon Press: Oxford, UK, 2004; pp. 573, 577–578.

107. Keppel v. Wheeler 1927 1KB 557. Available online: https://swarb.co.uk/keppel-v-wheeler-ca-1927/ (accessed on 25 February 2019).

108. Chaudry v.Prabhakar [1989] 1 WLR 29 (on reasonable diligence). Available online: https://www.lawteacher.net/cases/chaudhry-v-prabhakar.php (accessed on 25 February 2019).

109. Boardman v. Phipps [1966] UKHL 2 (no secret profits). Available online: http://www.trusts.it/admincp/UploadedPDF/200801171738350.jEnghouseBoardmanPhipps19661103.pdf (accessed on 25 February 2019).

110. Initial Services v Putterill [1968] 1 Q.B. 396. Available online: https://swarb.co.uk/initial-services-ltd-v-putterill-ca-1967/ (accessed on 25 February 2019).

111. Gartside v Outram (1856) 26 L.J. Ch. 113 (Duty Not to Divulge Confidential Information). Available online: https://swarb.co.uk/gartside-v-outram-1856/ (accessed on 25 February 2019).

112. Yasuda Fire & Marine Insurance Co of Europe v. Orion Marine Insurance Underwriting Agency 1995 2 Q.B. 174 (duty to account to principal). Available online: https://swarb.co.uk/yasuda-fire-and-marine-insurance-company-europe-ltd-v-orion-marine-insurance-underwriters-ltd-chd-27-oct-1994/ (accessed on 25 February 2019).

113. The spatial information standards and prescriptions provisions of the Spatial Data Infrastructure Act, No. 54 2003 *Government Gazette* Vol. 464 Cape Town 4 February 2004 No. 25973. Particularly S. 11-14 and 18. Available online: http://ggim.un.org/knowledgebase/Attachment234.aspx?AttachmentType=1 (accessed on 4 February 2004).

114. Cheng, B. The United Nations and Outer Space. Available online: https://academic.oup.com/clp/article-abstract/14/1/247/422457 (accessed on 1 December 1961).

115. Cowton, R. Satellite Brings Soviet Nuclear Carrier into Focus. *The Times* **1984**, *8*, 26.

116. Vujakovic, P. Soviet remote sensing: Commercial blueprint for planet management. *Geogr. Mag.* **1990**, *62*, 46–48.

117. Oduntan, G. The Never Ending Dispute: Legal Theories on the Spatial Demarcation Boundary Plane between Airspace and Outer Space. *Herts. Law J.* **2003**, *1*, 64–84.

118. Hoversten, M.R. U.S. National Security and Government Regulation of Commercial Remote Sensing from Outer Space. 50 *A.F. L. Rev.* 253 2001. pp. 262–263. Available online: https://heinonline.org/HOL/LandingPage?handle=hein.journals/airfor50&div=9&id=&page= (accessed on 26 February 2019).

119. Cheng op.cit (1965). pp. 134–136. Available online: https://academic.oup.com/clp/article-abstract/18/1/132/569648 (accessed on 25 February 2019).

120. Trail Smelter Arbitration (1935). Available online: legal.un.org/riaa/cases/vol_III/1905-1982.pdf (accessed on 25 February 2019).

121. Corfu Channel case (Merits) (1949). Available online: https://www.icj-cij.org/files/case-related/1/001-19 491215-JUD-01-00-EN.pdf (accessed on 25 February 2019).

122. Fisheries jurisdiction case (1974). Available online: https://www.icj-cij.org/files/case-related/56/056-19740 725-JUD-01-00-EN.pdf (accessed on 25 February 2019).

123. U. N-SPIDER. Space Applications. Available online: http://www.un-spider.org/space-application (accessed on 30 August 2018).

124. UN, Office for Outer Space Affairs, United Nations Platform for Space-Based Information for Disaster. Management and Emergency Response UN-SPIDER. pp. 5–7. Available online: http://www.unoosa.org/p df/publications/st_space_65E.pdf (accessed on 8 September 2018).

125. Chole, M.M.; Farago, M.E.; Mehra, A.; Vujakovic, P. Remote sensing imagery for the detection of stress in vegetation caused by drought and/or metal toxicity. In Proceedings of the 20th International Symposium on Remote Sensing of the Environment, Nairobi, Kenya, 4–10 December 1986; University of Michigan: Ann Arbor, MI, USA, 1987; pp. 889–898.

126. Vujakovic, P. The extent of the adoption of the Peters projection by 'Third World' organisations in the UK. *Bull. Soc. Univ. Cartogr.* **1987**, *21*, 11–16.

127. Vujakovic, P. Monitoring extensive 'buffer zones' in Africa: An application for satellite imagery. *Biol. Conserv.* **1987**, *39*, 195–208. [CrossRef]

128. Vujakovic, P. Rangeland management in Northern Botswana: The role of remote sensing in resource evaluation. In *African Resources (Vol. 1): Appraisal and Monitoring*; Millington, A.C., Mutiso, S.K., Binns, J.A., Eds.; Geographical Papers; University of Reading: Reading, UK, 1987; Volume 96, pp. 27–34.

129. Useful Space Application Matrix on the UNI-SPIDER Webpage. Available online: http://www.un-spider.or g/space-application-matrix (accessed on 31 August 2018).

130. Solberg, A.H.S. Oil Spill Detection by Satellite Remote Sensing. *Remote Sens. Environ.* **2005**, *95*, 1–13.

131. Brown, J.; Jenkerson, C.; Gu, Y. Using eMODIS Vegetation Indices for Operational Drought Monitoring. Available online: https://www.researchgate.net/profile/Jesslyn_Brown/publication/265190613_Using_ eMODIS_Vegetation_Indices_for_Operational_Drought_Monitoring/links/547f95380cf2ccc7f8b9716d.pdf (accessed on 8 February 2008).

132. Davies, A.G.; Tran, D.Q.; Mandrake, L.; Boudreau, K.; Cecava, J.; Vargas, A.M.; Behar, A.; Chien, S.; Castaño, R.; Frye, S.; et al. The Model-Based Volcano Sensor Web: Progress in 2007. Proceedings of the NASA Earth Science Technology Conference 2008, Paper A7P3. Available online: http://esto.nasa.gov/conf erences/estc2008/papers/Davies_Ashley_A7P3.pdf (accessed on 26 February 2019).

133. Hong, Y.; Adler, R.F.; Negri, A.; Huffman, G.J. Flood and Landslide Applications of Near Real-Time Satellite Rainfall Products. *Nat. Hazards* **2006**, *43*, 285–294. [CrossRef]

134. These Include MSG. Available online: http://oiswww.eumetsat.org/IPPS/html/MSG/RGB/DUST/ (accessed on 26 February 2019).

135. LIDAR Networks- France LIDAR Network (webpage available early 2016)—Germany DWD Ceilomap. Available online: http://www.dwd.de/ceilomap (accessed on 26 February 2019).

136. United Kingdom MO LIDARNET. Available online: http://www.metoffice.gov.uk/public/lidarnet/lcbr-ne twork.html (accessed on 30 October 2018).

137. EARLINET. Available online: http://earlinet.org/TOPROF - http://www.cost.eu/COST_Actions/essem/A ctions/ES1303 (accessed on 30 October 2018).

138. E-PROFILE. Available online: http://eumetnet.eu/e-profile (accessed on 12 October 2016).

139. EUR/NAT Office of ICAO, "Volcanic Ash Contingency Plan European and North Atlantic Regions" EUR Doc 019, Edition 2.0.0, NAT Doc 006, Part II. July 2016, p. 36. Available online: https://www.icao.int/EUR NAT/EUR%20and%20NAT%20Documents/EUR+NAT%20VACP.pdf (accessed on 8 September 2018).

140. Brannigan, V. Alice's Adventures in Volcano Land: The Use and Abuse of Expert Knowledge in Safety Regulation. *Eur. J. Risk Regul.* **2010**, *1*, 107–113. [CrossRef]

141. See "GMES info, "Space for a safer world: Reinforcing the European Emergency Response Capacity" No. 6. *Research EU*. June 2010, p. 18. Available online: https://ec.europa.eu/docsroom/documents/8646/attachm ents/1/translations/en/renditions/native (accessed on 25 February 2019).

142. Tim Jacobs and VGT4AFRICA Partners: VGT4 Distribution of Vegetation Data in Africa through EUMETCast, Final Publishable Activity Report. Available online: http://www.vgt4africa.org (accessed on 12 June 2010).

143. UN COPOUS Report of the Committee on the Peaceful Uses of Outer Space. 2009, pp. 16–17. Available online: www.unoosa.org/pdf/gadocs/A_55_20E.pdf (accessed on 25 February 2009).

144. Space Tourism's Toll on Climate. *Newscientist*. 2010. Available online: https://www.newscientist.com/artic le/dn19626-space-tourism-could-have-big-impact-on-climate/ (accessed on 26 October 2010).

145. Business Insider. Available online: https://www.businessinsider.com/ap-report-imagery-shows-north-ko rea-razes-missile-test-stand-2018-6?IR=T (accessed on 24 July 2018).

146. Washington (Reuters, AFP). Images Indicate North Korea Dismantling Facilities at Test Site: Report. *Strait Times*. Available online: https://www.straitstimes.com/asia/images-indicate-north-korea-dismantling-fa cilities-at-test-site-report (accessed on 24 July 2018).

147. Gomez, C. SAFER and GMES Emergency Response Core Service: One Step Closer to Full-Scale Operational Deployment. Available online: www.kpk.gov.pl/ (accessed on 10 June 2010).

148. United Nations. *Report of the Committee on the Peaceful Uses of Outer Space, General Assembly*; Official Records Sixty-fourth Session Supplement No. 20; United Nations: New York, NY, USA, 2009; Paras. 100–104, 117–125; p. 18.

149. NIGCOMSAT Info and Materials. Available online: http://www.nigcomsat.gov.ng/what-we-do.php (accessed on 13 December 2018).

geosciences

MDPI

Article

Implementing a Modern E-Learning Strategy in an Interdisciplinary Environment—Empowering UNESCO Stakeholders to Use Earth Observation

Tobias Matusch [1,*], Anne Schneibel [1], Lisa Dannwolf [1] and Alexander Siegmund [2]

[1] Research Group for Earth Observation, Department of Geography, Heidelberg University of Education, Czernyring 22/11-12, 69115 Heidelberg, Germany; schneibel@ph-heidelberg.de (A.S.); dannwolf@ph-heidelberg.de (L.D.)

[2] Research Group for Earth Observation, Department of Geography, Heidelberg University of Education & University Heidelberg, Czernyring 22/11-12, 69115 Heidelberg, Germany; siegmund@ph-heidelberg.de

* Correspondence: matusch@ph-heidelberg.de; Tel.: +49-6221-477-772

Received: 31 October 2018; Accepted: 20 November 2018; Published: 23 November 2018

Abstract: The Copernicus Program and the fleet of available Earth observation satellites provide valuable services in sectors such as agriculture, forestry, urban monitoring, and heritage management. However, drawbacks such as knowledge gaps by the user, limited technical and financial facilities or the lack of ready-to-use data, result in insufficient exploitation of these opportunities by heritage site managers and other relevant stakeholders. Based on an initial assessment of current threats, existing limitations and potential applications, we developed the e-learning module Space2Place. Through the use of the learning module, stakeholders get a substantial introduction into Earth observation and knowledge barriers that may exist are removed. For this purpose, we refined an existing e-learning platform, which was developed in close relation to an online remote sensing application and adapted to the needs of UNESCO site stakeholders. One of the main features is the personalization of the learning modules content depending on the abilities or interests of the user. The platform offers information with different levels of difficulty and adaptable learning paths. A graduation certificate and practical exercises in an online remote sensing application increase the specific added value for UNESCO site manager. By using the associated remote sensing application and its link to Space2Place, heritage site managers also improve their knowledge on image processing by working with original satellite imagery. Additional advantages of using the platform will be enhanced through the introduction of new learning modules, translation into other languages and accompanying scientific research.

Keywords: e-learning; Earth observation; education; capacity development; cultural and natural heritage; UNESCO

1. Introduction

Education and learning approaches are changing dynamically in the course of time. Since the development of computer systems in the mid last century and through the spread of the internet, broadband access, and mobile phones, education systems changed tremendously [1]. Besides the challenges of a continuously changing educational landscape, e.g., with mobile learning and electronical learning application (hereafter e-learning), new approaches and technical opportunities offer a variety of new options for educational experts, teachers, and providers of education [2].

E-learning raised considerably in the past few years in quantity and quality [3]. The market is still booming with a compound annual growth rate of 7.5% up to a market volume of 275 Billion USD in 2022 [4]. Promoting factors can be easily identified, ranging from flexibility in learning, accessibility and comparatively low costs [5]. Distance learning opportunities became more and more popular, also as a part of a whole distance learning program [1]. Nowadays, massive open online courses (MOOCs), often even free of charge, are accessible for an increasing global education market. Prominent MOOCs raised the attention of several thousand learners simultaneously, distributed all over the globe [6]. Free and open-source learning platforms for different topics were also established during the past years, altering academic structures and revenue models of the educational sector [7]. The large variety of available platforms and courses creates an offer, which might appear inscrutable for the respective learner. This is additionally challenging due to the lack of information about the effectivity and impact of an individual course. Many offers from the existing major platforms such as edX, Coursera or Udacity can provide useful information, but are not specifically tailored to the needs of heritage site managers and other relevant stakeholders. By providing concise e-learning modules adapted to the specific needs and in close relation with practical exercises in a simple remote sensing application, gained knowledge can be easily integrated in daily working routines. Additionally, a direct feedback and communication loop to the providers of the e-learning environment can mitigate inherent weaknesses of e-learning programs such as technical restrictions or limited interaction between teachers and peer-to-peers, which may have a negative impact on the effectiveness of e-learning courses [5].

Similar to the development in the field of computer systems and the e-learning sector, a boom in the field of remote sensing has emerged during recent years [8,9]. Satellite systems including GPS satellites, meteorological and communication satellite as well as Earth observation satellites constantly increase their socio-economic benefit for societies [10]. Geoinformatics and Earth observation applications are more and more part of our everyday life. Applications such as maps or navigation systems are, often unconsciously, even known by the youngest generation [11]. Benefit from Earth observation can be found in almost all areas of society and professions, including sectors such as agriculture, forestry, insurance, renewable energies, and air quality [10]. In addition, UNESCO stakeholders already benefit from the current development [12]. The magnitude of such a benefit depends considerably on the expertise, abilities, and received training on satellite image assessment of each individual [13]. UNESCO stakeholders are responsible for the management of designated UNESCO sites, which fall into three categories, including World Heritage Sites, Global Geoparks, and Biosphere Reserves, respectively. Currently, there are 1092 World Heritage Sites in total, 140 Global Geoparks, and 686 Biosphere Reserves [14–16], with varying cultural and natural assets, which currently also benefit quantitatively and qualitatively in very different ways from Earth observation. The work of UNESCO sites stakeholders can for example be facilitated by the monitoring of land cover changes [17], the surveillance of structural and ground deformation [18,19] or the assessment of risks of natural and anthropogenic hazards [20–22].

However, Earth observation is still a field for experts and professionals with specific knowledge and experience [23]. This creates an entrance barrier for untrained students, the general public or pupils. Therefore, specialists and educational institutions are challenged to set-up entry points and lower the entry barriers. Many Earth observation specialists and users are not specifically experienced in computer sciences, but rather focused on natural and social sciences. Several initiatives were established in recent years, providing a hub for valuable information, serve as a starting point and provide opportunities to build professional networks. This includes initiatives e.g., organized by the European Space Agency (ESA) and associated with conferences on Earth observation (e.g., 3rd ESA-EARSeL course on remote sensing for archaeology) as well as specific activities on conservation of cultural and natural heritage such as the Digital Belt and Road (DBAR) Initiative, launched by the Institute of Remote Sensing and Digital Earth, Chinese Academy of Sciences. Despite these promising activities, which function as impulses, capacity development activities need to be broadly diversified. Accordingly, Earth observation industry and educational institutions need to adapt to the interdisciplinary field and face this challenge of educating an interdisciplinary group of actors. The varying research goals and objectives of different disciplines cannot be addressed with one unique approach. Accordingly, educational programs have the challenging task to bridge the gap between the users' relevant knowledge and the desired skills and capacities in Earth observation.

The presented e-learning strategy with the various applications are especially designed for an interdisciplinary community, in this case based on the needs of pupils of different grades. Integrated geographical topics are linked with remote sensing applications. The presented paper gives an overview about an e-learning system, which was especially designed for pupils and teachers. By adapting the course content and related methods, the e-learning module Space2Place attempts to empower UNESCO sites stakeholders to incorporate Earth observation into their working routines. The paper offers a detailed insight into the educational concept of the systems, its developing stages, and the technical framework conditions. The given details are supposed to contribute to the current debate on e-learning in the field of Earth observation. Furthermore, other actors should be encouraged to set up similar open source and barrier-free systems and improve the capacities of UNESCO stakeholders and other target groups in this sector.

2. Materials and Methods

The basis of the current e-learning system geospektiv (www.geospektiv.de, see Appendix A for further details of access), was developed in the framework of the project "Space4Geography" carried out between 2013 and 2017. The e-learning system has especially been designed for pupils, adolescent, and teachers in the context of secondary education. Its structure and content are based on a nationwide curricula analysis of the subject Geography in German high schools and secondary schools (www.rgeo.de/cms/p/bpa/). The federal states in Germany are primarily responsible for education, resulting in regional differences, e.g., in terms of temporal variations between the various topics [24]. Each of the learning modules is based on recent didactic research, partly conducted within the mentioned research project [25]. The modules offer a broad range of topics, some with focus on Germany (e.g., the Wadden Sea, Elbe flooding), but also from an international perspective (e.g., deforestation in the tropics, food security in Africa). The practical functions of the e-learning environment are based on the online remote sensing application BLIF (www.blif.de, see Appendix A further details of access). This educational software has been designed to improve the basic competences in the use of Earth observation with optical and radar sensors. Both websites have been operating together for about two years, resulting in important statistics, which were tracked by the website provider on general parameters and internal statistics. Analyses of these statistics are included in the results.

The technical infrastructure was developed in close cooperation with an external IT company, specialized in the design of websites and online platforms. In general, the e-learning system geospektiv and the online remote sensing application BLIF are two different platforms, but were developed on a common PostgreSQL database, joining the user groups of both platforms (see Figure 1). PostgreSQL itself is the standard open source relational database management system, developed by an independent community and is available free of charge. Users of the geospektiv homepage thus only need to register once. In this framework, the platform geospektiv is used for authorization and the administration of rights and user settings. Furthermore, specific statistics are tracked, and learning modules developed and stored. BLIF, on the other hand, is used to track the common users' records on remote sensing tools. This platform enables the users to administrate satellite imagery, or to upload and implement external imagery resources. Users and administrators have direct access to the BLIF and geospektiv back end, developed as a Pyramid web framework. As universally usable, Pyramid is based on the programming language Python, providing manifold opportunities from simple applications to complex web projects. From the back end, direct communication to the front end is possible, developed with html (Hypertext Markup Language) and JavaScript. Therefore, we used AngularJS, a web application framework for the client-side generation of the html website and to extent the related html vocabulary. Access is granted after authentication, whereby the system distributes the rights of the various user groups. The back end is running on a central server and provides necessary services e.g., for administrators, the illustration and functions of the site. The same applies to the connection between the BLIF front end and the BLIF back end. Geospektiv and BLIF offer a browser compatibility that is able to deal with the diverse technical requirements of common browsers such as Firefox, Chrome, Opera, and Safari. All browsers connect to the web server via the web addresses, which are in this case provided by Caddy. The web server Caddy is a http/2 web server, which automatically produces https (Hypertext Transfer Protocol Secure) sites. The http/2 is a further development of the former Hypertext Transfer Protocol (http), to optimize and accelerate the transmission between the specific user browser and the internet website. The https standard for a tap-proof transmission of data is an important requirement, especially when working with German educational and governmental institutions. The web server requests the relevant information from the web application framework, provided by AngularJS.

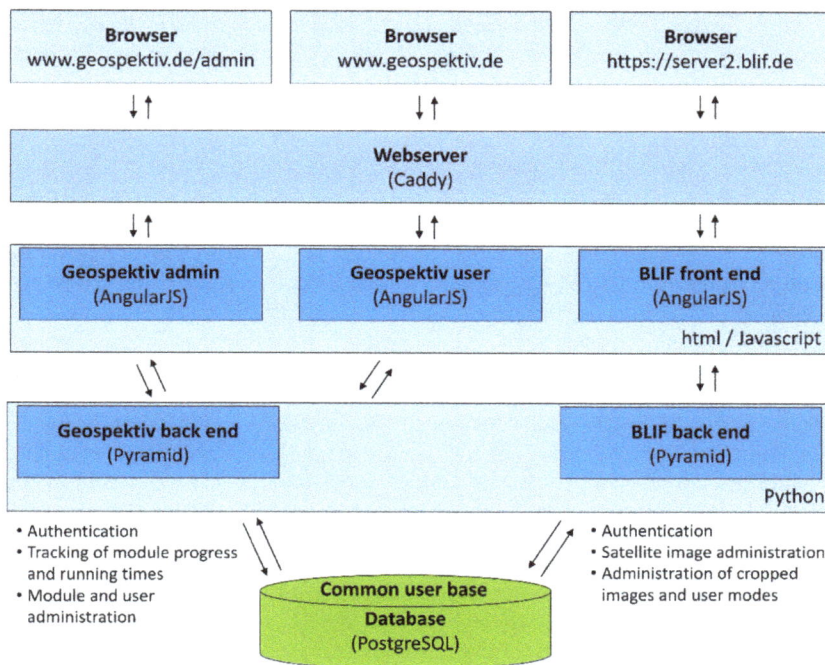

Figure 1. Technical structure of the e-learning environment with the most important functions and applications. An explanation of terms and acronyms can be found in Chapter 2.

The established software infrastructure is based on common, free accessible, and frequently used products. It is thus possible to copy the architecture and provide similar e-learning environments with own content or topics. However, the development relied on the strong technical expertise of the IT company. In particular, the fine-tuning and iterative adjustment steps increased additional costs and time expenditure considerably.

During an expert survey in 2017, the needs of UNESCO site managers regarding the potentials of Earth observation for the monitoring and management of UNESCO sites were recorded and will be presented in the results. By using Google forms, we received detailed feedback from 11 national and international experts. The majority of questions was posed in an open way, providing the opportunity for the experts to answer completely free and without limitations of specific categories. The results of the questionnaire were used for the design of Space2Place and the further development of geospektiv and BLIF. Of particular importance were the existing threats and the possible fields of application within the UNESCO sites.

The aforementioned Space2Place is the final outcome of our activities, offering an e-learning module for adults to get a fundamental introduction into Earth observation and potential fields of applications. Based on the developed learning materials and methods shown above, the structure of Space2Place is described in detail in Section 3.3.

3. Results

For a better understanding of the strategy to empower UNESCO stakeholders, three distinct levels of results can be differentiated. The first level of results deals with the technical implementation of the e-learning system geospektiv. On the second level, results about the expert survey offering insights in current Earth observation usage practices of various institutions, current challenges and future potential are discussed. The third level of results describes the final Space2Place module, especially designed for the needs of UNESCO stakeholders and similar external experts. By combining these three distinct levels, the approach becomes clear and future potentials of the e-learning environment are apparent.

3.1. Adaptive E-Learning System Geospektiv

As explained in the previous chapter, the e-learning system is designed for pupils, based on a nation-wide curricula assessment. The high variety of different education formats in Germany on which the almost 11 million pupils are distributed show the demand for specifically adapted learning resources [26]. The implementation of Earth observation in German curricula aims at acquiring the ability to independently obtain, analyze and evaluate geographical information from aerial and satellite imagery. Accordingly, a competent interaction with digital Earth observation data is a prerequisite [11]. During the development of the platform, all learning modules were intensively evaluated by students and teachers in Germany. The various learning modules of geospektiv are based on different geographical questions. All learning modules are implemented in a capacity development program of the GIS-Station. Everybody can use these modules free of charge. However, the GIS-Station also offers assistance to give teachers and students a greater understanding of the modules and to promote their usability. New partner schools and teachers in general book a guided exercise through a pre-selected module to test the application and functionality. This often takes place in the facilities of the GIS-Station but is also possible through a set of iPads or laptops within the school itself. After completing a first module, teachers tend to use the e-learning function to conduct their courses independently with their school classes, in the framework of their lectures.

Notwithstanding the arrangement of the courses, geospektiv offers a range of advantages for teachers and pupils besides the integration of new techniques and interactive learning methods: (I) teachers can track the activity and results of every single student, also during the course. Thereby, the teacher can immediately offer help and support when necessary. (II) The e-learning system offers the possibility of adaptive learning and various learning paths are implemented in different modules, creating the opportunity for self-guided and self-paced learning.

The concept of adaptivity is provided in three optional ways. The first option is based on the selection of the user and his preferences (see Figure 2, box 4). At certain stages of the module, pupils and teachers can choose different paths they want to follow. Whether students want to find out more about the successful Landsat programme or the new Copernicus programme depends on their own decision. The second option is based on the number of points achieved within the intermediate tests and quizzes (see Figure 2, box 2). Pupils with higher scores receive more complex learning units according to their abilities. Pupils with lower scores receive easier units to create the opportunity for repetition and comprehension. The third option is based on timing. If pupils need more time to complete a quiz compared to a predetermined threshold value, they will receive easier learning units. This makes it possible for pupils that require more learning time to completely pass learning modules and share a feeling of success. Since all learning paths meet at the end and each pupil ends the module with the same concluding topic, the teacher can directly start the follow-up. Accordingly, personalization of modules allows for the promotion of gifted students and the support of students with existing knowledge deficiencies at the same time.

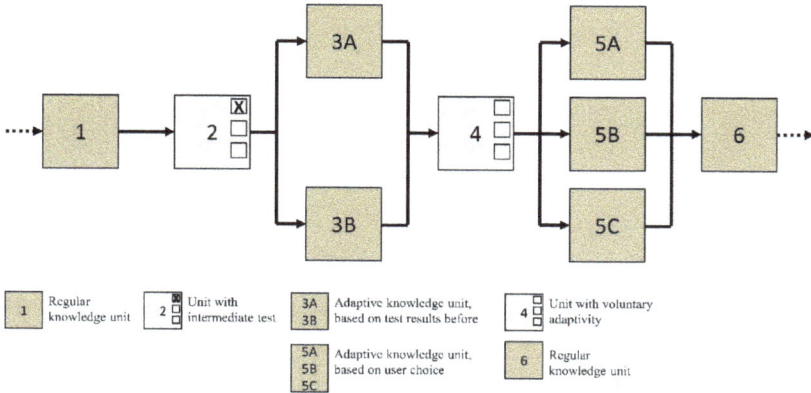

Figure 2. Adaptivity of the e-learning system with individual learning paths, based on score, time, and voluntary adaptivity.

Depending on the requirements and skills of the user, the e-learning system makes it possible to create modules with varying content and levels of difficulty. Currently available in geospektiv are four levels, which refer to German classes 5–6, 7–8, 9–10, and 11–13. The Organisation for Economic Co-operation and Development (OECD) offers an excellent overview of the different international education systems and various indicators within their reports through the interactive Education GPS [27,28]. Furthermore, there are numerous publications that present GIS technologies and their educational potentials in a pedagogical context. See for example Milson et al. (Eds.), who provide studies about the integration of GIS in education for several countries around the globe [29]. In geospektiv, an additional level has been designed for adults via Space2Place, creating off-the-job training opportunities. The duration of the modules is in general 90 min, which temporarily refers to a double lesson in the German school system. It is not intended to use the modules autonomously out of regular school lessons. In fact, they should be used accompanying a thematic unit to generate new impulses and perspectives as well as to introduce the pupils to meaningful Earth observation applications. To keep the motivation and attention high during this period, the modules were designed to be interactive and multimedia content is also provided. The adaptive design of the system provides the opportunity that learning modules can be used inside and outside of the classroom with external devices such as tablets and smartphones. Furthermore, users have the freedom to choose their favorite operating system or browser. An overview of the distribution of used browsers, operating systems, and end devices in 2018 is shown in Figure 3.

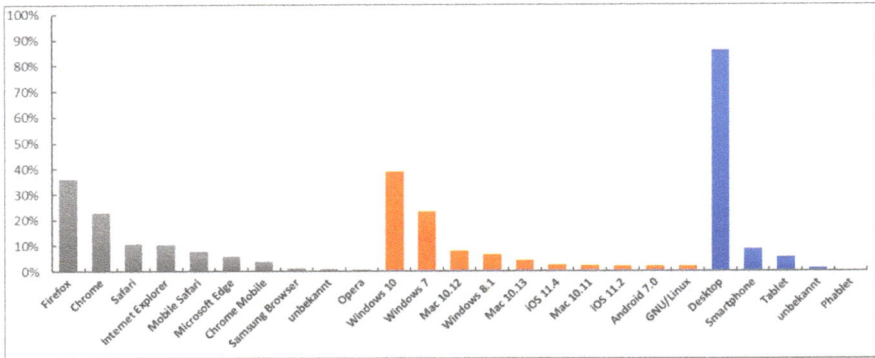

Figure 3. Percentage distribution of browsers (grey), operation systems (orange), and used end devices (blue) on the website geospektiv in 2018.

Results about the distribution partly reflect the current situation of IT infrastructure in German classrooms. Smartphones, tablet PCs, and wireless internet are only slowly being integrated [30]. The number of devices in schools is correspondingly low. The majority uses desktop PCs to access geospektiv (see Figure 4). Only a small proportion of participants use a smartphone or tablet to access geospektiv. However, the responsive web design was adopted to offer a flexible handling of the platform for the diverse devices. Both, smartphones and tablets, are represented with less than 10% of total users. This is in good accordance with the predestined use of the online remote sensing application BLIF, for which the major amount of information has been developed for desktop PCs.

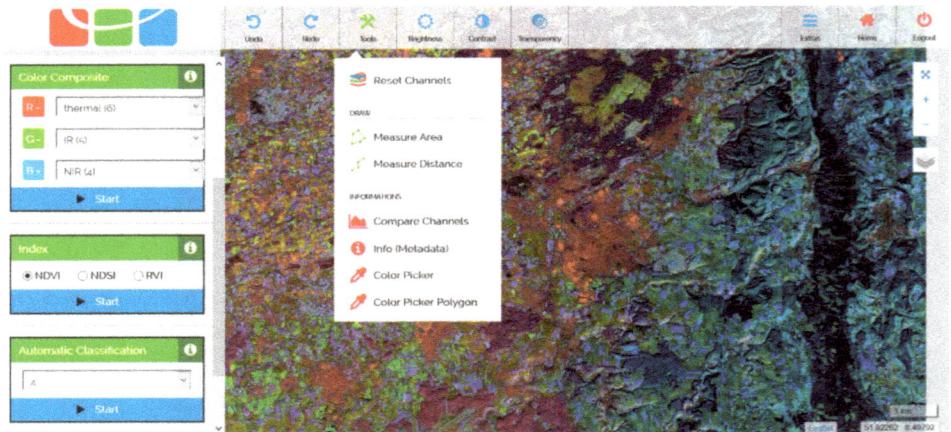

Figure 4. Graphical user interface (GUI) of the online remote sensing application BLIF, illustrating a false-color composite of a Landsat 5 image. On the left side, there are several available processing steps, above the picture some tools to evaluate the image.

Windows (incl. version 10 and 7) is the main operating system. However, also Mac, as operating system of Apple, with more than 10%, is partly represented and thus almost identical with the market shares of the leading operating systems in Germany [31]. These figures show that for extensive usage, interoperability needs to be provided by an educational learning website as a critical factor. Even more widely spread is the use of common browsers, which is in good agreement to the current market shares of leading browsers in Germany, where Chrome, Firefox and the Internet Explorer are the most used browser [32]. For geospektiv in 2018, Firefox is leading with about 36%, followed by Google Chrome with about 23%, and Safari with 11%. These three browsers cover only 70% of users. Especially the Internet Explorer as default browser on Microsoft operating systems and Microsoft Edge should not be neglected when dealing with educational websites for pupils and schools.

Independent of the used system or browser, all modules are based on a responsive design and provide immediate feedback to the user. Accordingly, pupils and teachers can use the feedback to reflect their learning progress and see what they have learned or where further repetition is necessary. The users can save their current status to interrupt and continue the module at any time. This allows, for example, the teacher to pause the module for additional interactions with the class. Of course, users also have the option of repeating certain learning units.

3.2. Evaluation of the Expert Survey

The expert survey, which was conducted before Space2Place was designed and established, gives a comprehensive insight into current applications of Earth observation as well as drawbacks in terms of the UNESCO-sites. Table 1 provides information about existing threats for UNESCO heritages sites, including biosphere reserves, main barriers for stakeholders to integrate Earth observation into working routines and how stakeholders perceive Earth observation to be beneficial for their respective individual organizations.

Table 1. Results of the expert survey, conducted at the beginning of 2017 with 11 experts coming from various institutions.

Organization of the Expert	Specific Threats	Main Barriers	Beneficial Applications
UNESCO-associated organization	Manmade destruction and war	Software usage	Management of UNESCO sites
Governmental organization	Urban development and urban growth	Lack of knowledge	Monitoring of changes, impacts, and interventions
Free consultant	Destruction and war, natural disasters, climate change, uncontrolled human development, tourism	Work overload, inability to handle data correctly	Monitoring of land cover changes, providing accurate mapping, detection of building movements
Intergovernmental organization	Climate change, shipping, oil and gas exploration, tourism	Unavailability of data, interpretation and ground-truth	Long-term monitoring of large areas
UNESCO site	Intensive land use	Lack of time	Updated information base
UNESCO site			Monitoring without aerial photography
UNESCO site	Intensification of agriculture, fragmentation and habitat loss	Unavailability of ready-to use data, already classified datasets	Cost-efficiency
NGO related to UNESCO	Urban sprawl, renewable energies	Lack of time, unavailability of data, hard- and software requirements, lack of knowledge and advantages	Landscape monitoring (i.e., urban sprawl or cultivation of fallow areas)

Table 1. *Cont.*

Organization of the Expert	Specific Threats	Main Barriers	Beneficial Applications
Scientific institute	Flooding and salt intrusion	Compromise between large area and high spatial resolution	Planning of interventions incl. vulnerable areas
Private company	Vegetation growth and vandalism	Coarse spatial resolution	Updated data availability and avoidance of field investigations at dangerous locations
Environmental protection agency	Climate change	Unavailability of ready to use data	Larger spatial footprint

The diversity of institutions reflects the wide range of potential applications in the field of Earth observation and related sciences as well as educational backgrounds. The list of specific threats for UNESCO sites includes global challenges such as climate change and urban sprawl up to local phenomena such as the impact of manmade destructions including vandalism. The information of current barriers is sobering and reflects known drawbacks well [33,34]. Several times, the lack of ready-to-use datasets, missing users' knowledge, and technical requirements was mentioned. Since Earth observation is often simply a tool for UNESCO stakeholders to improve the management effectiveness of their sites, additional time for further education, besides the regular workload, is limited. Accordingly, the lack of time to implement new tools and workflows was mentioned a couple of times in the survey.

While the spatially precise mapping of large areas is also one of the key demands of stakeholders, the choice of the sensor must be based on its capability to capture the desired information in the observed environment [35]. The large variety of almost 100 available operating Earth observation satellites by the end of 2013 [9] with their high variability in different sensor specifications are due to the trade-off that has to be made between spatial, temporal and spectral resolution [36]. One of the key tasks of capacity development initiatives is thus to enable UNESCO site stakeholders to choose the appropriate Earth observation sensor with regard to their specific demands, the study time, and the size and location of the site to be observed. This also meets the stated demand of the stakeholders to improve the detection of impacts, the evaluation of interventions and the early detection of critical changes by using up-to-date information with high cost efficiency (Table 1).

The Copernicus is the most ambitious Earth observation program and it will be soon fully operational [30]. The freely available and free of charge datasets can be used in numerous applications. Therefore, we asked the experts about their awareness of the Copernicus program. The results, illustrated in Figure 5, show that the majority of respondents were not aware of the program, but consider that it can be useful for their organization. Especially UNESCO site stakeholders themselves seem to be less informed about the available Copernicus Earth observation data. Only about 1/3 already uses Copernicus data, which includes in particular private companies and UNESCO associated organizations. Only one respondent, an intergovernmental organization, answered that they know the Copernicus program, but that they are not relevant for their specific tasks. However, the cause for this specific answer could be a lack of information.

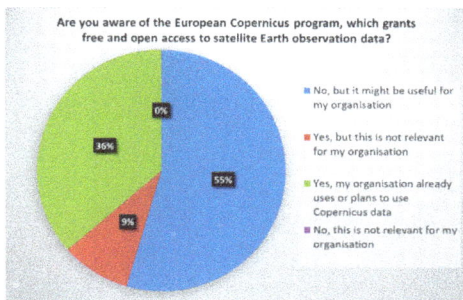

Figure 5. Awareness of the Copernicus program of interviewed experts at the beginning of 2017.

Accordingly, it is not surprising that Earth observation data is especially very often used in those organizations where the awareness about the Copernicus Program is equally high. About 60% of the interviewed experts use Earth observation data at least sometimes, reflecting the already high importance of these datasets in 2017. However, still four of eleven experts rarely use Earth observation data or not even at all (see Figure 6).

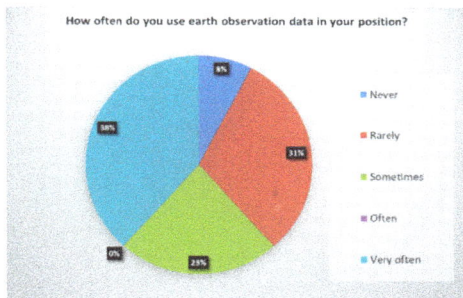

Figure 6. Frequency of EO data utilization of interviewed experts.

3.3. The Space2Place Module

The major aim of the learning module Space2Place is to empower UNESCO-site managers and planning authorities to incorporate data from the Copernicus programme into their daily working routines. Besides regular used very high resolution optical satellite images, current research also shows the potential of high resolution images such as Sentinel-2 images for the monitoring of cultural and archaeological sites [13]. By using practical examples and exercises, the capacities of stakeholders at UNESCO sites in the field of Earth observation should be enhanced. In addition to the general objective of integrating Earth observation applications into the daily working routines of UNESCO stakeholders, numerous intermediate objectives are useful. Space2Place should initially strengthen general knowledge and enhance the ability to communicate about Earth observation issues. This promotes the capacity to formulate needs and claim demands. The module was published online as an introductory course in Earth observation at the beginning of 2018.

The module consists of 21 learning units, including four intermediate tests and one final examination. Figures 7 and 8 give an impression of the usability of the GUI while Figures 9 and 10 show the usage of geospektiv and BLIF by students. Additionally, four links to the online remote sensing application BLIF are implemented. The module is designed for a duration of 120 min and registered users receive a graduation certificate at the end. The learning units are kept very broad to present an overview as large as possible.

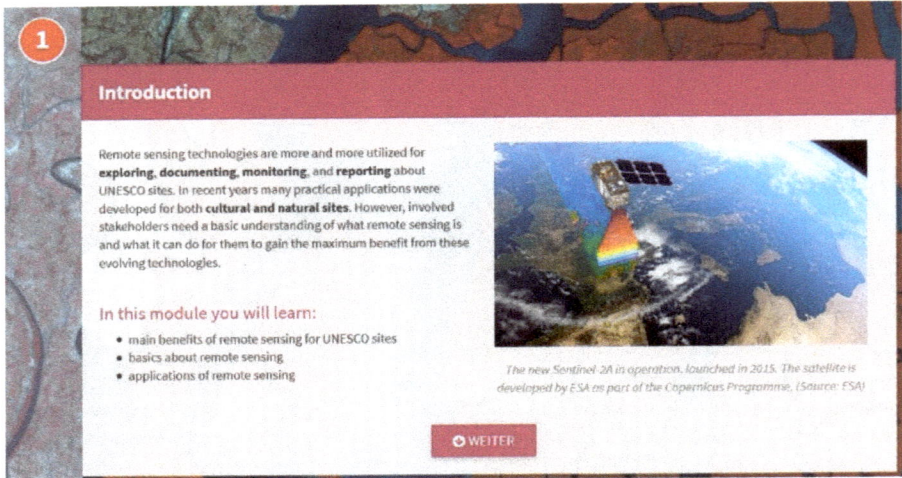

Figure 7. Front end of Space2Place and the first learning unit, introducing the learning module and giving a description of the learning objectives.

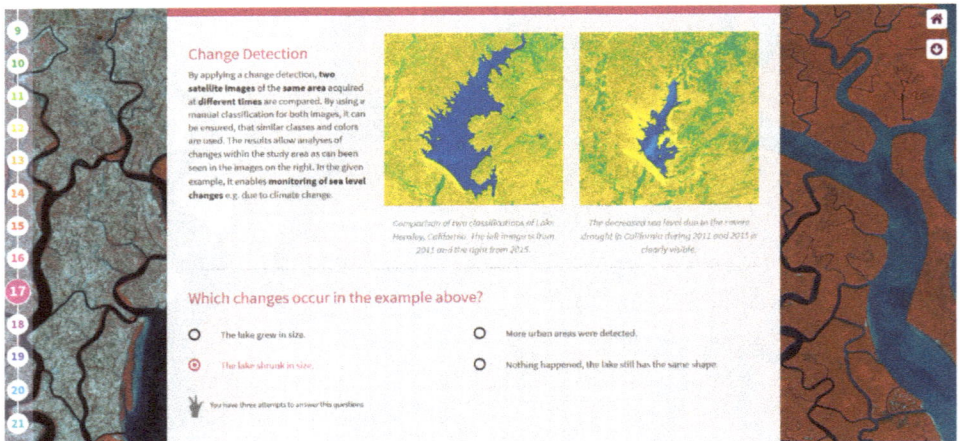

Figure 8. Graphical user interface of the learning module Space2Place, describing the change detection process with illustrations of the Lake Hensley in California. Below, a single choice test assesses the knowledge of participants. The results of this quiz and other quizzes are the basis for the evaluation on the final certificate.

Figure 9. Students testing the learning module Space2Place in the premises of the GIS-Station. Under the guidance of a lecturer, the various processing steps and tasks are evaluated from a didactical point of view and with regards to the content.

Figure 10. Students using the remote sensing application BLIF under the instruction of a lecturer. BLIF is integrated in all geospektiv modules and provides the basic functions to analyze remote sensing data.

Due to manifold applications and availability of data, a specific focus was placed on the Copernicus Program. Due to the different purposes of the satellites, various applications are possible, reflected by the Copernicus Services: Atmosphere, Marine, Land, Climate, Emergency, and Security. However, other data sources are also available within the module, providing the user with a wider spectrum of usable sensors such as Landsat and RapidEye. As a result, the integrated application examples are also kept without great detail, but will be specified for a deeper understanding by additional future learning modules.

4. Discussion

The developed learning module Space2Place is the result of ongoing activities and the high demand of capacity development and awareness raising for Earth observation applications in the context of UNESCO-sites and the SDGs [37]. Especially since the UNESCO signed various conventions on the use of space technologies in 2007 to strengthen monitoring activities of world heritage sites, these aspects have gained importance [38–40]. As shown above, various applications such as for example on natural disasters and natural hazards, like flooding already available to facilitate the work of UNESCO-site stakeholders [21,22]. The potential negative impacts of the predicted climate change will probably increase the necessity of an operational monitoring [41]. Applications on radar satellites to monitor movements of single buildings or small-scale unites in a range of cm or even mm will probably be introduced in the future [18,19]. Additionally, the use of Sentinel-2 images offers various possibilities for monitoring UNESCO sites, which are e.g., affected by armed conflicts and crises [13]. Another useful source of information are InSAR products and related knowledge, which can be utilized by the stakeholders to improve their management and monitoring strategies [42].

On the other hand, a variety of shortcomings limited the desired impact and widespread participation of UNESCO site stakeholders on the technological progress through Earth observation. Accordingly, educational programs are challenged by the variety of educational backgrounds and intended purposes by the UNESCO stakeholders. Clearly, such multifaceted challenges cannot be overcome by a standardized online education program alone. The developed Space2Place learning module and further modules which will follow should act as initial starting points for individual users, also to refer to other more advanced and specific offers. The setting in geospektiv and BLIF allows the handling of both, optical and radar satellite images, covering the needs of a broad range of users. A distinction into different e-learning modules for cultural heritage sites, often optimally studied with radar data and natural heritage sites, which are often best observed with optical data, is therefore not necessary. A smooth transition between different sensors, which can often be used in practice, is feasible and necessary for an educational program covering the diverse UNESCO heritage sites.

As other studies show, maintenance and technical support of the e-learning platform is of major importance [5]. According to the user statistics of geospektiv and written feedback, users are quickly frustrated and show little understanding, if technical errors appear regularly or even permanently and thus reduce the performance of the website. Furthermore, the rapid development in the Earth observation sector requires regular adjustments and adaptation to the changing data availability. Since the beginning of May 2018, BLIF offers the opportunity to incorporate Sentinel-2 satellite images, possible via a connection to the German Copernicus data and exploitation platform CODE-DE (https://code-de.org/), which is also the access point for information products of Copernicus services.

Ideally, e-learning offers should be supplemented by classroom events and face-to-face training. Especially during the winter season, an increase in e-learning activities as indoor classroom activities can be recorded. Particular emphasis should be placed on accompanying learning to enable direct interaction between pupils, tutors and teachers. This also promotes social skills and communication capabilities of pupils. By embedding external courses into classroom activities and the regular curricula, cross-thematic and interdisciplinary approaches are encouraged. Nevertheless, focusing on didactical aspects, distance and e-learning approaches also have numerous weaknesses. This includes e.g., often a limited interaction with the teacher and other participants, decreasing motivation and technical problems. There are also numerous obstacles in the field of didactics. This includes the disparate perception of male and female users in dealing with e-learning programs [43,44]. More research is necessary to implement these differences in current and future e-learning offers.

A blended learning strategy seems also more promising for UNESCO site stakeholders than mere distance learning. Such approaches provide the opportunity to implement interactive inputs in face-to-face trainings, opportunities for peer-to-peer networking, and the exchange on a peer-to-peer level. These networks are not only important to increase the effectivity of learning, but also for the daily working routines in managing UNESCO sites.

Various factors support the use of the e-learning modules and the remote sensing application BLIF:

1. Personalization of learning paths based on the adaptive design of the website is probably the most important factor for the application of geospektiv and Space2Place. By promoting fast learning users and support slow learners, more efficient learning outcomes can be realized [45].
2. A crucial point is the award in terms of a graduation certificate by the end of each learning module. In various countries, such certificates are part of the necessary prerequisites for job promotions or financial compensations. Regardless of the content, the expected final certificate additionally stimulates the brain's reward system, increases the motivation and reduces the rate of termination during the modules [46].
3. Another important factor is the security of the website, especially if it is used in an educational context in Germany. To control the access to forbidden websites or undesired changes in the system, considerable restrictions on computers and networks are often set by administrators. Accordingly, the website needs to be secured by https to be accessed, which is in the case of

geospektiv and BLIF automatically provided by Candy. However, this also requires that only secure content is implemented in the learning modules, otherwise no https security can be provided by the system.

4. The language barrier is another crucial factor, often resulting in a reduced understanding of the learning units. Space2Place is currently offered in English to meet the standard in the global education system. New courses will also initially be offered in English upon introduction. The online remote sensing application BLIF is additionally available in German, French and Italian as well as in Spanish soon. The e-learning platform geospektiv will be soon available in Spanish as well, next to the default languages of German and English.

UNESCO was an active partner during the adoption of the SDGs and creation of the sustainable development agenda [47]. UNESCO sites are important places for learning and creating opportunities to implement sustainable development. Earth observation and its application can significantly contribute to the 2030 Agenda as well as to monitor the progress for certain indicators [48]. Knowledge sharing and education activities on an international level are necessary to facilitate the process of implementing Earth observation data into working routines of heritage site stakeholders [48]. By trying to fill the existing gaps in knowledge of site managers and other relevant stakeholders related to the potentials of Earth observation, the offered e-learning environment can contribute to the preservation of our heritage sites. Despite useful operating instructions and the help of user descriptions from the internet, many of today's remote sensing software packages are still very complex and cluttered, especially for non-experts and users with less experience. The offered learning platform and online remote sensing application does not claim to be complete, but serves as an easy and barrier-free entry point into the field of Earth observation. Not all users will be transferred into Earth observation experts, but the gained knowledge will enhance understanding about the Earth observation technology and support the decision-making process.

By offering an e-learning platform, we can not only foster the use of Earth observation in general and impart knowledge to the different stakeholders, but also address various SDGs and contribute to the 2030 Agenda. One important aspect is the existing gender gap and under-representation of women in STEM (science, technology, engineering and mathematics). This aspect is closely related to the quality of education (SDG 4), gender equality (SDG 5) and the reduction of inequalities (SDG 10) as important objectives of the United Nations. Several studies show that distant learning approaches can minimize the existing gender gap and provide useful education opportunities for women [5,49]. Accordingly, there is an urgent need to implement results of known studies about the differences of male and female in participation of e-learning courses [43,44]. By integrating relevant results, geospektiv and Space2Place can also assist to close the gender gap and promote women in STEM fields.

5. Conclusions and Outlook

The e-learning environment including geospektiv and BLIF is constantly under development and expansion. New language packages will be added in the medium-term future. This includes the platform itself, but also the translation of learning modules into other languages. The focus is on translations into English, but courses with content in Spanish and French will follow. Accompanying scientific research and evaluations set the focus in two areas. The first one concentrates on the difference between pupils and adults, the second focus is on the mentioned gender gap and difference in learning types between male and female.

New learning content in line with state-of-the-art didactical research will be added and a practical introduction with didactic preparation will be made available. By using different cooperation, further contact persons can be introduced to refer the UNESCO site stakeholders accordingly. This will extend the network of stakeholders and lead to valuable linkages with Earth observation experts.

Author Contributions: The first author T.M. made the following contributions: Conceptualization; Formal analysis; Investigation; Methodology; Project administration; Software; Validation; Visualization; Writing—original draft; Writing—review & editing. A.S. (Anne Schneibel) as second author contributed to the paper as follows: Validation; Visualization; Writing—original draft; Writing—review & editing. As third author L.D. made significant contributions to the following fields: Methodology; Software; Validation Writing—review & editing. A.S. (Alexander Siegmund) as fourth author contributed to: Conceptualization; Funding acquisition; Supervision

Acknowledgments: This research was partially supported by German Aerospace Center (DLR), due to the project funding of Space4Geography (50 RO 1301). Furthermore, we would like to thank the GIS-Station, Klaus-Tschira-Centre of Competence for digital Geo-media, for providing insights and expertise that greatly assisted the research. Additionally, we like to thank ebene fünf GmbH for technical support.

Conflicts of Interest: The authors declare no conflict of interest.

Appendix A

Geospektiv (www.geospektiv.de) and BLIF (https://server2.blif.de/login) are freely available for everyone. Interested users can access the websites and create an account by registering. They automatically receive user rights as a teacher, which makes it possible to create group records for single classes and to monitor the learning progress of individual students. Any questions or feedback is possible through the contact e-mail address given in the websites. A member of the rgeo team takes care of processing the request. In future video tutorials, the user will be guided through the remote sensing software and the most important functions will be explained.

References

1. Arkorful, V.; Abaidoo, N. The role of e-learning, advantages and disadvantages of its adoption in higher education. *Int. J. Instr. Technol. Distance Learn.* **2015**, *12*, 29–42.
2. El-Hussein, M.O.M.; Cronje, J.C. Defining Mobile Learning in the Higher Education Landscape. *J. Educ. Technol. Soc.* **2010**, *13*, 12–21.
3. Fallon, C.; Brown, S. *e-Learning Standards—A Guide to Purchasing, Developing, and Deploying Standards-Conformant E-Learning*; CRC Press: Boca Raton, FL, USA, 2016.
4. Orbis Research. Global E-Learning Market Research Report and Forecast to 2017–2022 (Report). Retrieved from Statistics MRC. Available online: www.reuters.com/brandfeatures/venture-capital/article?id=11353 (accessed on 23 November 2018).
5. Welsh, E.T.; Wanberg, C.R.; Brown, K.G.; Simmering, M.J. E-learning: Emerging uses, empirical results and future directions. *Int. J. Train. Dev.* **2003**, *7*, 245–258. [CrossRef]
6. The 50 Most Popular MOOCs of All Time 2018, Online Course Report. Available online: https://www.onlinecoursereport.com/the-50-most-popular-moocs-of-all-time/ (accessed on 22 November 2018).
7. Peters, M.A. Technological unemployment: Educating for the fourth industrial revolution. *Educ. Philos. Theory* **2017**, *49*, 1–6. [CrossRef]
8. Schreier, G.; Dech, S. High resolution earth observation satellites and services in the next decade—A European perspective. *Acta Astronaut.* **2005**, *57*, 520–533. [CrossRef]
9. Belward, A.S.; Skøien, J.O. Who launched what, when and why; trends in global land-cover observation capacity from civilian earth observation satellites. *ISPRS J. Photogramm. Remote Sens.* **2015**, *103*, 115–128. [CrossRef]
10. PricewaterhouseCoopers International. Copernicus Market Report—November 2016. In *Study to Examine the Socio-Economic Benefits of Copernicus in the EU*; PricewaterhouseCoopers International: Luxembourg, 2016.
11. Ditter, R. *The Effectiveness of Digital Learning Paths in Remote Sensing—An Empirical Study of Secondary School Students' Learning based on Motivation and Self-Concept*; Pädagogische Hochschule Heidelberg: Heidelberg, Germany, 2014.
12. Negula, I.D.; Sofronie, R.; Virsta, A.; Badea, A. Earth Observation for the World Cultural and Natural Heritage. *Agric. Agric. Sci. Procedia* **2015**, *6*, 438–445. [CrossRef]
13. Tapete, D.; Cigna, F. Appraisal of Opportunities and Perspectives for the Systematic Condition Assessment of Heritage Sites with Copernicus Sentinel-2 High-Resolution Multispectral Imagery. *Remote Sens.* **2018**, *10*, 561. [CrossRef]

14. UNESCO. UNESCO World Heritage Centre—World Heritage List. Available online: http://whc.unesco.org/en/list/ (accessed on 24 January 2018).

15. UNESCO. UNESCO Global Geoparks | United Nations Educational, Scientific and Cultural Organization. Available online: http://www.unesco.org/new/en/natural-sciences/environment/earth-sciences/unesco-global-geoparks/ (accessed on 24 January 2018).

16. UNESCO. *Biosphere Reserves—Learning Sites for Sustainable Development 2017*; United Nations Educational, Scientific and Cultural Organization: Paris, France, 2017.

17. Giri, C.; Pengra, B.; Zhu, Z.; Singh, A.; Tieszen, L.L. Monitoring mangrove forest dynamics of the Sundarbans in Bangladesh and India using multi-temporal satellite data from 1973 to 2000. *Estuar. Coast. Shelf Sci.* **2007**, *73*, 91–100. [CrossRef]

18. Tapete, D.; Cigna, F. InSAR data for geohazard assessment in UNESCO World Heritage sites: State-of-the-art and perspectives in the Copernicus era. *Int. J. Appl. Earth Obs. Geoinf.* **2017**, *63*, 24–32. [CrossRef]

19. Zhou, W.; Chen, F.; Guo, H. Differential Radar Interferometry for Structural and Ground Deformation Monitoring: A New Tool for the Conservation and Sustainability of Cultural Heritage Sites. *Sustainability* **2015**, *7*, 1712–1729. [CrossRef]

20. Hadjimitsis, D.; Agapiou, A.; Alexakis, D.; Sarris, A. Exploring natural and anthropogenic risk for cultural heritage in Cyprus using remote sensing and GIS. *Int. J. Digit. Earth* **2013**, *6*, 115–142. [CrossRef]

21. Wang, J.-J. Flood risk maps to cultural heritage: Measures and process. *J. Cult. Herit.* **2015**, *16*, 210–220. [CrossRef]

22. Joyce, K.E.; Belliss, S.E.; Samsonov, S.V.; McNeill, S.J.; Glassey, P.J. A review of the status of satellite remote sensing and image processing techniques for mapping natural hazards and disasters. *Prog. Phys. Geogr. Earth Environ.* **2009**, *33*, 183–207. [CrossRef]

23. Wang, L.; Ma, Y.; Yan, J.; Chang, V.; Zomaya, A.Y. pipsCloud: High performance cloud computing for remote sensing big data management and processing. *Future Gener. Comput. Syst.* **2018**, *78*, 353–368. [CrossRef]

24. Fuchsgruber, V.; Schütt, F.; Viehrig, K.; Wolf, N.; Siegmund, A. Thematische und regionale Schwerpunkte in deutschen Bildungsplänen—Eine bundesweite Vergleichsstudie für das Fach Geographie an Gymnasien. *Prax. Geogr.* **2017**, *3*, 54–56.

25. Wolf, N.; Fuchsgruber, V.; Riembauer, G.; Siegmund, A. An Adaptive Web-based Learning Environment for the Application of Remote Sensing in Schools. *ISPRS Int. Arch. Photogramm. Remote Sens. Spat. Inf. Sci.* **2016**, *XLI-B6*, 53–56. [CrossRef]

26. Destatis. *Schülerzahlen im Schuljahr 2017/2018 um 0.4% Zurückgegangen*; Destatis—Statistisches Bundesamt: Wiesbaden, Germany, 2018.

27. OECD. *Education GPS*; OECD Publishing: Paris, France, 2018.

28. OECD. *Education at a Glance 2018: OECD Indicators*; OECD Publishing: Paris, France, 2018.

29. Milson, A.J.; Demirci, A.; Kerski, J.J. *International Perspectives on Teaching and Learning with GIS in Secondary Schools*; Springer: Dordrecht, The Netherlands, 2012; ISBN 978-94-007-2120-3.

30. Bos, W.; Lorenz, R.; Endberg, M.; Eickelmann, B.; Kammerl, R.; Welling, S. *Schule Digital—Der Länderindikator 2016 Kompetenzen von Lehrpersonen der Sekundarstufe I im Umgang Mit Digitalen Medien im Bundesländervergleich*; Waxmann Verlag: Münster, Deutschland, 2016; ISBN 978-3-8309-3540-7.

31. Destatis. *Marktanteile der Führenden Betriebssysteme in Deutschland von Januar 2009 bis September 2018*; Destatis—Statistisches Bundesamt: Wiesbaden, Germany, 2018.

32. Destatis. *Marktanteile der Führenden Browserfamilien an der Internetnutzung in Deutschland von Januar 2009 bis September 2018*; Destatis—Statistisches Bundesamt: Wiesbaden, Germany, 2018.

33. Rose, R.A.; Byler, D.; Eastman, J.R.; Fleishman, E.; Geller, G.; Goetz, S.; Guild, L.; Hamilton, H.; Hansen, M.; Headley, R.; et al. Ten ways remote sensing can contribute to conservation: Conservation Remote Sensing Questions. *Conserv. Biol.* **2015**, *29*, 350–359. [CrossRef] [PubMed]

34. Turner, W.; Spector, S.; Gardiner, N.; Fladeland, M.; Sterling, E.; Steininger, M. Remote sensing for biodiversity science and conservation. *Trends Ecol. Evol.* **2003**, *18*, 306–314. [CrossRef]

35. Woodcock, C.E.; Strahler, A.H. The factor of scale in remote sensing. *Remote Sens. Environ.* **1987**, *21*, 311–332. [CrossRef]

36. Townshend, J.R.G.; Justice, C.O. Selecting the spatial resolution of satellite sensors required for global monitoring of land transformations. *Int. J. Remote Sens.* **1988**, *9*, 187–236. [CrossRef]

37. Paganini, M.; Petiteville, I.; Ward, S.; Dyke, G.; Steventon, M.; Harry, J.; Kerblat, F. *Satellite Earth Observation in Support of the Sustainable Development Goals*; European Space Agency: Paris, France, 2018.

38. Hernandez, M.; Huth, U.; Schreier, G. Earth Observation from Space for the Protection of UNESCO World Heritage Sites: DLR Assisting UNESCO. In Proceedings of the International Archives of the Photogrammetry, Remote Sensing and Spatial Information Sciences, Beijing, China, 3–11 July 2008; pp. 643–646.

39. Stewart, C.; Rast, M.; Sarti, F.; Arino, O. ESA Activities in Earth Observation for Cultural Heritage Applications. *Int. J. Herit. Digit. Era* **2015**, *4*, 325–338. [CrossRef]

40. UNESCO. *Agreement between UNESCO and the United States National Aeronautics and Space Administration (NASA) for Cooperation on World Heritage Conservation, Monitoring of Biosphere Reserves and Natural Hazards, and Space Earth Science Education, Capacity-Building, and Outreach Activities 2005*; UNESCO: Paris, France, 2005.

41. Reimann, L.; Vafeidis, A.T.; Brown, S.; Hinkel, J.; Tol, R.S.J. Mediterranean UNESCO World Heritage at risk from coastal flooding and erosion due to sea-level rise. *Nat. Commun.* **2018**, *9*, 4161. [CrossRef] [PubMed]

42. Pastonchi, L.; Barra, A.; Monserrat, O.; Luzi, G.; Solari, L.; Tofani, V. Satellite Data to Improve the Knowledge of Geohazards in World Heritage Sites. *Remote Sens.* **2018**, *10*, 992. [CrossRef]

43. Wehrwein, E.A.; Lujan, H.L.; DiCarlo, S.E. Gender differences in learning style preferences among undergraduate physiology students. *Adv. Physiol. Educ.* **2007**, *31*, 153–157. [CrossRef] [PubMed]

44. Ong, C.-S.; Lai, J.-Y. Gender differences in perceptions and relationships among dominants of e-learning acceptance. *Comput. Hum. Behav.* **2006**, *22*, 816–829. [CrossRef]

45. Klašnja-Milićević, A.; Vesin, B.; Ivanović, M.; Budimac, Z.; Jain, L.C. Personalization and Adaptation in E-Learning Systems. In *E-Learning Systems*; Springer International Publishing: Cham, Switzerland, 2017; Volume 112, pp. 21–25. ISBN 978-3-319-41161-3.

46. Grimes, P.M.; Grimes, L.S.; Grimes, C.M. Method and System for Integrated Reward System for Education Related Applications. U.S. Patent 14/213,928, 19 September 2014.

47. UNESCO. *UNESCO Moving Forward the 2030 Agenda for Sustainable Development*; United Nations Educational, Scientific and Cultural Organization: Paris, France, 2017.

48. Anderson, K.; Ryan, B.; Sonntag, W.; Kavvada, A.; Friedl, L. Earth observation in service of the 2030 Agenda for Sustainable Development. *Geo Spat. Inf. Sci.* **2017**, *20*, 77–96. [CrossRef]

49. Cuadrado-García, M.; Ruiz-Molina, M.-E.; Montoro-Pons, J.D. Are there gender differences in e-learning use and assessment? Evidence from an interuniversity online project in Europe. *Procedia Soc. Behav. Sci.* **2010**, *2*, 367–371. [CrossRef]

MDPI

St. Alban-Anlage 66

4052 Basel

Switzerland

Tel. +41 61 683 77 34

Fax +41 61 302 89 18

www.mdpi.com

Geosciences Editorial Office

E-mail: geosciences@mdpi.com

www.mdpi.com/journal/geosciences

www.ingramcontent.com/pod-product-compliance
Lightning Source LLC
Chambersburg PA
CBHW051717210326
41597CB00032B/5513